MODALITIES AND MULTIMODALITIES

LOGIC, EPISTEMOLOGY, AND THE UNITY OF SCIENCE

VOLUME 12

Editors
Shahid Rahman, *University of Lille III, France*
John Symons, *University of Texas at El Paso, U.S.A.*

Editorial Board
Jean Paul van Bendegem, *Free University of Brussels, Belgium*
Johan van Benthem, *University of Amsterdam, the Netherlands*
Jacques Dubucs, *University of Paris I-Sorbonne, France*
Anne Fagot-Largeault *Collège de France, France*
Bas van Fraassen, *Princeton University, U.S.A.*
Dov Gabbay, *King's College London, U.K.*
Jaakko Hintikka, *Boston University, U.S.A.*
Karel Lambert, *University of California, Irvine, U.S.A.*
Graham Priest, *University of Melbourne, Australia*
Gabriel Sandu, *University of Helsinki, Finland*
Heinrich Wansing, *Technical University Dresden, Germany*
Timothy Williamson, *Oxford University, U.K.*

Logic, Epistemology, and the Unity of Science aims to reconsider the question of the unity of science in light of recent developments in logic. At present, no single logical, semantical or methodological framework dominates the philosophy of science. However, the editors of this series believe that formal techniques like, for example, independence friendly logic, dialogical logics, multimodal logics, game theoretic semantics and linear logics, have the potential to cast new light no basic issues in the discussion of the unity of science.

This series provides a venue where philosophers and logicians can apply specific technical insights to fundamental philosophical problems. While the series is open to a wide variety of perspectives, including the study and analysis of argumentation and the critical discussion of the relationship between logic and the philosophy of science, the aim is to provide an integrated picture of the scientific enterprise in all its diversity.

For other titles published in this series, go to
www.springer.com/series/6936

Walter Carnielli · Claudio Pizzi

Modalities and Multimodalities

With the assistance and collaboration
of Juliana Bueno-Soler

 Springer

Prof. Walter Carnielli
Centre for Logic, Epistemology
and the History of Science - CLE
State University of Campinas
P.O. Box 6133
13083-970 Campinas, SP
Brazil

Prof. Claudio Pizzi
Dipto. Filosofia e Scienze Sociali
Università Siena
Via Roma, 47
53100 Siena
Italy

Cover image: Adaptation of a Persian astrolabe (brass, 1712–13), from the collection of the Museum of the History of Science, Oxford. Reproduced by permission.

ISBN 978-90-481-7924-4   e-ISBN 978-1-4020-8590-1

All Rights Reserved
© 2008 Springer Science + Business Media B.V.
Softcover reprint of the hardcover 1st edition 2008
No part of this work may be reproduced, stored in a retrieval system, or transmitted in any form or by any means, electronic, mechanical, photocopying, microfilming, recording or otherwise, without written permission from the Publisher, with the exception of any material supplied specifically for the purpose of being entered and executed on a computer system, for exclusive use by the purchaser of the work.

Printed on acid-free paper

9 8 7 6 5 4 3 2 1

springer.com

With the assistance and collaboration
of Juliana Bueno-Soler

Preface

In the last two decades modal logic has undergone an explosive growth, to the point that a complete bibliography of this branch of logic, supposing that someone were capable to compile it, would fill itself a ponderous volume.

What is impressive in the growth of modal logic has not been so much the quick accumulation of results but the richness of its thematic developments. In the 1960s, when Kripke semantics gave new credibility to the logic of modalities – which was already known and appreciated in the Ancient and Medieval times – no one could have foreseen that in a short time modal logic would become a lively source of ideas and methods for analytical philosophers, historians of philosophy, linguists, epistemologists and computer scientists.

The aim which oriented the composition of this book was not to write a new manual of modal logic (there are a lot of excellent textbooks on the market, and the expert reader will realize how much we benefited from many of them) but to offer to every reader, even with no specific background in logic, a conceptually linear path in the labyrinth of the current panorama of modal logic. The notion which in our opinion looked suitable to work as a compass in this enterprise was the notion of multimodality, or, more specifically, the basic idea of grounding systems on languages admitting more than one primitive modal operator.

It cannot be denied that multimodality has been the focus of many fertile inquiries, inasmuch as it gave rise to progressive branches of modal logic such as temporal logics, epistemic logics, dynamic logics and so on. It is in the realm of multimodal logics that philosophers and mathematicians could find a common ground of interest, the former by gaining more powerful languages to be applied to philosophical analysis, the latter by reaching domains of highly abstract structures. One of the main results presented in our book (see Chapter 8) is a general completeness theorem for multimodal logics with full attention to all the essential ingredients. Though the theorem is not the most general possible, it opens a direct way of access to an arbitrarily wide class of specific concepts and sheds light to the interweaving between multimodalities and their algebraic trait.

The choice of a unifying subject has suggested, or indeed imposed, a drastic selection in the *mare magnum* of bibliographical material which is open to students working in this area. Some subjects which are currently treated in introductory textbooks, such as non-normal modal systems or conditional logic, had to be sacrificed to the dominant objective. By converse, normally neglected subjects such as, for instance, the logic of contingency and the logic of propositional quantifiers, receive some attention here.

The text of the book is an amplified and revised version of its Italian ancestor, "Modalità e Multimodalità", published in 2001 (Franco Angeli, Milan). In comparison to the Italian version, here we emphasize the existence of two levels of reading. At the first level, the book is proposed as a didactic tool, as one can see from Chapter 1 devoted to Standard Logic– and from the exercises placed at the end of every chapter. In this respect we advise that definitions, propositions, corollaries, lemmas and remarks are numbered consecutively to facilitate reference and localization: so, for instance, "Remark 2.3.7" means the seventh item in Section 3 of Chapter 2. Exercises are numbered by Chapter, so that "Exercise 2.4", for instance, means the fourth item in the exercise section of Chapter 2. Even if some exercises are optional, others are intended to integrate arguments introduced in the corresponding chapter.

However, at the second level, the book is proposed as a repertory of ideas for students interested in progressive inquiry in this field of logic. Readers with some experience in modal logic will find that every chapter is relatively independent and may be read as a short essay on the relevant topic. Furthermore, each chapter is followed by an abridged commented bibliographical note which is mainly thought for historically minded readers. The bibliography at the end of the book is not intended to be a complete repertoire of modal logic, but is restricted to the treated topics.

A terminological warning may be convenient. In the language employed by professional logicians there is an ambiguity in the use of the terms logic/logics and modality/modalities, which are actually used with more than one meaning. In the present book we have preserved such flexibility of uses, but respecting some conventions which are worthwhile making explicit:

(1a) The singular term "logic" indicates either logic as a science or a family of logical systems based on the same language (as in the expression "modal logic"); such family may also consists of only one system or of a class of equivalent systems (as in the case of the expressions "first order logic" or "standard logic").

(1b) The plural term "logics" stands for "logical systems" or for "families of logical systems of the same kind".

(2a) The singular term "modality" in technical sense indicates a string of negated or non-negated modal operators, while in non-technical sense it indicates a modal notion of some kind.

(2b) The plural term "modalities" refers to a plurality of objects of one of two kinds indicated at point (2a).

This book is the fruit stemming from a cooperation between a Brazilian and an Italian logician, each with an affinity to the philosophical side of modalities. The idea of writing a joint book would not have arisen without the first visit of Pizzi to Brazil in 1991, when he was invited to give a course in modal logic by the Centre for Logic, Epistemology and History of Science (CLE) of the State University of Campinas (UNICAMP). Various other stays of Pizzi in Brazil and of Carnielli in Italy have been financially supported by both Italian and Brazilian entities: CNR, MIUR and the Siena University from Italy and FAPESP, CNPq and UNICAMP from Brazil. Carnielli also acknowledges support from the Alexander von Humboldt Foundation in Germany for several stays in Europe which permitted sojourns in Italy, and from the Fonds National de la Recherche Luxembourg for a research stay at the University of Luxembourg, essential for polishing the final draft. Despite the amazing facility of telematic communications, living in two different continents did not make the work of the authors easier. As a reward for the discomfort, each one of the authors had the privilege to delve into the culture of a distant, but not (so) foreign country.

To conclude, it is a pleasure to express our warm thanks to the many colleagues who accepted to read and comment parts of the book, of course freeing them from any responsibility. The names we cannot avoid mentioning are Patrick Blackburn (Nancy), Marcelo Coniglio (Campinas), Giovanna Corsi (Firenze), Itala Loffredo D'Ottaviano (Campinas), Oscar Esquizabel (Buenos Aires), Maurizio Fattorosi-Barnaba (Roma), Shane Frank (Denver), Valentin Goranko (Johannesburg), Yuri Manin (Evanston), João Marcos (Natal), Antonio Marmo (Campinas), Maria da Paz Nunes de Medeiros (Natal), César Mortari (Florianópolis), Massimo Mugnai (Florence), Tanja Osswald (Bonn), Duccio Pianigiani (Siena), Vaughan Pratt (Stanford) and

Cristina Sernadas (Lisbon). Carnielli also thanks his graduate students at UNICAMP Anderson de Araújo, Samir Gorski, Bruno Jacinto, Alberto Batista Neto, Newton Peron and Luiz Henrique Silvestrini who read and criticized previous versions of the book.

The thanks we owe to Juliana Bueno-Soler (Campinas) are of a different order of importance. She provided not only a translation from Italian to English, but a revision and a composition in LATEX style of the whole text, along with the general bibliography. Many improvements and corrections are due to her suggestions, and it is difficult to say what the final result would have been without her intelligent cooperation.

Campinas and Milano, August 2008

Contents

Preface		**vii**
1	**Modal logic and standard logic**	**1**
	1.1 Modal notions and quantifiers	1
	1.2 A non-modal basis for modal logics	4
	1.3 The semantical analysis of PC	9
	1.4 Constructive completeness of PC	12
	1.5 Decidability of PC	14
	1.6 Post-completeness and other properties of PC	17
	1.7 Exercises	19
	1.8 Further reading	22
2	**The syntax of normal modal systems**	**25**
	2.1 The relationship among modal operators	25
	2.2 Minimal properties of modal systems	31
	2.3 Systems between K and S5	34
	2.4 Modalities in S5	43
	2.5 Exercises	46
	2.6 Further reading	48
3	**The semantics of normal modal systems**	**49**
	3.1 Matrices and Dugundji's Theorem	49
	3.2 Carnapian models and relational models	53
	3.3 Correspondence theory and bisimulations	64
	3.4 The method of relational tableaux	72
	3.5 Exercises	81
	3.6 Further reading	84

4 Completeness and canonicity — 87
- 4.1 The constructive completeness of K and KT … 87
- 4.2 Completeness by Henkin's method … 92
- 4.3 Completeness: models versus frames … 104
- 4.4 The logic of arithmetical provability … 107
- 4.5 Exercises … 114
- 4.6 Further reading … 115

5 Incompleteness and finite models — 117
- 5.1 An incompleteness result … 117
- 5.2 Finite model property and filtrations … 123
- 5.3 Exercises … 136
- 5.4 Further reading … 138

6 Temporal logics — 141
- 6.1 Logics with two primitive modal operators … 141
- 6.2 Completeness and incompleteness of *PF*-logics … 156
- 6.3 Monomodal fragments of *PF*-logics … 162
- 6.4 Other temporal systems … 166
- 6.5 *US*-logics, metric tense logics and hybrid logics … 174
- 6.6 Exercises … 178
- 6.7 Further reading … 180

7 Epistemic logic: knowledge and belief — 183
- 7.1 To know, to believe and their difficulties … 183
- 7.2 Knowledge, belief and agents … 187
- 7.3 The minimal logic of knowledge … 189
- 7.4 The systems K^m, KT^m, $S4^m$ and $S5^m$ … 194
- 7.5 Common knowledge and implicit knowledge … 196
- 7.6 The logic of belief … 200
- 7.7 Exercises … 202
- 7.8 Further reading … 203

8 Multimodal logics — 205
- 8.1 What are multimodalities? … 205
- 8.2 Multimodal languages … 206
- 8.3 The elementary multimodal systems … 209
- 8.4 Axioms for multimodal logics … 213
- 8.5 Multimodal systems and strict implication … 220
- 8.6 Multimodal models and completeness … 222

8.7	Exercises	235
8.8	Further reading	237

9 Towards quantified modal logic — 241
- 9.1 Propositional quantifiers — 241
- 9.2 Necessary and contingent identities — 250
- 9.3 The problem of completeness in first-order modal logic — 256
- 9.4 Inclusive domains and arbitrary domains — 263
- 9.5 Quantification and multimodalities — 267
- 9.6 Exercises — 270
- 9.7 Further reading — 271

Bibliography — 273

Index of names — 289

Index of notation — 293

Index of subjects — 297

Chapter 1
Modal logic and standard logic

1.1 Modal notions and quantifiers

Modal logics, the logics which study the modes of qualifying truth such as possibility and necessity, belong to the family of logics classified as *non-classical* or *non-standard* logics. The most studied non-standard logics may be conventionally grouped, for the sake of simplicity, into two distinct classes:

1. The class of logics which differ from standard logic for lacking some classical laws (e.g. intuitionistic logic, paraconsistent logics or quantum logics), and

2. The class of logics which differ from standard logic for being linguistic or axiomatic extensions of it.

Currently studied modal logics occupy a central place in the latter of the two mentioned classes, and this already offers an important preliminary information: modal logics are not essentially rivals of standard logic, but are linguistic and axiomatic enrichments of it. This means *inter alia* that the meaning of standard connectives – negation, conjunction, disjunction, etc. – is essentially preserved in modal systems, and that they need no deviation from the treatment they receive in standard logic.

The link between standard logic and modal logics is so strong that the dominant attitude shared by contemporary logicians, until the first decades of the 20th century, was to deny the legitimacy of an independent modal logic. The leading idea was, in fact, that modal notions might be reinterpreted as metasystematic notions or, alternatively, that modal statements could be paraphrased into the language of first-order standard logic. This reduction is *prima facie* plausible. On the one hand, the statement that "2+2 = 4 is necessary" may be claimed to have the same meaning of "2 + 2 = 4 is a theorem of Peano arithmetic", while on the other hand one could maintain that "Necessarily all cats are mortal" might be paraphrased into something like "for every x and every y, if x is a cat and lives in the spatio-temporal region y, then x is mortal".

Now, such phrases as "my cat lives in y" are not, strictly speaking, full-blooded statements but open statements or, in more mathematical jargon, *propositional functions*. Propositions are stably true or false, while propositional functions receive a truth value after the quantification of all their free variables or after the substitutions of nominal constants to their free variables.

However, already in the 19th century, one could recognize the germs of a different conception of the propositions and more generally of modal logic itself. Boole, for instance, treated propositions as the set of *the times* in which they are true. This approach of course implies the non-orthodox view that propositions are something which may have different truth values in different times.

Boole's interest in probability theory was strictly dependent on this basic intuition. Probability is, in fact, a function which associates to propositions real numbers between 0 and 1. If a proposition is treated as a set of times, its probability value is a measure of the size of such set of times which, in the case of tautologies, turns out to be exactly the universal set of times. The relevance of this view for modal logic should be obvious. The transition from probabilistic notions to modal notions is straightforward: in fact, "Possibly α" might be translated into "The probability of α is greater than 0", while "Necessarily α" might be translated into "The probability of α is 1".

Despite the original elaborations of such logicians as Hugh McCall (1837–1909), the pioneers of contemporary logic did not leave any space to modal logic, apparently forgetting that modal syllogistic was an essential part of ancient and medieval logic. Actually, the dominant connection between logic and mathematics seemed to encourage not only indifference towards modal logic, but hostility as well, an attitude which after the Second World War has been mainly represented by W. O. Quine's antimodalism.

Let us now examine the proposed reduction of modal operators to quantifiers. Even if such a reduction has some intuitive appeal, it has a certain amount of drawbacks which can be synthetically outlined in the following points:

(a) If we want to say that "Necessarily α" is to be paraphrased into something like "For all the reference points x, α is true with respect to x", propositions must be conceived as predicates of reference points. This

1.1. MODAL NOTIONS AND QUANTIFIERS

idea is highly problematic from the viewpoint of formalization: it suffices to remark that, if p is a propositional variable, "$\forall x p x$" is syntactically improper (i.e., it is not a well-formed formula). The only open alternative seems to be to introduce a two-place function R such that $R(x,p)$ is read as "Proposition p is realized in x". Then "Necessarily p" may be *prima facie* translated into $\forall x R(x,p)$. But the problem is that we should have axioms describing the properties of R, for instance, to determine which is the relation between $\forall x \neg R(x,p)$ and $\forall x R(x, \neg p)$, and such axioms do not follow from the first-order theorems for quantifiers.

(b) In standard first-order logic, $\forall x \alpha(x)$ implies $\exists x \alpha(x)$. This correlation suggests that a law of every modal logic should be that necessity implies possibility. If we use the symbol □ for necessity and ◊ for possibility, this would lead to the law $\Box \alpha \supset \Diamond \alpha$. This assumption is surely in line with the basic intuitions of Aristotle's *Organon* (see Section 2.1), but its universality is questionable if we observe that there are different notions of necessity (logical, physical, deontic etc.) endowed with distinguishable properties. Let us, for instance, endorse some notion of necessity by which to say that α is necessary means that there exists a certain background theory K such that, for any reference point x, if K is true and R is as in (a), then $R(x, \alpha)$ is also true. So "it is necessary that α" could be represented by $\forall x(K \to R(x, \alpha))$ (where $\to$ stands for some kind of implicative relation), while "it is possible that α" seems to be properly rendered by the dual formula $\neg \forall x(K \to R(x, \neg \alpha))$. Now, it is not generally the case that $\forall x(K \to R(x, \alpha))$ implies $\neg \forall x(K \to R(x, \neg \alpha))$. Indeed, if $\to$ is the material conditional (see Section 1.2) the inference is not valid. Suppose in fact that K is a false theory; then $K \to R(x, \alpha)$ and $K \to R(x, \neg \alpha)$ are both true for every x. So $\neg \forall x(K \to R(x, \neg \alpha))$ is false, and the implication is false. Thus in this interpretation, $\Box \alpha$ would not imply $\Diamond \alpha$.

(c) If it were true that modal operators are quantifiers in disguise and $P(x)$ is a propositional function containing x as the only variable, it should be indifferent to write $\forall x \Box P(x)$ in place of $\Box \forall x P(x)$, or $\exists x \Diamond P(x)$ in place of $\Diamond \exists x P(x)$. But Aristotle was the first to emphasize the distinction between what Medieval logicians called *de dicto* and *de re* modalities, i.e., between saying that a proposition is necessary or possible (as, for instance, in $\Box \forall x P(x)$ or $\Diamond \exists x P(x)$) and saying that everything or something has a necessary or a possible property (as, for instance, in $\forall x \Box P(x)$ or $\exists x \Diamond P(x)$).

In fact, there is a remarkable difference between saying:

(i) There is someone who has the possibility to become President of the Republic, and

(ii) It is possible that someone becomes President of the Republic.

The sentence (ii) appears to be true in circumstances in which (i) is false, for instance, for lack of eligible candidates (see Chapter 9).

(d) While it is obvious that $\Box p \supset \Box p$ is a logical truth, it is not obvious that $\Box p \supset \Box\Box p$ is a logical truth: for example, some philosophers find that necessity is something "created" by the rules of natural or artificial languages, but that the existence of such rules is itself something contingent, so essentially non-necessary. The latter formula, however, is granted by the interpretation of $\Box$ in terms of quantifiers, given that $\forall x \forall x \alpha(x)$ is equivalent to $\forall x \alpha(x)$ in standard quantificational logic.

1.2 A non-modal basis for modal logics

The preceding remarks are not meant to deny that some important relation exists between quantification and modality, but to deny that every kind of modal notion is trivially translatable using quantifiers as axiomatized in first-order logic. This result opens some interesting questions. Why exclude, for instance, that some modal notions may be represented not by first-order but by second-order quantifiers? Why exclude that the relationship between quantifiers and modal operators subsists in the reverse direction, i.e., that first-order quantifiers are treatable as a special kind of modal operators? The latter alternative (which will be formulated in Section 9.5) is suggested not only by the amazing plurality of modal notions which are used in common and scientific language, but by the possibility of treating them all together inside what we shall call a *multimodal* language (see especially Chapter 8).

An implicit suggestion from the above remarks is that modal systems are in the first place *propositional* systems which are extensions of propositional standard logic. It may be then useful to propose here a synthetical presentation of the standard Propositional Calculus (here named **PC**), whose properties will be used within subsequent chapters of this book. The most used approach to modal logic is the axiomatic approach, so **PC** will be also introduced in this book as an axiomatic system along with its semantics. There are many recommendable books dealing with **PC**, but the intention

1.2. A NON-MODAL BASIS FOR MODAL LOGICS

here is just to present the propositional calculus in a propedeutic way in view of our general treatment of modal logics.

Primitive symbols of the language of **PC** will be the *atomic variables* or *propositional variables* $p, q, r \cdots$ (with or without numerical subscripts), the constant $\bot$ for something which is always false (the *absurd*), and the symbol $\supset$ for the so-called *material conditional*. The *well-formed formulas* (*wffs*), apart from the atomic variables, will be either $\bot$ or formulas of the form $\alpha \supset \beta$, where α and β stand for any wffs. In this way, wffs are *inductively* defined, which permits us to prove facts about wffs by using induction. It is very useful, for such reason, to have a notion of *complexity* of wffs, which may be introduced by stipulating that atomic variables and $\bot$ have complexity zero, and that $\alpha \supset \beta$ has strictly higher complexity than both α and β. Usually, the *length* (number of symbols) of formulas is used as a natural way to measure complexity, and we will generally speak about a proof by "induction on the complexity", or more specifically, about "induction on the length" of formulas. This kind of proof by induction is typical in logic, and we will also use a lot of "induction on the length" of proofs, derivations, etc., as exemplified below.

Parentheses "(" and ")" will be used for grouping and will be eliminated according to standard conventions.

It is to be remarked that α, β, γ, etc. are not, strictly speaking, wffs, but metavariables which range over wffs. Thus, when we write $\alpha \supset \beta$, for instance, we are actually referring to an infinite number of wffs sharing the same logical form and using what is called a *formula schema*. Along the same line, when proposing an axiom written with metavariables we are, as a matter of fact, proposing an *axiom schema*. To avoid pedantic details, we shall not insist on this distinction unless necessary and will most of the times simply write "wff" or "formula" instead of "wff schema" or "formula schema".

The auxiliary symbols are defined as follows:

1. $\neg \alpha \stackrel{\text{Def}}{=} \alpha \supset \bot$

2. $\alpha \wedge \beta \stackrel{\text{Def}}{=} \neg(\alpha \supset \neg \beta)$

3. $\top \stackrel{\text{Def}}{=} \neg \bot$

4. $\alpha \vee \beta \stackrel{\text{Def}}{=} \neg \alpha \supset \beta$

5. $\alpha \equiv \beta \stackrel{\text{Def}}{=} (\alpha \supset \beta) \wedge (\beta \supset \alpha)$

We now define the notion of *deduction* in an arbitrary system **S** (consisting of axioms and inference rules) in the following way:

Definition 1.2.1 *Given a collection Γ of wffs, a deduction of α from Γ in S is a finite sequence of wffs $\alpha_1 \cdots \alpha_n$ where α is α_n and each one of the α_i ($1 \leq i \leq n$) satisfies one of the following properties:*

1. *α_i is an axiom of S.*

2. *α_i is an element of Γ.*

3. *α_i is a wff derived from the preceding wffs in the sequence by application of one of the inference rules of S.*

In this case we shall also say that α is *deducible* (or *derivable*) from Γ in **S**. If Γ is empty, we will say that α is a *theorem* of **S** and we simply write $\vdash_S \alpha$ to indicate that α has a *proof in* **S** (that is, that there is a deduction of α in **S** from an empty set of wffs).

In particular, we shall write $\Gamma \vdash_{PC} \alpha$ to indicate that there is a deduction (or derivation) of α from Γ in **PC**. The symbol $\vdash$ will be used in place of $\vdash_{PC}$ when this simplification will not yield equivocation, and axioms and theorems of **PC** will be called **PC**-*theses*.

There is also a natural way to refer to the notion of *length* of proofs, by saying that elementary proofs (constituted by axioms only) are proofs of length 1, and every application of rules increases proof length.

The notion of *substituting* a formula β for a variable p inside a formula α (notation: $\alpha[p/\beta]$) means the result of uniformly substituting β for every occurrence of the variable p in α. The case of simultaneous substitution of n variables will be noted by $\alpha[p_0/\beta_0, \cdots p_n/\beta_n]$.

An axiomatic basis for **PC** is as follows:
Axioms:

(**Ax1**) $p \supset (q \supset p)$

(**Ax2**) $(p \supset (q \supset r)) \supset ((p \supset q) \supset (p \supset r))$

(**Ax3**) $(\neg p \supset \neg q) \supset ((\neg p \supset q) \supset p)$

Rules:

(**US**) *Uniform Substitution* : If α is a **PC**-thesis and p is any atomic variable occurring in α, then the wff which results from uniformly substituting β for p (notation: $\alpha[p/\beta]$) is also a **PC**-thesis.

1.2. A NON-MODAL BASIS FOR MODAL LOGICS

(**MP**) *Modus Ponens*: β is deducible from α and $\alpha \supset \beta$ (in symbols, $\alpha, \alpha \supset \beta \vdash \beta$).

The rule (**US**) should not be confused with a derived rule, the rule of *Replacement of Proved Equivalents* (**Eq**), which is stated as follows:

Lemma 1.2.2 *If $\vdash_{PC} \alpha \equiv \beta$ then $\vdash_{PC} C(\alpha) \equiv C(\beta)$, where $C(\beta)$ is as $C(\alpha)$ with the only difference that β replaces α in one or more of its occurrences.*

Proof: Exercise 1.2. ♠

In what follows, we give a list of **PC**-theses which are of common use and are normally identified with a specific name:

Table 1.2.3 *Remarkable **PC**-theses*

1.	$p \supset p$	(Identity)
2.	$p \vee \neg p$	(Excluded Middle)
3.	$\neg \neg p \equiv p$	(Double Negation)
4.	$(p \supset q) \supset (\neg q \supset \neg p)$	(Contraposition)
5.	$(p \supset q) \supset ((q \supset r) \supset (p \supset r))$	(Transitivity)
6.	$(p \wedge q) \supset r \equiv (p \supset (q \supset r))$	(Importation–Exportation)
7.	$(p \supset (q \supset r)) \supset (q \supset (p \supset r))$	(Permutation of Antecedents)
8.	$\bot \supset p$	(Pseudo-Scotus)
9.	$(p \supset q) \supset ((p \wedge r) \supset q)$	(Weakening)
10.	$p \supset (p \vee q)$	(Disjunctive Expansion)
11.	$(p \wedge q) \supset p$	(Simplification)
12.	$p \vee (q \wedge r) \equiv (p \vee q) \wedge (p \vee r)$	(Distribution of $\vee$ over $\wedge$)
13.	$p \wedge (q \vee r) \equiv (p \wedge q) \vee (p \wedge r)$	(Distribution of $\wedge$ over $\vee$)
14.	$(p \wedge q) \equiv \neg(\neg p \vee \neg q)$	(De Morgan's Law I)
15.	$(p \vee q) \equiv \neg(\neg p \wedge \neg q)$	(De Morgan's Law II)
16.	$((p \supset q) \wedge (\neg p \supset q)) \equiv q$	(Proof by Cases)
17.	$((p \supset q) \wedge (p \supset r)) \supset (p \supset (q \wedge r))$	(Composition of Consequents)
18.	$(p \wedge (p \supset q)) \supset q$	(Conditional Elimination)

Starting from **PC**-theses and using (**MP**), one may obtain several useful derived rules and meta-rules. Some remarkable derived (meta)rules are the following:

Table 1.2.4 *Remarkable derived (meta)rules*

1.	$\alpha \vdash \alpha$	(Identity)
2.	$\vdash \alpha \supset \beta, \vdash \beta \supset \gamma$ implies $\vdash \alpha \supset \gamma$	(Transitivity)
3.	$\alpha \vdash \beta$ implies $\Gamma, \alpha \vdash \beta$	(Monotonicity)
4.	$\alpha, \beta \vdash \alpha \wedge \beta$	(Adjunction)
5.	$\alpha, \neg \alpha \vee \beta \vdash \beta$	(Disjunctive Syllogism)
6.	$\alpha \equiv \beta \vdash \neg \alpha \equiv \neg \beta$	(Negation of Equivalents)
7.	$\alpha \wedge (\alpha \supset \beta) \vdash \beta$	(Conditional Elimination)
8.	$\vdash \alpha, \vdash \alpha \supset \beta$ implies $\vdash \beta$	(Secondary Modus Ponens)
9.	$\alpha \vdash \gamma, \beta \vdash \gamma$ implies $\alpha \vee \beta \vdash \gamma$	(Disjunction Introduction)

There is, indeed, an infinite number of derived rules (it should be clear that virtually any **PC**-thesis will give rise to one of such rules). In most cases, we shall refer informally to them, while in other cases we shall represent them in fractional style (e.g. in the case of Negation of Equivalents above, by $\frac{\alpha \equiv \beta}{\neg \alpha \equiv \neg \beta}$).

The proof procedures in **PC** may be highly simplified by using various devices provided by some metatheorems. The first of them is the *Syntactical Deduction Metatheorem* (**SDM**), which asserts what follows:

Proposition 1.2.5 (SDM) *If Γ is a set of wffs and $\Gamma, \alpha \vdash \beta$, then $\Gamma \vdash \alpha \supset \beta$.*

Proof: The proof is by induction on the length of derivations. Let $\beta_1 \cdots \beta_n$ be a derivation of β_n from $\Gamma \cup \{\alpha\}$, where $\beta = \beta_n$. We want to prove that $\Gamma \cup \{\alpha\} \vdash \beta_i$ for every i such that $1 \leq i \leq n$.

- If $i = 1$, then β_i must be either an axiom or a member of Γ or α itself. In each case, it is easy to check that $\Gamma \vdash \alpha \supset \beta_i$ by making use of the axioms of **PC** and the law of Identity $\alpha \supset \alpha$.

- By induction hypothesis, suppose that $\Gamma \vdash \alpha \supset \beta_k$ for every $k < i$. We have to prove $\Gamma \vdash \alpha \supset \beta_i$.
  Now there are only the following possibilities for deriving β_i:
  1. β_i is an axiom.
  2. β_i is a member of Γ.
  3. β_i is α itself.

4. β_i follows via (**MP**) from two premises β_j and $\beta_j \supset \beta_i$.
5. β_i follows via (**US**) from a previous theorem $\vdash \beta_j$.

In the first three cases the argument runs as before.

In case (4) we have, by induction hypothesis, that for some j both derivations hold: $\Gamma \vdash \alpha \supset \beta_j$ and $\Gamma \vdash \alpha \supset (\beta_j \supset \beta_i)$. Thanks to (**Ax2**) and Monotonicity, we have $\Gamma \vdash (\alpha \supset (\beta_j \supset \beta_i)) \supset ((\alpha \supset \beta_j) \supset (\alpha \supset \beta_i))$ and by (**MP**) it follows $\Gamma \vdash \alpha \supset \beta_i$.

In case (5) β_i has been obtained by applying (**US**) to $\vdash \beta_j$, and thus β_i is also a theorem (of the form $\beta_j[p/\gamma]$ for some p and γ). Thus, by (**Ax1**), $\vdash \alpha \supset \beta_i$ holds, and by Monotonicity it follows $\Gamma \vdash \alpha \supset \beta_i$.

♠

The Syntactical Deduction Metatheorem is the key of the syntactical treatment of propositional logic. Given a theorem in implicational form $\vdash \alpha \supset \beta$, via *Modus Ponens* we may conclude $\alpha \vdash \beta$. The (**SDM**) grants that the converse derivation holds. It follows that every axiom in implicative form, for instance $p \supset (q \supset p)$ or the equivalent $p \supset (\neg q \vee p)$, may in principle be replaced by a rule of inference, such as for instance $\alpha \vdash \neg \beta \vee \alpha$.

Given this possibility, there is ground to build a presentation of **PC** by using an empty set of axioms and a suitable stock of inference rules, as it happens in the so-called "Natural Deduction Calculus".

1.3 The semantical analysis of PC

Up to now we have given a syntactical connotation of logical laws, identifying them with provable formulas of the formal system **PC**. But another approach, the semantical one, has been considered by many logicians to be more intuitive. In this alternative approach, logical laws are seen as *tautologies*, namely as formulas whose truth is invariant under any possible value assignment to their propositional atomic variables. This simple idea may receive an exact meaning in the following way.

Let v be a function which assigns to every atomic variable of the language a member of the set $\{0, 1\}$, where 0 and 1 may be read as "False" and "True". The properties of v are defined as follows:

1. For every p, $v(p) = 1$ or $v(p) = 0$.
2. $v(\bot) = 0$.
3. $v(\alpha \supset \beta) = 0$ iff $v(\alpha) = 0$ or $v(\beta) = 1$.

Given that $\neg\alpha$ is defined as $\alpha \supset \bot$, then $\neg\alpha$ has the same semantic conditions of $\alpha \supset \bot$; on the other hand, it is easy to check that $\bot$ has the same semantic conditions of $\neg(\alpha \supset \alpha)$, so we could also take $\neg$ and $\supset$ as primitives and define their truth conditions by the following truth-tables:

$\supset$	1	0
1	1	0
0	1	1

$\neg$	
1	0
0	1

Other derived truth-tables might be formulated for the auxiliary connectives. For instance, $v(\alpha \wedge \beta)$ is 1 if and only if $v(\alpha)$ and $v(\beta)$ are both 1, $v(\alpha \vee \beta)$ is 0 if and only if $v(\alpha)$ is 0 or $v(\beta)$ is 0, $v(\alpha \equiv \beta)$ is 1 if and only if $v(\alpha)$ and $v(\beta)$ are both 1 or both 0.

A *tautology* α is a wff such that $v(\alpha) = 1$ for every value assignment to its atomic variables. We shall write $\vDash \alpha$ to mean that α is a tautology. The notation $\Gamma \vDash \alpha$ means that any value assignment which assigns value 1 to all formulas of Γ also assigns value 1 to α. Of course, $\vDash \alpha$ is the special case of $\Gamma \vDash \alpha$ in which Γ is empty.

The relationship between theorems and tautologies is governed by the following two (meta)theorems, which establish the properties that we call *soundness* and *strong soundness*. Soundness is formulated as follows:

Proposition 1.3.1 *If α is a **PC**-theorem, then α is a tautology.*

Proof: By induction on the length of proofs.
Since axioms are theses whose proofs are of length 1, we have to prove that:

(a) All **PC**-axioms are tautologies.

(b) The **PC**-inference rules preserve tautologicity (i.e., the property of being a tautology).

Step (a) is performed by checking (using truth-tables) that every axiom of **PC** is a tautology. This is a tedious but unproblematic task.
Step (b) is performed by a *Reductio ad Absurdum* reasoning. First, suppose that *Modus Ponens* does not preserve tautologies. Then, for some value assignment to atomic variables we should have α true, $\alpha \supset \beta$ true, and β false, which is impossible. Suppose also by *Reductio* that α is a tautology and some wff $\alpha[p/\beta]$, in which some atomic variable p is uniformly substituted by a wff β, is not a tautology. This is impossible since the truth tables for α and for $\alpha[p/\beta]$ are structurally identical. So, rule (**US**) preserves tautologicity. ♦

1.3. THE SEMANTICAL ANALYSIS OF PC

A variant of soundness is *strong soundness*: if Γ is a set of wffs and α is a single wff, then $\Gamma \vdash \alpha$ implies $\Gamma \vDash \alpha$. **PC** may be proved to be strongly sound.

This proof presupposes the proof of two important properties of **PC** which are *consistency* and *syntactical compactness*.

In general, to say that a (finite or infinite) set of formulas Γ is *consistent* with respect to a logical system **S** (or **S**-consistent) means that **S** does not exclude (from the point of view of deduction) any conjunction $\alpha_1 \wedge \cdots \wedge \alpha_n$ of wffs in Γ: in symbols, $\nvdash_S \neg(\alpha_1 \wedge \cdots \wedge \alpha_n)$ for any $\alpha_1, \cdots, \alpha_n$ in Γ.

Conversely, to say that Γ is *inconsistent* with **S** means to say that $\vdash_S \neg(\alpha_1 \wedge \cdots \wedge \alpha_n)$ for some $\alpha_1, \cdots, \alpha_1$ in Γ.

It can be easily seen (cf. Exercise 1.12) that a set Γ is **S**-consistent if and only if, for any arbitrary wff γ, it does not happen that $\Gamma \vdash_S \gamma$ and $\Gamma \vdash_S \neg \gamma$, or equivalently, $\Gamma \nvdash_S \bot$, which is the same as saying that Γ is deductively non-trivial with respect to **S** (i.e., there exists at least a γ such that $\Gamma \nvdash_S \gamma$).[1] In particular, if **S** is **PC** and Γ is empty, it does not happen, for any γ, that γ and $\neg \gamma$ are both **PC**-theses.

A logical system **S** is *syntactically compact* iff, for every set Γ of formulas, the following holds: $\Gamma \vdash \alpha$ iff there exists a finite $\Delta \subseteq \Gamma$ such that $\Delta \vdash \alpha$.

It is easy to prove that **PC** is consistent: if, for some α, both α and $\neg \alpha$ were **PC**-theorems, they would be both **PC**-tautologies by Proposition 1.3.1, which is impossible. On the other hand, **PC** is also syntactically compact. In fact, every proof of α from a set Γ consists in a finite number of steps, so it involves only a finite number of premises $\Delta \subseteq \Gamma$.

Proposition 1.3.2 *PC is a strongly sound system.*

Proof: Let us suppose $\Gamma \vdash \alpha$. Then by syntactical compactness, there is a finite subset Γ' of Γ such that $\Gamma' \vdash \alpha$. Thus, by the Syntactical Deduction Metatheorem (Proposition 1.2.5), if $\alpha_1 \cdots \alpha_n$ are the only members of Γ', we obtain $\vdash \alpha_1 \supset \cdots (\alpha_n \supset \alpha)$. By the Soundness Theorem (Proposition 1.3.1), this means $\vDash \alpha_1 \supset \cdots (\alpha_n \supset \alpha)$, and by Importation-Exportation $\vDash (\alpha_1 \wedge \cdots \wedge \alpha_n) \supset \alpha$. This means that all value assignments to atomic variables which assign value 1 to members of Γ' also assign value 1 to α, i.e., that $\Gamma' \vDash \alpha$. *A fortiori* this means that all value assignments which assign value 1 to members of Γ also assign value 1 to α, i.e. that $\Gamma \vDash \alpha$. ♠

---

[1] The identification between consistency, non-contradictoriness and deductive non-triviality does not necessarily hold for every logic system. The paraconsistent logics, in particular the *logics of formal inconsistency*, support contradictions without falling into deductive triviality, and retain most properties of classical reasoning (cf. W. A. Carnielli and J. Marcos [CM02] and W. A. Carnielli, M. E. Coniglio and J. Marcos [CCM07]).

1.4 Constructive completeness of PC

An important (meta)theorem is the *completeness* of **PC**, i.e. a result establishing that all **PC**-tautologies are **PC**-theses. Soundness and completeness then jointly state that the semantical and syntactical definition of logical laws identify the same class of wffs. The following proof of completeness makes essential use of (**SDM**). As every wff is equivalent to a wff containing only $\neg$ and $\supset$, we shall presuppose that any α contains only these two connectives; this strategy makes the proof more natural than using $\bot$ and $\supset$ as primitives.

The theorem needs a special lemma (Lemma 1.4.1), for which we introduce the following transformation of an arbitrary formula α, called *rectification under a valuation* (notation: α^*). Given a valuation v for α, α^* is defined as follows:

$$\alpha^* = \begin{cases} \alpha & \text{if } v(\alpha) = 1 \text{ or} \\ \neg \alpha & \text{if } v(\alpha) = 0 \end{cases}$$

The lemma shows that the rectification of the atomic variables of any formula α under a valuation v entails the rectification of α under the same valuation.

Lemma 1.4.1 *Let v be any valuation and $p_1 \cdots p_n$ be the atomic variables occurring in α. Then $p_1^*, \cdots, p_n^* \vdash \alpha^*$*

Proof: By induction on the length of α.

- Suppose $\alpha = p$, for p an atomic variable;
  – If $v(p) = 1$, then $\alpha^* = p^* = p$. By Identity $p \vdash p$, thus $p^* \vdash \alpha^*$.
  – If $v(p) = 0$, then $\alpha^* = p^* = \neg p$ and similarly $p^* \vdash \alpha^*$.

- Induction hypothesis;
  – Suppose that $p_1^*, \cdots, p_n^* \vdash \beta^*$ holds for all formulas β of length $j < n$.

- Suppose $\alpha = \neg \beta$;
  – If $v(\alpha) = 1$, then $\alpha^* = \alpha$, and since $v(\beta) = 0$, $\beta^* = \neg \beta$. By induction hypothesis, $p_i^*, \cdots, p_n^* \vdash \beta^*$. Therefore $p_i^*, \cdots, p_n^* \vdash \alpha^*$ as $\alpha^* = \neg \beta = \beta^*$.
  – If $v(\alpha) = 0$, then $\alpha^* = \neg \alpha$, and as $v(\beta) = 1$, $\beta^* = \beta$. By induction hypothesis $p_i^*, \cdots, p_n^* \vdash \beta^*$. By Double Negation and Monotonicity $p_i^*, \cdots, p_n^* \vdash \beta \supset \neg\neg \beta$, hence by (**MP**) $p_i^*, \cdots, p_n^* \vdash \neg\neg \beta$. since $\alpha^* = \neg \alpha = \neg\neg \beta$, then $p_i^*, \cdots, p_n^* \vdash \alpha^*$.

- Suppose $\alpha = \beta \supset \gamma$;
  – If $v(\alpha) = 1$, then $\alpha^* = \alpha = \beta \supset \gamma$. Consequently, $v(\beta) = 0$ or $v(\gamma) = 1$. In the case $v(\beta) = 0$, then $\beta^* = \neg\beta$. By induction hypothesis, $p_i^*, \cdots, p_n^* \vdash \beta^*$. From **PC**-thesis and Monotonicity, $p_i^*, \cdots, p_n^* \vdash \neg\beta \supset (\beta \supset \gamma)$. Therefore, by (**MP**), $p_i^*, \cdots, p_n^* \vdash \alpha^*$.
  In case $v(\gamma) = 1$, then $\gamma^* = \gamma$. By induction hypothesis, $p_i^*, \cdots, p_n^* \vdash \gamma^*$, that is, $p_i^*, \cdots, p_n^* \vdash \gamma$. By (**Ax1**) and Monotonicity, $p_i^*, \cdots, p_n^* \vdash \gamma \supset (\beta \supset \gamma)$, and by (**MP**) $p_i^*, \cdots, p_n^* \vdash \beta \supset \gamma$. Therefore, $p_i^*, \cdots, p_n^* \vdash \alpha^*$.
  – If $v(\alpha) = 0$, then $v(\beta) = 1$ and $v(\gamma) = 0$, thus $\beta^* = \beta$ and $\gamma^* = \neg\gamma$. By induction hypothesis, $p_i^*, \cdots, p_n^* \vdash \beta^*$ and $p_i^*, \cdots, p_n^* \vdash \gamma^*$. By the derived rule of Composition of Consequents, we have that $p_i^*, \cdots, p_n^* \vdash \beta \wedge \neg\gamma$. Since $v(\alpha) = 0$, then $\alpha^* = \neg\alpha = \neg(\beta \supset \gamma)$, but by definition $\beta \wedge \neg\gamma$ is $\neg(\beta \supset \gamma)$. Therefore, $p_i^*, \cdots, p_n^* \vdash \alpha^*$.

♠

The completeness theorem is then proved as follows:

Proposition 1.4.2 *If α is a tautology, then α is a thesis of* **PC**.

Proof: Let α be a tautology and $p_1 \cdots p_n$ be the atomic variables occurring in α. Then α^* is α. By Lemma 1.4.1, $p_1^* \cdots p_n^* \vdash \alpha$ for any valuation v. Let v_1 and v_2 be two valuations which agree on all atomic variables but not on p_n, that is $v_1(p_n) = 1$ and $v_2(p_n) = 0$. This means that $p_1^* \cdots p_{n-1}^*, p_n \vdash \alpha$ and $p_1^* \cdots p_{n-1}^*, \neg p_n \vdash \alpha$. But by the Deduction Theorem (Proposition 1.2.5), $p_1^* \cdots p_{n-1}^* \vdash p_n \supset \alpha$ and $p_1^* \cdots p_{n-1}^* \vdash \neg p_n \supset \alpha$. So by Proof by Cases $((p \supset q) \wedge (\neg p \supset q)) \equiv q$ we reach the conclusion $p_1^* \cdots p_{n-1}^* \vdash \alpha$. By applying the same argument a finite number of times and by discharging the n atomic variables one by one we obtain α as a **PC**-theorem. ♠

Just as we distinguished a variant of soundness which we called "strong soundness", we may distinguish a variant of completeness which may be called *strong completeness*:

Proposition 1.4.3 $\Gamma \vDash \alpha$ *only if* $\Gamma \vdash \alpha$

Proof: Exercise 1.4. ♠

So, **PC** is a strongly complete system. The reader should be advised that strong completeness does not coincide with another property called Post-completeness (see Section 1.6), which is also enjoyed by **PC**.

1.5 Decidability of PC

The propositional calculus **PC** has the remarkable property of being decidable. In other words, there exists a mechanical procedure that establishes, for every **PC**-wff α, whether α is a **PC**-theorem or not (or equivalently, in view of completeness of **PC**, whether α is a tautology or not).

The standard procedure is called the truth-table method: it consists in writing all the truth-value assignments to the atomic variables of α in tabular form and in calculating the value of the subformulas of α moving from the most simple ones to the most complex ones, until ending with α itself. If α receives value 1 for every assignment, α is a tautology, and by the Completeness Theorem (Proposition 1.4.2), this means that α is a **PC**-thesis. If some truth assignment gives value 0 to α, then α is not a tautology and consequently is not a **PC**-thesis.

The so-called *semantic tableaux* method is derived from the truth-table method, but normally is a handier procedure. Let α be again a wff under test (call it an *input formula*) and let us suppose by *Reductio ad Absurdum* that α is not a tautology, which means that we are allowed to assign truth-value 0 to α as as result of some possible value assignment to its variables. As a consequence of this assignment, we are able to give (consequential) assignments to all subformulas of α, say $\beta_1 \cdots \beta_n$, beginning from the longest ones (the so-called immediate subformulas) and going on with the same method.[2] It is understood that $\perp$ always receive value 0.

In calculating the corresponding values for some subformula γ, there are two possibilities:

1. The consequential assignment is univocal (e.g. $v(\alpha \supset \beta) = 0$ implies $v(\alpha) = 1$ and $v(\beta) = 0$), or

2. The consequential assignment is non-univocal (e.g. $v(\alpha \supset \beta) = 1$ implies either $v(\alpha) = v(\beta) = 0$, $v(\alpha) = 0$ and $v(\beta) = 1$, or $v(\alpha) = v(\beta) = 1$. This obviously amounts to $v(\alpha) = 0$ or $v(\beta) = 1$).

In case every assignment is univocal, the tableau is called *sequential* while, in the second case, it yields two or more parallel *alternatives*. From the graphical viewpoint, the case of alternatives implies that different diagrams are to be worked out with all possible consequential assignments.[3]

---

[2] For a rigorous definition of immediate subformula see Definition 2.3.2.
[3] Semantic tableaux may also be visualized as *semantic trees*. The graphical representation used in this book is however more suitable to be extended to modal logics, as we will see in Section 3.4.

1.5. DECIDABILITY OF PC

The procedure runs until one of the following two situations is reached:

(a) In all the examined alternatives, it so happens that some subformula of the input formula α receives incompatible truth-values (or $\bot$ receives value 1). This proves that the supposition that α is not a tautology is mistaken and, consequently, α is a tautology.

(b) At least one of the examined alternatives does not yield any contradictory assignment. This implies that the supposition is correct and that α is not a tautology.

Given that α has a finite number of subformulas and that there is a finite number of alternatives, in both cases the procedure ends in a finite number of steps.

In order to apply the tableau method, it is better to follow some heuristic rules of thumb. For instance:

1. In assigning truth-values, always proceed from left to right.

2. If you have to consider more than one alternative, i.e. more than one consequential value, choose the ones which have univocal consequences first, and then the others.

3. After attributing a value to a subformula β, check whether β occurs in other parts of the formula. If so, report the same value given to β in all of its occurrences.

4. Check the compatibility (consistency) of the obtained assignments at any step.

5. Stop the procedure when the full analysis of some alternative does not yield an inconsistent assignment. It is useless, in fact, to check any other alternative if the wff under test has already been proved not to be a tautology.

It is useful to give graphical evidence of the wff (or wffs) where an inconsistent assignment appears by underlying them inside the diagram. It should be clear that the same method of proof, formulated for wffs containing only $\supset$ and $\bot$, will apply *mutatis mutandis* to wffs containing $\neg, \wedge, \vee$ and $\equiv$.

Example 1.5.1 *Consider the following formula $(p \vee q) \supset (p \wedge q)$ with truth-value 0, noticing that the consequential assignment to the immediate subformulas is univocal:*

$$\mathcal{T} : \boxed{\begin{array}{ccc} 1 & 0 & 0 \\ (p \vee q) & \supset & (p \wedge q) \end{array}}$$

Here, given that the consequential assignments of value 1 to $p \vee q$ and value 0 to $p \wedge q$ are not univocal, we have to examine three alternative diagrams – let us call them $\mathcal{T}_1, \mathcal{T}_2, \mathcal{T}_3$ – which are built by making explicit the three different assignments compatible with $v(p \vee q) = 1$. In performing the next steps, we will try to apply the rules of thumb.

- $v(p) = 1, v(q) = 1$:

$$\mathcal{T}_1 : \boxed{\begin{array}{ccccccc} 1 & 1 & 1 & 0 & 1 & 0 & 1 \\ (p & \vee & q) & \supset & (p & \wedge & q) \end{array}}$$

Here the wff $(p \wedge q)$ receives values 0 while its conjuncts receive value 1, a contradiction.

- $v(p) = 1, v(q) = 0$:

$$\mathcal{T}_2 : \boxed{\begin{array}{ccccccc} 1 & 1 & 0 & 0 & 1 & 0 & 0 \\ (p & \vee & q) & \supset & (p & \wedge & q) \end{array}}$$

The consequential values assigned in this case do not yield any contradiction. So the wff under test is not a tautology, as there is a value assignment to the variables which makes it false. This conclusion makes useless to consider the third diagram $\mathcal{T}_3$ in which $v(p) = 0$ and $v(q) = 1$.

Example 1.5.2 *Consider the following formula $(p \wedge q) \supset (p \vee q)$:*

$$\mathcal{T} : \boxed{\begin{array}{ccc} 1 & 0 & 0 \\ (p \wedge q) & \supset & (p \vee q) \end{array}}$$

Here the consequential assignments are univocal, giving rise to:

$$\mathcal{T}' : \boxed{\begin{array}{ccccccc} 1 & 1 & 1 & 0 & 1 & 0 \\ (p & \wedge & q) & \supset & (p & \vee & q) \end{array}}$$

1.6 Post-completeness and other properties of PC

The value 1 *to p in p* ∧ *q is reported inside p* ∨ *q. This ends the procedure as p* ∨ *q has both values* 0 *and* 1, *a contradiction.*

1.6 Post-completeness and other properties of PC

Among the many theorems which describe the properties of **PC**, a very peculiar one is the theorem of *Reduction to Conjunctive Normal Form*. Let us call *literal* an atomic variable or the negation of an atomic variable, and *atomized disjunction* a disjunction whose disjuncts are literals. We will say that a wff α is in *Conjunctive Normal Form* (CNF) when it is a conjunction of atomized disjunctions. α is then in CNF if and only if α has the form $\beta_1 \wedge \beta_2 \wedge \cdots \wedge \beta_n$, where each β_i has the form $\beta_i = l_1 \vee l_2 \vee \cdots \vee l_m$, where each l_j for $1 \leq j \leq m$ is a literal. The statement of the theorem is as follows:

Proposition 1.6.1 *Every wff α is equivalent in **PC** to a wff α' in CNF.*

Proof: We just sketch the main argument here, considering that a rigorous proof of this theorem may be completed by the reader by using the following remarkable **PC**-theses: De Morgan's laws, Distributivity of ∧ on ∨ and of ∨ on ∧ and Double Negation.

Furthermore, we may perform replacement of equivalent wffs by using the rule (**Eq**). The procedure to reduce any α into CNF consists of the following steps:

(a) All the connectives are eliminated by suitable definitions with the exception of ¬, ∨, ∧.

(b) Whenever $2n + m$ negations occur at the left of some wff, they are reduced to m negations by applying Double Negation laws, i.e., by iterating the application of $\neg\neg\alpha \equiv \alpha$.

(c) By using De Morgan and Distributive laws, we find that conjunctions within disjunctions are transformed into disjunctions within conjunctions, and vice versa.

By iterating this procedure, in a finite number of steps we reach the required form. ♠

An example of the procedure may be useful here.

Example 1.6.2 *The wff* $\neg((\neg p \supset \neg q) \supset (\neg\neg p \wedge q)) \vee r$ *is transformed in CNF thanks to the following steps:*

(a) $\neg((\neg p \supset \neg q) \supset (\neg\neg p \wedge q)) \vee r$

(b) $(\neg(\neg p \wedge \neg\neg q) \wedge \neg(\neg\neg p \wedge q)) \vee r$ \hfill *(Eliminating $\supset$)*

(c) $(\neg(\neg p \wedge q) \wedge \neg(p \wedge q)) \vee r$ \hfill *(Eliminating double negations)*

(d) $((\neg\neg p \vee \neg q) \wedge (\neg p \vee \neg q)) \vee r$ \hfill *(Using De Morgan laws)*

(e) $((p \vee \neg q) \wedge (\neg p \vee \neg q)) \vee r$ \hfill *(Eliminating double negations)*

(f) $(p \vee \neg q \vee r) \wedge (\neg p \vee \neg q \vee r)$ \hfill *(Distributivity)*

It is worth remarking that the theorem of reduction to CNF grants another completeness proof which, differing from the proof of Proposition 1.4.2, does not depend on the Syntactical Deduction Metatheorem. The basic idea of the argument may be sketched as follows.

Let α be a tautology. Then Proposition 1.6.1 implies that α is equivalent to a tautology α' which is in CNF, i.e., it is a conjunction of atomized disjunctions γ_i which must, in turn, be tautologies (as, for every α and β, $\vDash \alpha \wedge \beta$ iff $\vDash \alpha$ and $\vDash \beta$). Now we can see that the general form of every such tautological disjunction γ_i is $p \vee \neg p \vee \delta$, where p is some atomic variable and δ is an atomized subdisjunction of γ_i. It is easy, in fact, to prove that if all atomic variables p in γ_i were such that p (but not $\neg p$) is among the disjuncts of γ_i, then there is some assignment to them which gives value 0 to γ_i, so that γ_i would not be a tautology.

Now, given that we have every instance of axiom (**Ax1**) $p \supset (p \vee q)$ among the theorems of **PC** and that $p \vee \neg p$ is a **PC**-theorem, then every such disjunction γ_i, being of form $p \vee \neg p \vee \delta$, is provable as a **PC**-theorem. By Adjunction, then, $\gamma_i \wedge \gamma_j$ is also a **PC**-theorem for every γ_i and γ_j. Since α is equivalent to a conjunction of disjunctions $\gamma_1, \cdots, \gamma_n$, then α is also provable as a **PC**-theorem. This will establish the proposed completeness argument.

Let us conclude this chapter by considering a property which positively characterizes **PC**. This property, called *Post-completeness*, is defined in terms of the notion of extension. A system **S'** is an *extension* of **S** if all theses of **S** are theses of **S'**, and **S'** is a *proper extension* of **S** if it is an extension of **S** and does not coincide with **S**. An extension is consistent (as already defined) if it does not contain contradictory theses.

Definition 1.6.3 *A system S is Post-complete iff it has no proper consistent extensions.*

Otherwise said, any proper axiomatic extension of a Post-complete system **S** is necessarily inconsistent. We may now prove:

Proposition 1.6.4 *PC is a Post–complete system.*

Proof: Let α be any wff which is not a **PC**-thesis. By the semantic completeness of **PC**, α is not a tautology. As α is equivalent to a wff α' which is in CNF, this means that at least a conjunct γ_j of α' is not a tautology. If α were added to **PC** as an axiom, then γ_j could be derived from α as a theorem of the extended system. But γ_j is an atomized disjunction whose form is $p \vee q \vee \neg r \vee \cdots \vee s \vee \neg t$, where no disjunct is the negation of any other disjunct.

Suppose then, by *Reductio ad Absurdum*, that such a disjunction were a **PC**-thesis. As **PC** is closed under (**US**), then we might put a variable, say p, in place of all the non negated atomic wffs, and its negation $\neg p$ in place of the negated atomic wffs, so as to obtain a theorem $p \vee p \vee \neg\neg p \vee \cdots \vee p \vee \neg\neg p$. However, any disjuction with this form is equivalent to p, and from p the absurd formula $\bot$ may be derived by (**US**). It is thus proved that **PC**, extended with any non-tautology α, yields an inconsistent system. ♠

It is interesting to remark that Post-completeness is a property of **PC** but not of first-order logic and, as we shall show in Section 3.4, it is neither a property of any non-trivial modal logic. This helps to identify an aspect in which propositional reasoning differs from modal and quantificational reasoning. The next chapters will enlighten several converging or diverging features of modal and quantificational reasoning.

1.7 Exercises

1. Some of the following intuitively valid arguments are not representable in the propositional language of **PC**. Which?

    (a) All cats are sly, so my cat is sly.
    (b) If it necessarily rains, it necessarily rains.
    (c) If necessarily every material thing is heavy, then every material thing is heavy.
    (d) If it is allowed to rob, it is allowed to rob or it is allowed to be silent.
    (e) If it is allowed to rob, it allowed to rob or to be silent.

(f) All philosophers are donkeys, Socrates is a philosopher, so Socrates is a donkey.

(g) It is possible that only two distinct objects exist in the universe.

(h) If some angel can dance on the point of a pin, every angel can do it.

2. Prove Lemma 1.2.2. (Hint: induction on the complexity of wffs – the proof given in Proposition 2.3.10 can be easily adapted).

3. Give a syntactic proof for the **PC**-theses in Table 1.2.3 from axioms (**Ax1**)-(**Ax3**). (Hint: use the Syntactical Deduction Metatheorem proved in Proposition 1.2.5).

4. Prove the Strong Completeness Theorem of **PC** formulated as Proposition 1.4.3. (Hint: use compactness).

5. Prove the derived rules listed at Table 1.2.4 by deriving them from some corresponding **PC**-theorem.

6. Use the tableaux method to test the **PC**-validity of the following wffs:

    (a) $p \supset ((p \vee q) \wedge (r \vee q))$
    (b) $((p \supset q) \supset q) \supset q$
    (c) $(p \wedge r \wedge s) \supset (p \vee q)$
    (d) $((p \vee r \vee s) \supset q) \supset (p \supset q)$
    (e) $(r \supset s) \supset ((r \wedge (q \vee t)) \supset s)$

7. Identify the **PC**-valid wffs of the list reported in Exercise 1.6 and convert them into CNF.

8. Given a truth-table for some arbitrary wff α with atomic variables $p_1, \cdots, p_n$; suppose that α has value 1 in one and only one of the rows of the table. Then construct a conjunction $p_1^* \wedge \cdots \wedge p_n^*$ where p_i^* is the rectification of p_i, that is, p_i^* is p_i if p_i receives value 1 in that row, and $\neg p_i$ if p_i receives value 0. Prove that $p_1^* \wedge \cdots \wedge p_n^*$ is true if and only if α is true.

9. Generalize the result of Exercise 1.8 to the case in which α has value 1 in more than one row of the table. Show that α is equivalent to a formula which is the disjunction of all the wffs α built with the method exposed in Exercise 1.8 and in correspondence with the rows in which α has value 1.

1.7. EXERCISES

10. The wff $\alpha : (p \vee q \vee \neg r) \wedge (p \vee r)$ is in CNF. By applying Double Negation, De Morgan and Distribution Theorem, convert α into DNF, i.e. into a disjunction of conjunctions of literals.

11. Besides the usual connectives, other connectives can be defined starting from truth-tables. Let NOR be the connective whose semantic properties are defined by the following truth-table:

NOR	1	0
1	0	0
0	0	1

    Show that NOR defines all other connectives, and can be defined in terms of $\neg$ and $\vee$.

12. Prove that the following properties are equivalent for any **S** which extends **PC**:

    (a) Γ is **S**-consistent.
    (b) There is no wff γ such that $\Gamma \vdash_S \gamma$ and $\Gamma \vdash_S \neg \gamma$.
    (c) $\Gamma \nvdash_S \bot$.
    (d) Γ is deductively non-trivial with respect to **S**, i.e., there exists at least a wff γ such that $\Gamma \nvdash_S \gamma$.

13. A set X of connectives is a *complete set* if all other connectives can be defined from the elements of X (for instance, $X = \{\neg, \vee\}$ is a complete set, cf. Exercise 1.11). A connective is a *Sheffer connective* if it defines all other connectives, that is, if the singleton whose only element is this connective is a complete set (for instance, NOR is a Sheffer connective, cf. Exercise 1.11). Let $NAND$ be the connective whose semantic properties are defined by the following truth-table:

NAND	1	0
1	0	1
0	1	1

    (a) Show that $NAND$ is a Sheffer connective.
    (b) Show that NOR and $NAND$ are the only Sheffer connectives.
    (c) Show that the following pairs are complete sets of connectives: $\{\neg, \wedge\}, \{\neg, \supset\}$.

(d) Show that there are no other pairs of complete sets of connectives besides $\{\neg, \vee\}, \{\neg, \wedge\}$ and $\{\neg, \supset\}$.

14. Show that our axioms (**Ax1**)–(**Ax3**) can be replaced by the following, due to Jan Łukasiewicz, preserving rules (**MP**) and (**US**) (Hint: for one direction, use the Syntactical Deduction Metatheorem as in Proposition 1.2.5).

    L1. $(\neg p \supset p) \supset p$
    L2. $p \supset (\neg p \supset q)$
    L3. $(p \supset q) \supset ((q \supset r) \supset (p \supset r))$

15. Prove the Craig Interpolation Lemma (propositional form), i.e, prove what follows. Let φ and ψ be any **PC**-formulas; if $\varphi \vdash \psi$, φ is not a contradiction and ψ is not a tautology, then there exists a formula ρ (called *interpolation formula*) such that $\varphi \vdash \rho$ and $\rho \vdash \psi$, and such that all propositional variables occurring in ρ occurs simultaneously in φ and in ψ. Hint: since φ is not a contradiction and ψ is not a tautology, given that $\varphi \vdash \psi$ show that:

    (a) If $Var(\varphi)$ and $Var(\psi)$ are the sets of propositional variables of φ and ψ respectively, $Var(\varphi)$ and $Var(\psi)$ are not disjoint.

    (b) Consequently, as $Var(\varphi) \cap Var(\psi) \neq \emptyset$, let $p_1, p_2, \cdots, p_n$ be the propositional variables that occur in $Var(\varphi) \cap Var(\psi)$, and let $p_{n+1}, p_{n+2}, \cdots, p_m$ be the variables of $Var(\varphi)$ which do not occur in $Var(\psi)$ (if any). Consider the contradiction $\bot = p_1 \wedge \neg p_1$ and the tautology $\top = p_1 \vee \neg p_1$ (notice that at least p_1 belongs to $Var(\varphi) \cap Var(\psi)$).

    (c) Now construct an interpolation formula ρ as:
    $\rho = \bigvee \varphi[x_1/p_{n+1}, x_2/p_{n+2}, \cdots, x_m/p_m]$
    for every substitution where $x_i \in \{\bot, \top\}$. Show that $Var(\rho) \subseteq Var(\varphi) \cap Var(\psi)$ and that $\varphi \vdash \rho$ and $\rho \vdash \psi$.

1.8 Further reading

What is called classical propositional logic is a research subject which is 24 centuries old, and which has proved to be surprisingly fecund even in recent years. **PC**, in fact, has strong connections to algebra, a route started by George Boole in the 19th century, and also has sharp relations to computer

1.8. FURTHER READING

science: in fact, the seminal work of S. Cook in [Coo71] showed that the problem (called *SAT* problem) of determining whether a **PC**-formula α is satisfiable (that is, if there exists a valuation v such that valuation $v(\alpha) = 1$) is *NP*-complete, i.e. it can represent the entire class of *NP*-problems.

The term *NP* ("nondeterministic polynomial")-problems denotes the class of problems which could be solved by a nondeterministic computer in an amount of time which varies as a polynomial function in the size of the input.

If a method could be devised to solve *SAT* in (deterministic) polynomial time, then every problem in *NP* could be solved in polynomial time. Thus, *SAT* is the "hardest" *NP*-problem: if *SAT* can be solved in polynomial time, then everything in *NP* can be solved in polynomial time. To decide whether or not such an algorithm for *SAT* exists is one of the most difficult problems in computer science.

For other features of **PC**, with regard to recursive functions and decidability, see R. L. Epstein and W. A. Carnielli [EC00].

A complete or selected bibliography about classical propositional logic would be of little use in the present context. It suffices to recall that the axiomatic method employed here is inspired by B. Russell's and A. Whitehead's *Principia Mathematica* (1910–1913) and has been dominating until recent times (see for instance E. Mendelson [Men64]).

As non-axiomatic methods are concerned, for the so called natural deduction method an important reference is still D. Kalish and R. Montague [KM64], while for so-called analytic tableaux method, not to be confused with the "semantic tableau" procedure used in the present book, the standard reference is R. Smullyan [Smu68].

Natural deduction is at the root of so-called Proof Theory, for which a basic reference is D. Prawitz [Pra65]. For a useful history of this trend of inquiry see F. J. Pelletier [Pel00].

The system of natural deduction used by G. Gentzen to prove the basic theorem known as Gentzen's Hauptsatz (1934) was actually different from the standard one, being based on the notion of a sequent, i.e. of a syntactic object of form $A_1, \cdots, A_n \vdash B_1, \cdots, B_k$. The sequent calculus is still a useful tool in advanced research, for instance in mechanical theorem proving. For a textbook of logic based on sequent calculus see A. B. Manaster [Man75].

The completeness result proved in Section 1.4 is derived from the original proof given by L. Kalmár in [Kal35].

The relation between standard and modal operators has been an object of both technical and philosophical investigations. A rigorous treatment of classical logic seen as a foundation to modal logic is offered by K. Segerberg [Seg82], while for a discussion of the subject in a philosophical framework a useful reference is still S. Haack [Haa74].

The controversy about the legitimacy of modal logic was originated in the 1940s by W. v. O. Quine's provocative claim that modal logic is both technically useless and philosophically suspicious (since, in his view, it is intertwined with pre-scientific essentialism). The literature on this topic is variegated. Among the many papers devoted by Quine to this subject, the most known is "Three grades of modal involvement" of 1953 (now in [vOQ66]). For the replies to Quine's skepticism, it is enough to mention two collections: L. Linsky [Lin71] and P. W. Humphreys and J. Fetzer [HF98]. Since a central question under discussion is the relation between quantifiers and modal operators, the mentioned essays could be very useful when read jointly with the content of Chapter 9.

Chapter 2
The syntax of normal modal systems

2.1 The relationship among modal operators

Aristotle's *Organon*, the book that prevailed in the logical culture until the Modern Age, contains the first known treatment of modal logic. In that work, modal syllogisms receive as much attention as the non-modal ones, which Aristotle calls *categorical*. The distinction between the modal and the categorical syllogisms is that the former are valid not only due to the meaning of terms like "every" and "some", but also because of the meaning of terms which refer to the basic modal notions: *necessary, possible, impossible* and *contingent*. It has been established that Aristotle's interest for modalities was motivated by his metaphysical convictions. In particular, modal logic allowed him to analyze the double distinction which will have been inherited by all philosophers who were inspired by his philosophy: the distinction between actual (real) and potential (possible) properties on the one hand, and the distinction between essential (necessary) and accidental (contingent) properties on the other.

In the Middle Ages, the logical relationship between the modal notions was visualized by means of a didactic device called *Aristotle's square* (see Figure 2.1).

In the diagonally opposite vertices of the square, one finds mutually contradictory statements (that is, statements such that if one statement is true, the other is false); in the collinear upper vertices one finds contrary statements (that is, statements such that they cannot be jointly true, but can be

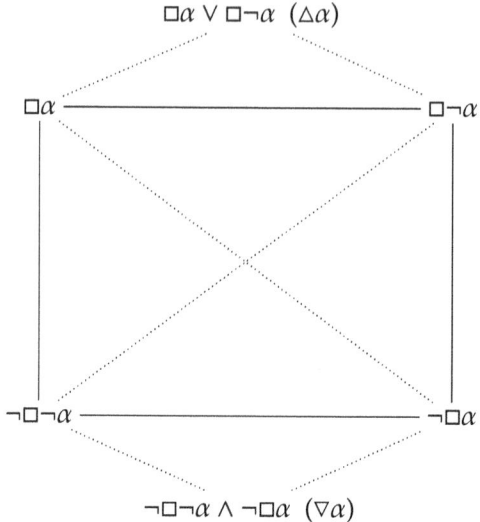

Figure 2.1: Aristotle's square

jointly false), whereas in the collinear lower vertices one finds subcontrary statements (those that cannot be jointly false, but can be jointly true).

The possible, in intuitive terms, corresponds to what is not necessarily false. Employing the already introduced symbol ◊, the notion of possibility can be defined as follows:

Definition 2.1.1 *(Def◊)*: $\quad \Diamond \alpha \stackrel{Def}{=} \neg \Box \neg \alpha.$

As anticipated in Chapter 1, modal logics are conveniently seen as linguistic and axiomatic extensions of classical propositional logic, **PC**, introduced in Section 1.2. Let us call **PC**$^\Box$ the linguistic extension of **PC** resulting from the addition of the operator □ to the language of **PC** and consequently, admitting □α, □□α, □α ⊃ □β, ¬□β ∧ α etc., as new well-formed formulas. If we further assume *(Def◊)*, it is straightforward to see that ◊ and □ are *dual* operators, i.e. that the relation that holds between them is analogous to the one that holds between the existential and the universal quantifiers in the usual predicate logic. In fact, on the grounds of the laws of truth-functional logic and *(Def◊)*, ¬◊α is equivalent to ¬¬□¬α and, by the laws of double

2.1. THE RELATIONSHIP AMONG MODAL OPERATORS

negation (see Section 1.2), it is also equivalent to $\Box\neg\alpha$. Therefore, the following equivalence holds for every α, where $\vdash_{PC^\Box}$ denotes theoremhood in $PC^\Box$:

1. $\vdash_{PC^\Box} \Box\neg\alpha \equiv \neg\Diamond\alpha$.

By applying the truth-functional laws of Section 1.2, we also obtain the following equivalences for every α:

2. $\vdash_{PC^\Box} \Box\alpha \equiv \neg\Diamond\neg\alpha$,   by (1), $\vdash_{PC^\Box} \neg\neg\alpha \equiv \alpha$ and (**Eq**).

3. $\vdash_{PC^\Box} \neg\Box\alpha \equiv \Diamond\neg\alpha$,   by the rule $\frac{\alpha \equiv \beta}{\neg\alpha \equiv \neg\beta}$ in (2) and $\vdash_{PC^\Box} \neg\neg\alpha \equiv \alpha$.

The equivalence $\Diamond\alpha \equiv \neg\Box\neg\alpha$ deduced from $(Def\Diamond)$ and the equivalences (1)–(3) are called *laws of $\Box\Diamond$-interchange*. It should be clear that, provided we have at our disposal the ordinary truth-functional connectives, we would obtain the same laws taking the modal operator $\Diamond$ for possibility as primitive, and defining the operator $\Box$ for necessity by $\neg\Diamond\neg\alpha$. Therefore, it is indifferent to take either the necessity or the possibility modal operator as primitive.

The notion of *contingency* deserves a special treatment. In Aristotle's works this notion is sometimes identified with the notion of possibility and sometimes with the notion of non-necessity, but in the philosophic tradition it is considered to be contingently true what is neither necessarily true nor necessarily false (in our language, $\neg\Box\alpha \land \neg\Box\neg\alpha$, equivalent to $\Diamond\neg\alpha \land \Diamond\alpha$). Strictly speaking, this notion and its negation (that is, non-contingency) do not belong to Aristotle's square, but to a hexagonal extension of it (see Figure 2.1). It is convenient to introduce two appropriate symbols for contingency and for non-contingency:

Definition 2.1.2 $(Def\nabla)$: $\nabla\alpha \stackrel{Def}{=} \Diamond\alpha \land \Diamond\neg\alpha$

Definition 2.1.3 $(Def\triangle)$: $\triangle\alpha \stackrel{Def}{=} \neg\nabla\alpha$

From $(Def\triangle)$, it follows that $\triangle\alpha$ is equivalent to $\Box\alpha \lor \Box\neg\alpha$. The most remarkable property of contingency is that the equivalence $\triangle\alpha \equiv \triangle\neg\alpha$ holds, and therefore $\nabla\alpha \equiv \nabla\neg\alpha$ also holds. As it can be seen from Figure 2.1, $\nabla\alpha$ implies both formulas in the lower vertices of the square, whereas $\triangle\alpha$ is implied from the formulas in the upper vertices of the square.

It is useful at this point to define the concepts of *modal function* and *modal degree*. Each formula containing a modal operator is said to be a modal function of its atomic variables. For example, the formula $q \lor \Diamond p$ is a modal function of the atomic variables p and q. The modal degree of a formula is the maximum number of iterated modal operators (or operators in the scope of modal operators) found in any subformula of that formula. For example, $\Diamond\Box(p \lor \Diamond r) \supset \Box t$ is a modal function of the variables p, r and t, which contains a subformula with three iterated operators, i.e., $\Diamond\Box(p \lor \Diamond r)$. As any other subformula has a lower modal degree, the formula itself has modal degree 3. Considering $\bot, \supset$ and $\Box$ as primitives, the formal definition of *modal degree* dg is the following:

(i) If α is a propositional variable or $\bot$, $dg(\alpha) = 0$

(ii) $dg(\alpha \supset \beta) = max\{dg(\alpha), dg(\beta)\}$

(iii) $dg(\Box\alpha) = dg(\alpha) + 1$

As $\neg\alpha$ is defined as $\alpha \supset \bot$, then $dg(\neg\alpha) = dg(\alpha)$.

Coming back to contingency, we can say that the formula $\Diamond\alpha \land \Diamond\neg\alpha$ (equivalent to $\nabla\alpha$) has the same modal degree as $\Diamond\alpha$ and as $\Diamond\neg\alpha$. However, the notions of contingency and non-contingency, even if displaying the same modal degree of others, are more complex than the notions which occur in the Aristotle's square. Considering that each formula in the square has minimal complexity, the concepts of contingency and non-contingency, being defined upon them, have thus a higher degree of complexity. In principle, it is indeed possible to define an arbitrary number of modal notions starting from (and thus being more complex than) the ones in the square, even if some of them lack a corresponding name in natural language. We could define, for example, a new modal operator as:

Definition 2.1.4 *(Def⊟):* $\boxminus\alpha \stackrel{Def}{=} \Diamond\alpha \lor \Diamond\neg\alpha \lor \Box\alpha \lor \Box\neg\alpha$.

This operator corresponds intuitively to the notion of "having one of the four possible modal status".[1] Another definition expressing a notion of necessity stronger than the usual one is the following:

Definition 2.1.5 *(Def⊞):* $\boxplus\alpha \stackrel{Def}{=} \Box\alpha \land \Diamond\alpha$.

The interdefinability of $\Box$ and $\Diamond$ suggests the following problem: is it possible to consider ∇ and $\triangle$ as primitive operators, and to define $\Box$ in terms of ∇ or $\triangle$? In order to discuss this problem, we have to go beyond **PC$^\Box$**. Let us take

---
[1] Note that $\boxminus\alpha$ is a **PC$^\Box$**-tautology, for every α.

2.1. THE RELATIONSHIP AMONG MODAL OPERATORS

the system $\mathbf{PC^\square}$ as our basis and accept for sake of discussion the following principle according to which "Every α which is necessarily true is true":

(T) $\square\alpha \supset \alpha$

As (T) is not a theorem of $\mathbf{PC^\square}$, it can be taken as a new axiom schema, and what results is a new system to be called $\mathbf{PC^\square}$+(T). As (T) is equivalent to $\neg(\square\alpha \wedge \neg\alpha)$, the negation of (T), $\square\alpha \wedge \neg\alpha$, is inconsistent with $\mathbf{PC^\square}$+(T). Thus, by standard propositional reasoning, $\square\alpha \wedge \neg\alpha$ can be proven to be equivalent to the contradiction $\alpha \wedge \neg\alpha$, or, which is the same, equivalent to $\bot$ (the reader is invited to go back to Section 1.2). As $\square\alpha \supset \alpha$ is, for every α, a thesis of $\mathbf{PC^\square}$+(T), from the laws of $\mathbf{PC^\square}$ (using here $\vdash_{\mathbf{PC^\square}} (\alpha \supset \beta) \supset (\alpha \supset (\alpha \wedge \beta))$ and $\vdash_{\mathbf{PC^\square}} (\alpha \wedge \beta) \supset \alpha$), it turns out that the following equivalent formula is also a thesis of $\mathbf{PC^\square}$ + (T) for every α:

(T') $(\square\alpha \wedge \alpha) \equiv \square\alpha$

Therefore, supposing that in $\mathbf{PC^\square}$+(T) the rule of Replacement of Proved Equivalents (Eq) still holds, we can deduce in $\mathbf{PC^\square}$+(T) a sequence of equivalences whose last step is $\square\alpha \equiv (\alpha \wedge \triangle\alpha)$. As usual, justifications appear at the right side:

1. $\square\alpha \equiv (\top \supset \square\alpha)$ $[\mathbf{PC^\square}]$
2. $\square\alpha \equiv (\top \supset (\alpha \wedge \square\alpha))$ $[(\mathbf{T'}), (\mathbf{Eq})$ in 1$]$
3. $\square\alpha \equiv (\bot \vee (\alpha \wedge \square\alpha))$ $[\mathbf{PC^\square}, (\mathbf{Eq})$ in 2$]$
4. $(\alpha \wedge \square\neg\alpha) \equiv \bot$ $[\mathbf{PC^\square}+ (\mathbf{T})]$
5. $\square\alpha \equiv ((\alpha \wedge \square\neg\alpha) \vee (\alpha \wedge \square\alpha))$ $[(\mathbf{Eq})$ in 4 and 3$]$
6. $((\alpha \wedge \square\neg\alpha) \vee (\alpha \wedge \square\alpha)) \equiv (\alpha \wedge (\square\alpha \vee \square\neg\alpha))$ $[\mathbf{PC^\square}]$
7. $\square\alpha \equiv (\alpha \wedge (\square\alpha \vee \square\neg\alpha))$ $[(\mathbf{Eq})$ in 6 and 5$]$
8. $\square\alpha \equiv (\alpha \wedge \triangle\alpha)$ $[\text{Def. }\triangle, (\mathbf{Eq})$ in 7$]$

Of course, by applying the definition of possibility and by trivial transformations, we shall also obtain the equivalence $\Diamond\alpha \equiv (\alpha \vee \nabla\alpha)$. Also, from line 8 we can obtain (T) again simply as:

1. $\square\alpha \supset (\alpha \wedge \triangle\alpha)$ $[\text{Line 8}, \mathbf{PC^\square}]$
2. $(\alpha \wedge \triangle\alpha) \supset \alpha$ $[\mathbf{PC^\square}]$
3. $\square\alpha \supset \alpha$ $[\mathbf{PC^\square}$ in 1,2$]$

It follows then from the above results that by subjoining to any calculus containing $\mathbf{PC^\square}$ the axiom (T) or $\square\alpha \equiv (\alpha \wedge \triangle\alpha)$ yields the same results on the light of the definition of $\triangle\alpha$. It is to be noted that if we had extended the language of **PC** with △ instead of with □, the following definition could be introduced in the resulting system **PC**$^\triangle$:

Definition 2.1.6 *(Def □):* $\Box\alpha \stackrel{Def}{=} \alpha \wedge \triangle\alpha$

and (T) could be obtained as a theorem.

The preceding arguments suggest, nonetheless, that there is a clear reason by which contingency cannot be used as primitive on the same grounds as necessity and possibility. In fact, ◊ and □ are definable in terms of contingency, given that we have at our disposal the principle (T), but this result is not granted in absence of (T).[2] It happens, however, as already suggested in Section 1.1, that albeit (T) is certainly a plausible principle of what have been called *alethic* (i.e. logical) modalities and has been widely accepted in ancient and Medieval philosophical tradition, there exist certain interpretations of the necessity operator which do not validate (T). In some contexts, for instance, "it is necessary that" conveys the meaning of "it is obligatory that", and that "it is possible that" conveys the meaning of "it is permissible that". Therefore, if □α is read as "it is obligatory that α", we cannot accept as valid a principle which asserts that □α implies the truth of α, as it is plainly false that any moral or legal obligation entails the realization of it. What can be accepted as a law of the logic of so-called *deontic* modalities, at most, is the weaker principle which holds for every α:

(D) $\Box\alpha \supset \Diamond\alpha$

(D) is a consequence of (T): from (T) in fact we obtain $\Box\neg\alpha \supset \neg\alpha$ by instantiation and this is equivalent, by contraposition and □◊-interchange, to $\neg\neg\alpha \supset \Diamond\alpha$, that is, to $\alpha \supset \Diamond\alpha$ (let us call this formula (T*)), and from $\Box\alpha \supset \alpha$ and (T*) it follows $\Box\alpha \supset \Diamond\alpha$. On the other hand (D) is equivalent to $\neg(\Box\alpha \wedge \Box\neg\alpha)$, so it simply describes the contrariety of the wffs located at the upper vertices of Aristotle's square. The implication described in (D), inside the square, is called a relation of *subalternance*.

While (T) is equivalent to (T'): $\Box\alpha \equiv (\alpha \wedge \Box\alpha)$, (D) is equivalent to the weaker (D'): $\Box\alpha \equiv (\Diamond\alpha \wedge \Box\alpha)$. Thus, the proof sketched in the steps (1)–(8) above cannot be performed, and the reader can experience the failure of any effort of defining □ in terms of △ on the grounds of the latter equivalence.

We realize then that the modal notions are so ambiguous that they admit at least two different axiomatizations: a stronger one (containing (T)) and

---
[2] Actually, it may be rigorously proved that the removal of (T) has the effect of making such a definition impossible in all interesting modal systems. See Further reading.

2.2 Minimal properties of modal systems

a weaker one (containing (**D**)) with different properties. As we shall see, as a matter of fact, there are various intuitively plausible meanings for the necessity operator, each one characterized by a different axiomatic behavior that can be associated to it. It is then useful to assume from the beginning that there is an unlimited number of possible axiomatizations for the modal notions and, consequently, an infinite number of modal systems.

The phenomenon of the plurality of logics is not only foreign to the spirit of ancient and medieval logic, but also to the rigid logical monism that characterized the initial progress of mathematical logic in the second half of the 19th century. Logical pluralism, on the other hand, is well represented in the book which gave birth to contemporary modal logic, Lewis and Langford's *Symbolic Logic* (see Further reading).

That book introduces the axioms of five different systems of modal logic of increasing strength (**S1** to **S5**). Lewis and Langford intended to emphasize the distinction between implication in the strict sense (the so-called *strict implication*, symbolized by "$\prec$") and material implication (symbolized by "$\supset$"). As we know, the latter can be reduced to a disjunction due to the truth-functional equivalence $\alpha \supset \beta \equiv \neg \alpha \vee \beta$. This allows us to derive certain logical laws considered to be paradoxical, such as the law of Pseudo-Scotus $\neg \alpha \supset (\alpha \supset \beta)$. All Lewis' systems from **S1** to **S5** have the expression of a fundamental intuition about strict implication in common: such implication holds when it is impossible that the antecedent is true and the consequent is false. In the weakest of Lewis' systems, **S1**, and, consequently, in all the stronger ones from **S2** to **S5**, the following definition can be introduced:

Definition 2.2.1 ($Def_{\prec}$): $\quad \alpha \prec \beta \stackrel{Def}{=} \neg \Diamond (\alpha \wedge \neg \beta)$

Thus, by the law of $\Box \Diamond$-interchange and truth-functional laws, one obtains the equivalence: $\alpha \prec \beta \equiv \Box(\alpha \supset \beta)$. Therefore, to say that α strictly implies β means to say that the corresponding material conditional is necessary. Another useful definition is:

Definition 2.2.2 ($Def_{\asymp}$): $\quad \alpha \asymp \beta \stackrel{Def}{=} (\alpha \prec \beta) \wedge (\beta \prec \alpha)$

C. I. Lewis also accepted, even for his weakest system **S1**, the validity of the law (**T**) (i.e., $\Box \alpha \supset \alpha$, for every α). But as we have seen, this law should be excluded from any system which intends to grasp notions of necessity which have nothing to do with logical necessity. On the other side, in the weaker Lewis' systems **S1**, **S2** and **S3**, the highly desirable Necessitation Rule (i.e, the rule that says that if α is a theorem of a modal system **S**, then $\Box\alpha$ is also a theorem of **S**) is not a valid rule. Without denying valid motivations in favor of weak modal systems, in our analysis we will consider the Necessitation Rule as an inherent component of a minimal modal system. The class of systems that we will examine, therefore, will not coincide with the class of Lewis' systems.

A second feature of necessity that may be considered to be minimal is the following: by analogy with Modus Ponens, necessity should distribute over clauses of material conditional, so that if $\Box\alpha$ and $\Box(\alpha \supset \beta)$ hold, then so holds $\Box\beta$. For this purpose, we shall include the following schema in all lists of modal axioms: $(\Box(\alpha \supset \beta) \wedge \Box\alpha) \supset \Box\beta$ or its equivalent variant $\Box(\alpha \supset \beta) \supset (\Box\alpha \supset \Box\beta)$. The systems containing such schema and the Necessitation Rule are called *normal systems*, and we will follow this terminology here. The weakest normal modal system, composed by **PC** extended with the above principles, is called **K** in honor of Saul Kripke and will be rigorously treated in the next chapter.

Beyond such principles, as our intuitions about modal notions are uncertain, we will not put any further constraints on the properties of normal modal systems. Anyway, even if it is difficult, in general, to decide which properties normal systems should enjoy, it is easy to state which properties they should exclude. Firstly, it is obvious that no modal system should contain among its theorems a formula of the form $\alpha \wedge \neg\alpha$, nor should it contain both α and its negation, $\neg\alpha$, among its theorems. In other words, the modal systems should be free of contradictions, or *consistent*.

Now, in modal logics, given that they are extensions of **PC**, any contradiction implies deductive triviality (i.e., the derivability of every wff) and vice versa (consistency is identified with deductive non-triviality). So in order to show that a modal system **S** is consistent, it is enough to show that **S** is deductively non-trivial, that is, that there exists at least a formula which is not deducible in **S** (see Section 1.3).

Still, there are other formulas that a system should exclude in order to be characterized as a legitimate modal system. It is intuitive that no modal system which could be interesting for philosophical aims may have among its theses the formula $p \supset \Box p$, whose meaning is that whatever is true is necessarily true. If we were to accept such a statement, we would lose the ability to express the distinction between what is necessarily true and what is factually true, and thus any construction of a modal logic would turn out to be superfluous. The formula (**Ban**): $p \supset \Box p$ will be called

2.2. MINIMAL PROPERTIES OF MODAL SYSTEMS

modal banalization formula, while the stronger equivalence (**Triv**): $p \equiv \Box p$ will be called *modal trivialization formula*. Any system containing (**Ban**) will be said to be *modally banal*; so the system **PC**$^\Box$+(**Ban**) (and any of its extensions) although consistent, is a degenerate modal system, and is sometimes said to express the *collapse of modalities*.

But there is also another sense in which a modal system may be said to be modally banal: we may say that a modal system **S** is such if it contains the wff $\Box p$ for every p. If $\Box p$ is a theorem in **S**, it is easy to see that this amounts to having the wff $\Box\bot$ as a theorem of this system. Henceforth, the wff $\Box p$ will be called *Verum* (**Ver**). We shall assume that no meaningful modal system can be either modally banal or inconsistent, even if it is sometimes technically convenient to admit modally banal systems as limit cases of modal logics. The next proposition shows that subjoining (**Ban**), (**Triv**) and (**Ver**) to **PC**$^\Box$ yields consistent systems which are distinct from **PC**$^\Box$.

Proposition 2.2.3 *(i) **PC**$^\Box$+(**Ban**) is a proper extension of **PC**$^\Box$, i.e. (**Ban**) is not provable in **PC**$^\Box$.*
*(ii) **PC**$^\Box$+(**Triv**) is a proper extension of **PC**$^\Box$, i.e (**Triv**) is not provable in **PC**$^\Box$.*
*(iii) **PC**$^\Box$+(**Ver**) is a proper extension of **PC**$^\Box$, i.e (**Ver**) is not provable in **PC**$^\Box$.*
*(iv) **PC**$^\Box$+(**Ban**), **PC**$^\Box$+(**Triv**) and **PC**$^\Box$+(**Ver**) are consistent systems.*

Proof: Consider the following tables, which will provide three distinct semantic interpretations of $\Box$ to be added to the standard truth-tables of **PC**:

p	$\Box p$
1	1
0	1

Table 1

p	$\Box p$
1	0
0	0

Table 2

p	$\Box p$
1	1
0	0

Table 3

1. (i) To see that (**Ban**) is not provable in **PC**$^\Box$, interpret $\Box$ as in Table 2. All value assignments v to atomic variables granted by Table 2 give value 1 to all theorems of **PC**$^\Box$, but there is a v' such that $v'(p) = 1$ and $v'(\Box p) = 0$, and, consequently, $v'(p \supset \Box p) = 0$, that which would not happen if (**Ban**) were a **PC**$^\Box$-theorem.

2. (ii) As (**Triv**) implies (**Ban**), if (**Triv**) were provable in **PC**$^\Box$ so would be (**Ban**), contrary to (i).

3. (iii) Every assignment v to atomic variables granted by Table 3 gives value 1 to all theorems of **PC**$^\Box$, but there is an assignment v' such that $v'(\Box p) = 0$. This would be impossible if (**Ver**) were a **PC**$^\Box$-theorem.

4. (iv) Let us call 1-*valid* (3-*valid*) any wff that receives value 1 by standard truth tables and Table 1 (Table 3). It is easy to check that all **PC**-theses, $\Box p$ and $p \supset \Box p$ are 1-valid, and that all **PC**-theses along with $p \equiv \Box p$ are 3-valid. Furthermore, it may be easily proved that what is derived in such systems by applying *Modus Ponens* and Uniform Substitution to their theorems is either 1-valid or 3-valid. Suppose then that $\bot$ were a theorem of $\mathbf{PC}^\Box$+(**Ban**), $\mathbf{PC}^\Box$+(**Triv**) or $\mathbf{PC}^\Box$+(**Ver**), i.e., that they were inconsistent. So $(p \supset \Box p) \supset \bot$, $(p \equiv \Box p) \supset \bot$ and $\Box p \supset \bot$ would also be theorems in such systems by standard logic. But if this were true, then $\bot$ would turn out to be 1-valid or 3-valid, since *Modus Ponens* preserves 1-validity or 3-validity in each system, which is impossible.

♠

It is to be noted that $\mathbf{PC}^\Box$+(**Ver**) extended with the axiom (**T**) yields an immediate inconsistency as from $\Box\bot \supset \bot$ (an instance of (**T**)) and $\Box\bot$ one derives $\bot$, thus leading to inconsistency. However, Proposition 2.2.3 can be almost straightforwardly extended to the systems **K**+(**Ban**), **K**+(**Triv**) and **K**+(**Ver**). Such systems will be called simply **Ban**, **Triv** and **Ver** (see Exercise 2.12).

2.3 Systems between K and S5

We can now take into consideration a rigorous definition of a modal language. A *modal language* is defined as a quadruple $\mathbf{ML} = \langle Var, \bot, \supset, \Box \rangle$, where *Var* is a set of symbols called *propositional variables* as defined in Section 1.2 and $\bot, \supset$ and $\Box$ are the operators[3] already introduced in Chapter 1 and in Section 2.1.

Definition 2.3.1 *The collection WFF of the well-formed formulas is defined as follows, where $\alpha, \beta, \gamma, \cdots$ will be metavariables ranging over formulas:*

(i) $\bot \in WFF$ and, for each α, if $\alpha \in Var$, then $\alpha \in WFF$.

(ii) *If* $\alpha, \beta \in WFF$, *then* $(\alpha \supset \beta) \in WFF$.

(iii) *If* $\alpha \in WFF$, *then* $(\Box \alpha) \in WFF$.

(iv) *No other sequence of symbols belongs to WFF.*

---
[3] It is worth noting that the constant $\bot$ can be seen as a 0-ary operator.

2.3. SYSTEMS BETWEEN K AND S5

As in the case of **PC**, any well-formed formula in *WFF* will be referred to as a *wff*. The same informal use of the term "formula" for "formula schema" that we adopted for **PC** (see Section 1.2) will be maintained for modal language. Also, from now on we will make use of parentheses only when necessary, using standard conventions to delete external parentheses.

The *auxiliary connectives* in a language **ML** are $\neg$, $\wedge$, $\vee$, $\top$ and $\equiv$, as defined in Section 1.2 plus $\lozenge$, $\triangledown$, $\triangle$, $\prec$ and $\asymp$, already defined in Sections 2.1 and 2.2.

The abbreviations $\square^n$ and $\lozenge^n$ mean that the symbols $\square$ and $\lozenge$ are iterated n times. Useful concepts are the following:

Definition 2.3.2 *The subformulas and immediate subformulas of a wff are simultaneously defined by the following clauses:*

(i) *Every formula α is a subformula of α.*

(ii) *No wff is an immediate subformula of $\bot$ and α, for every $\alpha \in Var$.*

(iii) *α is the only immediate subformula of $\square \alpha$.*

(iv) *α and β are the only immediate subformulas of $\alpha \supset \beta$.*

(v) *If α is a subformula of β and β is an immediate subformula of γ, then α is a subformula of γ.*

(vi) *No other sequence of symbols are subformulas.*

Definition 2.3.3 *A propositional normal modal logic S is any subset of WFF which contains:*

(i) *All the theorems of the Propositional Calculus* **PC**

(ii) *The axiom* (**K**): $\square(p \supset q) \supset (\square p \supset \square q)$

and is closed under the following rules:

(**US**) *Uniform Substitution: for each $p \in Var$ and $\beta \in WFF$, if $\vdash_S \alpha$ then $\vdash_S \alpha[p/\beta]$; (for the notation used see Section 1.2);*

(**MP**) *Modus Ponens: β is deducible from α and $\alpha \supset \beta$;*

(**Nec**) *Necessitation: if $\vdash_S \alpha$, then $\vdash_S \square \alpha$.*

We recall that the weakest normal modal system is called **K**. It is also to be noted that all normal systems could be reformulated by using axiom schemas, with the advantage of not introducing a rule of Uniform Substitution.

Remark 2.3.4 *The Lewis' systems **S1**, **S2** and **S3** are closed under the following rule (**NecR**), but not under (**Nec**):*

(NecR) For $S \in \{S1, S2, S3\}$, if $\vdash_{PC} \alpha$, then $\vdash_S \Box \alpha$.

(**NecR**) expresses the idea that the truth-functional tautologies (that is, the **PC**-theorems) are all necessary, while (**Nec**) extends this property to every provable formula of the reference system. Systems which are closed under (**NecR**) but not under (**Nec**) are called non-normal systems.

The notion of deduction for an arbitrary system **S**, given in Definition 1.2.1, is general enough to be applied to any normal modal logic, since it refers to applications of inference rules of **S** in general. The already known notations $\Gamma \vdash_S \alpha$ and $\vdash_S \alpha$ will be used for any normal modal system **S**. However, there are crucial complications when new rules are added to a logic; in our case, the addition of (**Nec**) fortunately preserves the validity of the important **SDM** (see Proposition 1.2.5). In fact, we are able to prove the Modal Syntactical Deduction Metatheorem (**MSDM**):

Proposition 2.3.5 (MSDM) *If $\Gamma \cup \{\alpha, \beta\}$ is a set of wffs and $\Gamma, \alpha \vdash_S \beta$ where S is a modal system axiomatized with **PC** axioms, (**US**), (**MP**) and (**Nec**), then $\Gamma \vdash_S \alpha \supset \beta$.*

Proof: By induction on the length of derivations, as in Proposition 1.2.5, adding an extra case corresponding to the application of rule (**Nec**):

6. β_i follows via (**Nec**) from a previous theorem $\vdash \beta_j$.

In all other cases, the argument is precisely the same as in Proposition 1.2.5. In the new case, β_i is $\Box \beta_j$ and has been obtained by applying (**Nec**) to $\vdash \beta_j$, so we have $\vdash \beta_i$. But then by (**Ax2**) $\vdash \alpha \supset \beta_i$ holds, and it follows by Monotonicity $\Gamma \vdash \alpha \supset \beta_i$. ♠

An immediate consequence of (MSDM) is the following theorem, which is usually taken as a definition of derivation in modal logics:

Proposition 2.3.6 (*Finitary character of modal proofs*) *Let S be a modal system under the conditions of Proposition 2.3.5; then $\Gamma \vdash_S \beta$ iff there is a finite subset $\{\alpha_1, \cdots, \alpha_n\} \in \Gamma$ such that $\vdash_S (\alpha_1 \wedge \cdots \wedge \alpha_n) \supset \beta$.*

2.3. SYSTEMS BETWEEN K AND S5

Proof: Exercise 2.13. ♠

Remark 2.3.7 *It is to be noted that normal modal systems are syntactically compact (see Section 1.3) due to such finitary character of modal proofs. The relevance of this fact is emphasized in Section 4.1, where we discuss the notion of strong completeness for modal logics.*

The modal system that results from **K** by eliminating both the axiom (**K**) and the rule (**Nec**) coincides with the already mentioned system **PC**$^\Box$ and can be called the *degenerate modal system*. **K** can be extended with an arbitrary number of axioms or axiom schemas, yielding the so-called family of normal modal systems. If $(\mathbf{X}_1)\cdots(\mathbf{X}_n)$ are the labels of such axioms, then $\mathbf{KX}_1\cdots\mathbf{X}_n$ will be the name of the corresponding system which results from extending **K** with such axioms. Some of these axioms, already introduced in Chapter 1, are (**D**) and (**T**). A table with the best known axioms in modal logic is the following:

Table 2.3.8 *Some modal axioms*

(K)	$\Box(p \supset q) \supset (\Box p \supset \Box q)$
(D)	$\Box p \supset \Diamond p$
(T)	$\Box p \supset p$
(4)	$\Box p \supset \Box\Box p$
(B)	$p \supset \Box\Diamond p$
(5)	$\Diamond p \supset \Box\Diamond p$

In the literature the systems **KT** and **KTB** are frequently called **T** and **B**, respectively, while the systems **KT4** and **KTB4** are respectively equivalent to Lewis' **S4** and **S5**. Note that different combinations of axioms may yield systems that are deductively equivalent: for example, it can be proven (see Exercise 3.15) that **KT5**=**KTB4**=**KDB4** =**KDB5**. Therefore, a same modal system can receive different names according to its axiomatization.

The relationship between the most important systems obtained by combining the above axioms is schematized below, where the arrow represents proper inclusion:

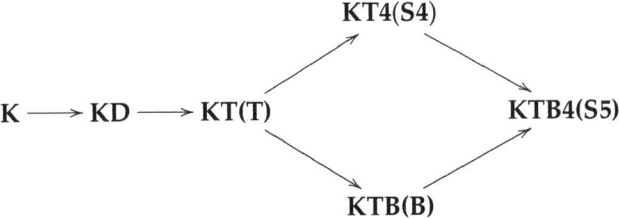

As axiom (**D**) follows from axiom (**T**), it is straightforward to see that every system includes the preceding one in the schema. It is a bit harder to see, however, that all the systems in the diagram are distinct, that is, that no system is included in the preceding one. To show this fact, we have to prove that axiom (**T**) is not derivable in **KD**, that (**4**) is not derivable in **KT**, and so on. The proof of non-derivability cannot be carried out by syntactical methods only, and this makes a non-trivial task to investigate the problem of independence among the various modal systems.

S5, the strongest system, has a special position in the family of modal normal systems, and this is why we shall treat it in the first place here. Some properties of **S5** hold for all normal systems, while others hold for **S5** alone.

Examples of useful properties which hold in **S5** and in all modal normal systems are the following derived inference rules that can be used in many contexts:

Lemma 2.3.9 *Let S be an arbitrary normal modal system. Then:*

(i) (**DR1**): *If* $\vdash_S \alpha \supset \beta$ *then* $\vdash_S \Box\alpha \supset \Box\beta$

(ii) (**DR2**): *If* $\vdash_S \alpha \supset \beta$ *then* $\vdash_S \Diamond\alpha \supset \Diamond\beta$

Proof:
  (i)
  1. $\vdash_S \alpha \supset \beta$ [**Hyp.**]
  2. $\vdash_S \Box(\alpha \supset \beta)$ [(**Nec**) in 1]
  3. $\vdash_S \Box(\alpha \supset \beta) \supset (\Box\alpha \supset \Box\beta)$ [(**K**)]
  4. $\vdash_S \Box\alpha \supset \Box\beta$ [(**MP**) in 2,3]
  (ii) Using (i), contraposition and Definition 2.1.1.   ♠

As another example, it can be also proven that **S5** and other normal modal systems admit the rule of Replacement of Proved Equivalents (**Eq**), already proved in the case of **PC** in Lemma 1.2.2. The proof of this fact can be given for a generic normal modal system **S**.

Proposition 2.3.10 (**Eq**) *Let S be an arbitrary normal modal system, and let us suppose* $\vdash_S \alpha \equiv \beta$. *Let* γ' *be obtained from* γ *by replacing some occurrences of* α *by* β; *then* $\vdash_S \gamma \equiv \gamma'$.

Proof: We prove the result by induction on the length of γ. Suppose $\vdash_S \alpha \equiv \beta$. We have to consider the result of the replacement for the cases where γ is, respectively: (i) the atomic formula $\bot$ or an atomic variable, (ii) $\delta \supset \epsilon$, or (iii) $\Box\delta$.

2.3. SYSTEMS BETWEEN K AND S5

(i) Suppose γ is atomic; there are two subcases to be analyzed:

   (a) No occurrences are replaced, in which case γ is coincident with γ' and obviously $\vdash_S \gamma \equiv \gamma'$ (note that this covers the case in which γ is $\bot$).

   (b) An occurrence of α has been replaced by β, but, as γ is atomic, an occurrence of a formula in γ is all of γ. Therefore, γ is α and γ' is β and by hypothesis $\vdash_S \alpha \equiv \beta$.

(ii) Suppose γ is $\delta \supset \epsilon$. If no occurrences have been replaced, then the result is trivially true. Otherwise, γ' is $\delta' \supset \epsilon'$, and, by induction hypothesis, $\vdash_S \delta \equiv \delta'$ and $\vdash_S \epsilon \equiv \epsilon'$. Therefore, by standard propositional reasoning, $\vdash_S \gamma \equiv \gamma'$.

(iii) Suppose γ is $\Box\delta$. If no occurrences have been replaced, then the result is trivially true. Otherwise, γ' is $\Box\delta'$ and, by induction hypothesis, $\vdash_S \delta \equiv \delta'$. As $\vdash_S \delta \equiv \delta'$ is equivalent to $\vdash_S (\delta \supset \delta') \land (\delta' \supset \delta)$, it is enough to apply (**DR1**) twice to obtain $\vdash_S \gamma \equiv \gamma'$.

♠

Just as the rule (**Eq**) holds in all normal modal systems, the following wffs are theorems of **S5**, including all other normal modal systems as well.

The proofs of the theorem schemas below are left to the reader with the exception of (v)[4] (see Exercise 2.9):

Proposition 2.3.11 *The following schemas are K-theorems:*

(i) $\Box(\alpha \land \beta) \equiv (\Box\alpha \land \Box\beta)$ *Distributivity of $\Box$ over conjunction*

(ii) $\Diamond(\alpha \lor \beta) \equiv (\Diamond\alpha \lor \Diamond\beta)$ *Distributivity of $\Diamond$ over disjunction*

(iii) $\neg\Diamond\alpha \supset (\alpha \strictif \beta)$ *Paradox of strict implication*

(iv) $\Box\alpha \supset (\beta \strictif \alpha)$ *Paradox of strict implication*

(v) $(\Box\alpha \lor \Box\beta) \supset \Box(\alpha \lor \beta)$ *Semi-distributivity of $\Box$ over disjunction*

(vi) $\Diamond(\alpha \land \beta) \supset (\Diamond\alpha \land \Diamond\beta)$ *Semi-distributivity of $\Diamond$ over conjunction*

(vii) $\Delta(\alpha \lor \neg\alpha)$ *Non-contingency of tautologies*

---

[4] When the reference system is explicit or when there is no danger of misunderstanding, we write simply $\vdash$ instead of $\vdash_K$, $\vdash_{KT5}$, $\vdash_{KTB4}$, etc.

Proof: We prove only item (v). The proof uses the derived rule of Lemma 2.3.9 (i):
1. $\alpha \supset (\alpha \vee \beta)$ [PC]
2. $\Box\alpha \supset \Box(\alpha \vee \beta)$ [(DR1) in 1]
3. $\beta \supset (\alpha \vee \beta)$ [PC]
4. $\Box\beta \supset \Box(\alpha \vee \beta)$ [(DR1) in 3]
5. $(\Box\alpha \vee \Box\beta) \supset \Box(\alpha \vee \beta)$ [PC in 2,4]

♠

Although **S5** is the name usually given to the system **KTB4**, the simplest axiomatization of it is obtained by adding axiom (**5**) to **KT**: in the presence of axioms (**K**) and (**T**), the conjunction of axioms (**B**) and (**4**) turns out to be, in fact, equivalent to (**5**), as shown in the next propositions:

Proposition 2.3.12 *KT5 contains KTB4.*

Proof: (i) Firstly, let us show that (**B**) is a theorem of **KT5**:
1. $\Diamond p \supset \Box\Diamond p$ [(5)]
2. $\Box\neg p \supset \neg p$ [(T)]
3. $\neg\neg p \supset \neg\Box\neg p$ [PC in 2]
4. $p \supset \Diamond p$ [(Eq) in 3]
5. $p \supset \Box\Diamond p$ [PC in 4,1]

(ii) Secondly, we show that (**4**) is a theorem of **KT5**:
1. $\Diamond\neg p \supset \Box\Diamond\neg p$ [(5)]
2. $\neg\Box\Diamond\neg p \supset \neg\Diamond\neg p$ [PC in 1]
3. $\Diamond\Box p \supset \Box p$ [(Eq) in 2]
4. $\Box\Diamond\Box p \supset \Box\Box p$ [(DR1) in 3]
5. $\Box p \supset \Box\Diamond\Box p$ [(B)]
6. $\Box p \supset \Box\Box p$ [PC in 5,4]

♠

Proposition 2.3.13 *KTB4 contains KT5.*

Proof: We show that **KTB4** deduces the formula (**5**)

1. $\Box\neg p \supset \Box\Box\neg p$ [(4)]
2. $\neg\Box\Box\neg p \supset \neg\Box\neg p$ [PC in 1]
3. $\Diamond\Diamond p \supset \Diamond p$ [(Eq) in 2]
4. $\Box\Diamond\Diamond p \supset \Box\Diamond p$ [(DR1) in 3]
5. $\Diamond p \supset \Box\Diamond\Diamond p$ [(B)]
6. $\Diamond p \supset \Box\Diamond p$ [PC in 5,4]

♠

2.3. SYSTEMS BETWEEN K AND S5

Concentrating on the properties of **S5** has many advantages, because some of its properties can be transferred to its subsystems. The most important among the transferrable properties are consistency and modal non-banality. We first define the useful concept of a translation between logics:

Definition 2.3.14 *A translation from a system **S** into a system **S′** is a function f from the language of **S** into the language of **S′** such that $\mathbf{S} \vdash \alpha$ implies $\mathbf{S'} \vdash f(\alpha)$.*[5] *A translation is said to be a strong translation when "implies" is replaced by "if and only if".*

Proposition 2.3.15 *S5 is consistent.*

Proof: Let **QL** be a system for quantificational logic (that is, for first-order logic) axiomatized in a standard way.[6] Let $f : \mathbf{S5} \longrightarrow \mathbf{QL}$ be a translation from **S5** into **QL** defined as follows:

(i) $f(p_n) = P_n(x)$, where $P_n(x)$ is a monadic predicate of **QL**

(ii) $f(\bot) = \bot$

(iii) $f(\alpha \supset \beta) = f(\alpha) \supset f(\beta)$

(iv) $f(\Box \alpha) = \forall x f(\alpha)$

Obviously, the translation of $f(\Diamond \alpha)$ is the existential formula $\exists x f(\alpha)$. It is then easy to see that the translation of the axioms of **S5**, i.e. (**K**), (**T**) and (**5**) are theorems of **QL**. Then, for every atomic variable p_1:
$f(\mathbf{K}) = f(\Box(p_1 \supset p_2) \supset (\Box p_1 \supset \Box p_2)) = \forall x(P_1(x) \supset P_2(x)) \supset (\forall x P_1(x) \supset \forall x P_2(x))$
$f(\mathbf{T}) = f(\Box p_1 \supset p_1) = \forall x P_1(x) \supset P_1(x)$
$f(\mathbf{5}) = f(\Diamond p_1 \supset \Box \Diamond p_1) = \exists x P_1(x) \supset \forall x \exists x P_1(x)$
As the rule (**MP**) holds both in **S5** and in **QL**, the only rule specifically modal in **S5** is (**Nec**), but its translation holds in **QL**: in fact (**Nec**) in **S5** amounts to "If $\vdash_{S5} \alpha$ then $\vdash_{S5} \Box \alpha$", which is translated by "If $\vdash_{QL} f(\alpha)$ then $\vdash_{QL} \forall x f(\alpha)$". By induction on the length of proofs, we conclude that the image under the f-translation of any theorem in **S5** is a theorem of **QL**.

Now, suppose by *Reductio* that **S5** is inconsistent, i.e. that for some formula α, α and its negation $\neg \alpha$ are both theorems of **S5**. This is equivalent

---
[5]For this notion of translation see, for instance, W. A. Carnielli e I. M. L. D'Ottaviano [CD97].
[6]The reader unfamiliar with first-order standard logic is invited to consult any good manual on the subject, or to go to Section 9.1.

to having a single atomic variable p_n as a theorem. In this case, the formula $f(p_n)$ (i.e., $P_n(x)$) would be a theorem of **QL**, which is known to be false. ♠

Proposition 2.3.16 *S5 is modally non-banal.*

Proof: Suppose, by way of contradiction, that the formula (**Ban**), and so in particular $p_1 \supset \Box p_1$, is a theorem of **S5**; then $P_1(x) \supset \forall x P_1(x)$ would be a theorem of **QL**, which is known to be false. ♠

Corollary 2.3.17 *For every arbitrary normal system S, if $S \subseteq S5$, then S is consistent and modally non-banal.*

Proof: If **S** were to contain a contradiction, then such contradiction would be derivable in **S5**; also, if (**Ban**) were a theorem of **S**, then (**Ban**) would be a theorem of **S5**. But such conclusions are incompatible with Propositions 2.3.15 and 2.3.16, respectively. ♠

In general terms, the advantages of studying the system **S5** include the fact that if a certain formula α is not a theorem of **S5**, it will obviously not be a theorem of either of its subsystems. Relying on the fact that **S5** is consistent and modally non-banal, these kinds of proofs can be made even easier. If we suppose that a certain formula α is a thesis of **S5**, and from such supposition, it follows $\bot$ or the formula (**Ban**), then we conclude that α cannot be a thesis of **S5** nor of any of its subsystems.

As an example, it can be shown that $\Diamond p \supset \neg \Box \Diamond p$ is not provable in **S5**, hence in none of its subsystems. Suppose, by *Reductio*, that we could prove $\Diamond p \supset \neg \Box \Diamond p$. Then we could perform the following proof:
1. $\vdash_{S5} \Diamond p \supset \neg \Box \Diamond p$ [Hyp.]
2. $\vdash_{S5} \Diamond p \supset \Box \Diamond p$ [(5)]
3. $\vdash_{S5} \Diamond p \supset (\neg \Box \Diamond p \land \Box \Diamond p)$ [PC in 1,2]
4. $\vdash_{S5} \Diamond p \supset \bot$ [(Eq) in 3]

Therefore, by $(Def_\neg)$, we obtain $\vdash_{S5} \neg \Diamond p$ and also $\vdash_{S5} \Box \neg p$. This is, however, impossible, because $\Box \neg p[p/\top]$ yields $\vdash_{S5} \Box \neg \top$, and considering that an instance of (**T**) is $\Box \neg \top \supset \neg \top$ it follows by (**MP**) $\vdash_{S5} \neg \top$, a contradiction with Proposition 2.3.15.

This method of refutation can be applied to a much discussed formula, the so-called Gödel-Löb formula (see Section 4.4), where the necessity operator can be interpreted as "provable in Peano arithmetic":

(**GL**): $\Box(\Box p \supset p) \supset \Box p$

2.4. MODALITIES IN S5

If, by *Reductio* (**GL**) were a theorem of **S5**, then we would obtain p for any formula p. In fact, $\Box(\Box p \supset p)$ is a theorem of **S5** by (**T**) and (**Nec**), hence (**MP**) would yield $\Box p$, therefore by (**T**) also the atomic wff p, which is impossible given that **S5** is consistent. Therefore (**GL**) cannot be a theorem of **S5** nor of any of its subsystems.

The above result is obtained by using the consistency of **S5**, while other results can be obtained by appealing to its modal non-banality. For example, let us remark that the converse of semi-distributivity of **S5** (items (v) and (vi) of Proposition 2.3.11) does not hold in **S5**. Here is the proof, by *Reductio*, for the impossibility of the converse of item (v):

1. $\vdash_{S5} \Box(p \vee q) \supset (\Box p \vee \Box q)$ [Hyp.]
2. $\vdash_{S5} \Box(p \vee \neg p) \supset (\Box p \vee \Box \neg p)$ [[$q/\neg p$] in 1]
3. $\vdash_{S5} \Box(p \vee \neg p)$ [PC, (**Nec**)]
4. $\vdash_{S5} \Box p \vee \Box \neg p$ [(**MP**) in 2,3]
5. $\vdash_{S5} \Diamond p \supset \Box p$ [PC, Eq in 4]
6. $\vdash_{S5} \Box \neg p \supset \neg p$ [(**T**)[$p/\neg p$]]
7. $\vdash_{S5} p \supset \Diamond p$ [PC, (**Eq**) in 6]
8. $\vdash_{S5} p \supset \Box p$ [PC in 7,5]

Therefore, if (**1**) were an **S5**-thesis, we would derive the formula (**Ban**), which we know to be underivable in **S5**.

The above results of course show that $\Box(p \vee q) \supset (\Box p \vee \Box q)$ is underivable in all subsystems of **S5**, including the minimal system **K**. By applying the same method, it can be shown, for instance, that $\Box \Diamond p \supset \Box p$ is not a thesis of **K** as it is not a thesis of **S5**. If it were, since $\Diamond p \supset \Box \Diamond p$ is a thesis of **S5** (in fact, it is axiom (**5**)) then we would have $\Diamond p \supset \Box p$ and $p \supset \Diamond p$ by transitivity (by (**T**) and contraposition), and we would again obtain (**Ban**) as a thesis.

2.4 Modalities in S5

The system **S5** has some features which make it a special system in the panorama of modal logics. By a *modality* we will mean every (possibly empty) finite sequence of symbols $\Box$, $\Diamond$ and $\neg$. It is convenient to realize that any modality in **S5** is equivalent to some one which either has no occurrences of negation symbols, or has just one occurrence of a negation symbol before some propositional variable. To see this, note that each formula $\neg \Box \alpha$ or $\neg \Diamond \alpha$ can be replaced by $\Diamond \neg \alpha$ and by $\Box \neg \alpha$, thanks to applications of the $\Box\Diamond$-interchange rules (see Section 2.1). By applying this procedure a finite number of times we obtain a finite sequence of $\Box$ and $\Diamond$ at the left side of any formula, to the right of which it will appear a finite number of negation symbols. The number of negation symbols can be reduced to zero (if the number of negations is even) or to one (if the number of negations is odd) by applying the equivalence $\neg\neg\alpha \equiv \alpha$ and the rule (**Eq**) as many times as needed. Hence, in this section, we will use the term "modality" to denote sequences of $\square$ and $\lozenge$, and will define its *length* as the number of modal symbols in the string. We will call *Reduction Theorem* any equivalence of the form $M\alpha \equiv M'\alpha$, where M is a modality of length k, M' is a modality of length l, and $k \neq l$.

The following reduction meta-theorem can be proven.

Proposition 2.4.1 *If M is a modality and α is a well-formed formula in S5, then $M\square\alpha$ is equivalent to $\square\alpha$ and $M\lozenge\alpha$ is equivalent to $\lozenge\alpha$.*

Proof: By induction on the length of modalities (for readiness we will denote by M^n a modality of length n).

- For $n = 1$, it can be easily seen that both $M^n\square\alpha \equiv \square\alpha$ and $M^n\lozenge\alpha \equiv \lozenge\alpha$ are **S5**-theorem schemas.

    As a matter of fact, M^1 can only be $\square$ or $\lozenge$, and then there are only four cases to analyze:
    TR1: $\square\square\alpha \equiv \square\alpha$ (which follows from axioms (**T**) and (**4**)).
    TR2: $\lozenge\lozenge\alpha \equiv \lozenge\alpha$ (which follows from axioms (**T**) and (**4**)).
    TR3: $\square\lozenge\alpha \equiv \lozenge\alpha$ (which follows from axioms (**5**) and (**T**)).
    TR4: $\lozenge\square\alpha \equiv \square\alpha$ (which follows from axioms (**5**) and (**T**)).

- By induction hypothesis suppose, for $n = k$, that $M^k\square\alpha \equiv \square\alpha$ and $M^k\lozenge\alpha \equiv \lozenge\alpha$.

- We must show then that (i) $M^{k+1}\square\alpha \equiv \square\alpha$ and (ii) $M^{k+1}\lozenge\alpha \equiv \lozenge\alpha$.

    (i) $M^{k+1}\square\alpha$ can be either $M^k\square\square\alpha$ or $M^k\lozenge\square\alpha$
    In the first case, by **TR1** and (**Eq**), one obtains $M^k\square\alpha$. Now, by applying the induction hypothesis, $M^k\square\alpha$ is proved to be equivalent to $\square\alpha$.
    In the second case, one uses **TR4** instead and obtains, by the same argument, a wff which is equivalent to $\square\alpha$.

    (ii) $M^{k+1}\lozenge\alpha$ can be either $M^k\square\lozenge\alpha$ or $M^k\lozenge\lozenge\alpha$.
    In the first case, by **TR3** and (**Eq**), one obtains $M^k\lozenge\alpha$. Now, by applying the induction hypothesis, $M^k\lozenge\alpha$ is proved to be equivalent to $\lozenge\alpha$.
    In the second case, one uses **TR2** instead and obtains the equivalence between $M^k\lozenge\lozenge\alpha$ and $\lozenge\alpha$.

♠

2.4. MODALITIES IN S5

In **S5**, every modality is then equivalent to some modality which belongs to the Aristotle's square. **S5** has yet another special feature: the following theorem of **S5**, which we shall call *Absorption Theorem*, shows that not only do we prove the equivalence between formulas of arbitrarily distinct modal length in **S5**, but also the equivalence between formulas of distinct modal degrees.

Proposition 2.4.2 *(Absorption Theorem) The following absorption laws hold in S5.*
AT1: $\Box(p \vee \Box q) \equiv \Box p \vee \Box q$
AT2: $\Box(p \vee \Diamond q) \equiv \Box p \vee \Diamond q$
AT3: $\Diamond(p \wedge \Diamond q) \equiv \Diamond p \wedge \Diamond q$
AT4: $\Diamond(p \wedge \Box q) \equiv \Diamond p \wedge \Box q$

Proof: We only prove **AT1** here; the others are left to the reader (see Exercise 2.4).

1. $\Box(\neg\Box q \supset p) \supset (\Box\neg\Box q \supset \Box p)$ [(**K**)[$p/\neg\Box q, q/p$]]
2. $\Box(\Box q \vee p) \supset (\Diamond\Box q \vee \Box p)$ [**PC**, $Def_\Diamond$ in 1]
3. $\Box(p \vee \Box q) \supset (\Box p \vee \Box q)$ [**TR4**, (**Eq**) in 2]
4. $(\Box p \vee \Box\Box q) \supset \Box(p \vee \Box q)$ [Proposition 2.3.11 (v)]
5. $(\Box p \vee \Box q) \supset \Box(p \vee \Box q)$ [**TR1**, (**Eq**) in 4]
6. $\Box(p \vee \Box q) \equiv \Box p \vee \Box q$ [**PC** in 3, 5]

♠

The following theorem is a consequence of the Absorption Theorem:

Proposition 2.4.3 *Every formula α can be reduced in S5 to an equivalent formula α' of first degree.*

Proof: It is enough to illustrate a procedure of reduction to the first degree. Instead of dealing with the primitive connectives of the basic modal language **ML**, here we use a more handy set of operators ($\Box, \Diamond, \neg, \vee,$ and $\wedge$). Let α be a second degree formula.

1. By using the adequate equivalences, eliminate all modal operators and connectives, except $\Box, \Diamond, \neg, \vee,$ and $\wedge$.

2. Apply De Morgan's laws and $\Box\Diamond$-interchange in order to place all negations in the most internal position, and then eliminate the double negations until at most one negation symbol at the left of a propositional variable remains.

3. Reduce the iterated modalities by means of the equivalences **TR1–TR4** stated in the proof of Proposition 2.4.1.

4. If α is still a second degree formula, there is a subformula of α which has the form $\Box\beta$ or $\Diamond\beta$, where β is a conjunction or a disjunction of some first degree formulas. Let us examine the case of $\Box\beta$, leaving to the reader the case of $\Diamond\beta$ (Exercise 2.14). Three cases have to be examined:

   (a) If β is a conjunction of the form $\gamma \wedge \delta$, apply Proposition 2.3.11, item (i), to obtain $\Box\gamma \wedge \Box\delta$ and reduce the modal degree by using the absorption laws of Proposition 2.4.2.

   (b) If β is a disjunction of the form $\gamma \vee \delta$, where γ or δ begin with a modal operator, apply again the absorption laws of Proposition 2.4.2.

   (c) If β is of the form $\gamma \vee \delta$ where neither δ nor γ begins with a modal operator, this means that one of the disjuncts of $\gamma \vee \delta$ is of form $\rho \wedge \epsilon$, where $\rho \wedge \epsilon$ contains a modal operator. In this case, apply the **PC** law $\gamma \vee (\rho \wedge \epsilon) \equiv (\gamma \vee \rho) \wedge (\gamma \vee \epsilon)$, and replace the disjunction with the equivalent conjunction. Subsequently, apply the distributivity law of $\Box$ over the conjunction and the absorption theorems whenever possible.

If α has degree higher than the second, the procedure is firstly applied to second degree subformulas of α, so that, if α has degree n, the resulting wff α' has degree $n-1$. By reiterating the procedure, α is then transformed into a first–degree formula in a finite number of steps. ♠

2.5 Exercises

1. If $\mathbf{PC}^\triangle$ is as defined at Section 2.1, page 30, show that $\mathbf{PC}^\triangle + \triangle\alpha \equiv \triangle\neg\alpha + Def_\Box$ is equivalent to $\mathbf{PC}^\Box + (\mathbf{T}) + Def_\triangle$.

2. (i) Show that an alternative axiomatization for **K** is obtained by extending $\mathbf{PC}^\Box$ with the following axioms and rules:
   1. $\Box p \wedge \Box q \equiv \Box(p \wedge q)$.
   2. $\Box(p \supset p)$.
   3. If $\vdash_K \alpha \supset \beta$, then $\vdash_K \Box\alpha \supset \Box\beta$.

   (ii) Show that the following is a **K**-theorem:
   $\vdash_K (\Box p \wedge \Diamond q) \supset \Diamond(p \wedge q)$. (Hint: observe that $\vdash_K \Box(q \supset \neg p) \supset (\Box p \supset \Box\neg q)$.

2.5. EXERCISES

3. Show that **K** can be axiomatized by extending **PC**$^\Box$ with the following rule:
   If $\alpha_1, \cdots \alpha_n, \vdash_K \beta$, then $\Box\alpha_1, \cdots, \Box\alpha_n \vdash_K \Box\beta$, for $n \geq 0$.

4. Prove in **S5** the absorption laws stated at Proposition 2.4.2.

5. Call Denecessitation (**Den**) and Possibilitation (**Pos**) the following rules:
   (**Den**) If $\vdash \Box\alpha$, then $\vdash \alpha$.
   (**Pos**) If $\vdash \alpha$, then $\vdash \Diamond\alpha$.
   Show that (**Den**) and (**Pos**) hold in every system that extends **KT**, but are inconsistent with the systems **PC**$^\Box$+(**Ver**) and **Ver**.

6. Show that each normal system contains the following axiomatic basis for non-contingency:
   K$\triangle$1. $\triangle p \equiv \triangle \neg p$
   K$\triangle$2. $(\triangle p \wedge \triangle q) \supset \triangle(p \wedge q)$
   K$\triangle$3. $(\triangle p \wedge \nabla(\neg p \vee r)) \supset \triangle(p \vee q)$

7. Reduce the following formulas of **S5** to first degree and show that the reduced formulas are theorems of **S5**.
   1. $\Box(p \supset \Box(q \vee \neg q))$
   2. $\Box((\Box p \wedge \Box q) \supset (\Box\Box p \wedge \Box\Box q))$
   3. $(\Diamond\Diamond(r \wedge \Box r) \vee \Box r) \supset \Box(\Diamond\Box p \supset (\Box p \vee r))$

8. Show that **KT5**=**KDB4**=**KDB5**.

9. Prove the items (i)–(iv) and (vi)–(vii) of Proposition 2.3.11.

10. Let us call **P4** the system which is **PC**$^\Box$ + (**T**) + $\Box(p \supset q) \supset \Box(\Box p \supset \Box q)$ and the rules (**US**), (**Nec**) and (**MP**). Prove that **P4** and **S4** are equivalent systems.

11. Show that **S5** can be axiomatized by adding to **PC** the following rules:
    1. If $\vdash \alpha \supset \beta$, then $\vdash \Box\alpha \supset \beta$.
    2. Under the proviso that all variables of α are in the scope of $\Box$ or $\Diamond$, if $\vdash \alpha \supset \beta$ then $\vdash \alpha \supset \Box\beta$.
    3. Under the proviso that all variables of β are in the scope of $\Box$ or $\Diamond$, if $\vdash \alpha \supset \beta$ then $\vdash \Diamond\alpha \supset \beta$.
    4. If $\vdash \alpha \supset \beta$, then $\vdash \alpha \supset \Diamond\beta$.

12. Show that:
    (a) **Ban**, **Triv** and **Ver** are consistent systems.
    (b) **Ver** is a proper extension of **K**, i.e (**Ver**) is not provable in **K**.
    (c) **Ban** is a proper extension of **K**, i.e (**Ban**) is not provable in **K**. (Hint: for (a) and (b) use the same tables as in Proposition 2.2.3; for (c) use Table 3 if p is a **K**-thesis and Table 2 otherwise).

13. Prove Proposition 2.3.6.

14. Complete the proof of Proposition 2.4.3 by showing that every second degree formula of the form $\Diamond\beta$ may be reduced to a first-degree formula.

15. Show that **Ver** and **Triv** are Post-complete systems.

2.6 Further reading

See W. Kneale and M. Kneale [KK62] for a qualified exposition of the history of modal logic in the Ancient and Middle Ages, and I. M. Bochenski [Boc61] for the general history of formal logic.

For the relationships between necessity and contingency, see M. J. Cresswell [Cre88], which proves the essentiality of (**T**) for the definition of $\Box$ in terms of $\triangle$. C. I. Lewis and C. H. Langford [LL32] is one of the great classics of logical literature: it contains a seminal treatment of strict implication, for which J. J. Zeman [Zem73] and N. M. Martin [Mar89] are also good recent references. For a study of the system **S5**, a basic reference is R. Carnap [Car47], while the first systematic treatment of Lewis' systems and their neighbors is due to R. Feys [Fey65]. The latter book also contains an almost complete bibliography up to 1965. For a detailed exposition of the syntax of the systems between **KT** and **S5**, see G. E. Hughes and M. J. Cresswell [HC68].

For a natural deduction formulation of basic modal systems see M. Ohnishi and L. Matsumoto [OM57a, b, c] and K. Schütte [Sch68]
.

Chapter 3
The semantics of normal modal systems

3.1 Matrices and Dugundji's Theorem

The problem of endowing modal systems with appropriate semantics has been considered to be a difficult, or even impossible task, for decades. In their *Symbolic Logic*, Lewis and Langford did not formulate truth conditions for modal propositions, nor did they propose any decision procedure for their five systems. They were able to prove, however, in spite of **S1** to **S5** being ordered by an inclusion relation, that these systems are all distinct. The strategy they used to establish this fact was the matrix method. A *matrix* $\mathcal{M}$, as defined below (see Definitions 3.1.1 and 3.1.2), consists of a set of objects (usually sets, natural or rational numbers) inside which some special objects are selected as "distinguished" objects (which play the role of the value "true" in the class of truth-values).

A logical system **S** can be *interpreted* in a matrix $\mathcal{M}$ if we take the propositional variables of the wffs from this logic as ranging over the elements of the matrix and interpret the connectives and modal operators of **S** as operations in $\mathcal{M}$. A wff α is then said to be *verified* by $\mathcal{M}$ if, for all interpretations of the atomic variables in α (also called *valuations*), the value of α is a distinguished value of $\mathcal{M}$. Otherwise, we say that α is *falsified* by the matrix. We also say that a system **S** is *verified* by a matrix $\mathcal{M}$ (or that $\mathcal{M}$ is a *model for* **S**) when all the theorems of **S** are verified by $\mathcal{M}$. We say that **S** is *characterized* by $\mathcal{M}$ if the set of wffs of **S** verified by the matrix coincides with the set of theorems of **S**.

The many-valued truth-tables are prototypical examples of matrices: in particular, in the familiar two-valued matrix for **PC**, the objects are the standard truth-values {0, 1}, 1 is the distinguished truth-value and the operations are the usual interpretations of the connectives.

Lewis and Langford were able to prove the independence of the systems **S1** to **S5** by defining a class of matrices $\mathcal{M}_1$ that verify **S1** but not the axioms of **S2**, a class of matrices $\mathcal{M}_2$ that verify **S2**, but not the axioms of **S3**, and so on. However, Lewis and Langford were unable to define *characteristic matrices* for their systems: in other words, they did not provide matrices assigning distinguished values to all theorems of a given system and *only* to them. In 1940, James Dugundji (cf. [Dug40]) gave, in a sense, a justification for this omission: in fact, he showed that no characteristic matrix for **S5** or any of its subsystems can have a finite number of truth-values.

To prove his theorem, Dugundji followed the strategy used by K. Gödel to prove that Heyting's formulation of intuitionistic logic cannot be characterized by finite-valued matrices (cf. [Göd32]).

Dugundji's argument consists of the following steps:

(a) It is shown that, for each n-valued matrix, there exists a (modal) disjunction with $n + 1$ variables (call it *Dugundji's formula*) which takes distinguished values.

(b) It is possible to define an infinite matrix that assigns a distinguished value to every theorem of **S5**, and thus it verifies (i.e. it is a model of) **S5** and of all of its subsystems.

(c) It is proved then that this infinite matrix falsifies all Dugundji's formulas.

(d) It follows that no modal system **S** $\subseteq$ **S5** can be characterized by matrices with finitely many values, as by (a) any finite-valued matrix verifies some Dugundji's formula which, by (c), is falsified by the infinite **S5**-matrix.

To carry out the argument, some formal definitions and two propositions are in order:

Definition 3.1.1 *A matrix $\mathcal{M}$ is a triple $\mathcal{M} = \langle M, D, O \rangle$, where:*
- *$M \neq \emptyset$.*
- *$D \subseteq M$ is a set of distinguished values.*
- *O is a set of operations over M.*

3.1. MATRICES AND DUGUNDJI'S THEOREM

Definition 3.1.2 *A matrix $\mathscr{M}$ characterizes a logical system S iff all theorems of S, and only them, receive distinguished values when S is interpreted in $\mathscr{M}$. A matrix $\mathscr{M}$ is a model for a logical system S if all theorems of S (but not necessarily only them) receive distinguished values when S is interpreted in $\mathscr{M}$.*

Proposition 3.1.3 *For each finite matrix with n truth-values which is a model for S5, there exists a formula D_n containing $n+1$ variables (Dugundji's formula), such that this formula receives a distinguished truth-value when S5 is interpreted in $\mathscr{M}$.*

Proof: Define the following Dugundji's formulas D_n, where, for $1 \leq i \leq n+1$ and $1 \leq j \leq n+1$:

$$D_n \stackrel{\text{Def}}{=} \bigvee_{i \neq j} (p_i \asymp p_j)$$

recalling from Definition 2.2.2 that $p_i \asymp p_j$ means $\Box(p_i \supset p_j) \wedge \Box(p_j \supset p_i)$.

As an example, $D_2 \stackrel{\text{Def}}{=} (p_1 \asymp p_2) \vee (p_1 \asymp p_3) \vee (p_2 \asymp p_3)$, written in the variables p_1, p_2, and p_3.

Suppose that there exists a finite matrix $\mathscr{M}$ with $n \geq 2$ values which is a model for **S5** and a formula D_n having $n+1$ variables. Provided that we have n values of the matrix being assigned to $n + 1$ variables, then two distinct variables p_i and p_j of D_n will necessarily receive the same value. Therefore, the value that is attributed to the equivalence $p_i \asymp p_j$ is the same value attributed to $p_i \asymp p_i$, which obviously is a distinguished value (as $p_i \asymp p_i$ is a theorem of **S5**). Moreover, $(p_i \asymp p_i) \vee \beta$, where β is any formula, also receives a distinguished value in any model for **S5**, as it is clearly a theorem of **S5**. Therefore $(p_i \asymp p_j) \vee \beta$ receives a distinguished value when interpreted in any matrix that identifies the values of p_i and p_j. But Dugundji's formula D_n is of the form $(p_i \asymp p_j) \vee \beta$, and because it has $n + 1$ variables, it receives a distinguished value when **S5** is interpreted in $\mathscr{M}$. ♠

In order to prove the second proposition, we first define an infinite matrix $\mathscr{M}_\infty$, and subsequently show that it assigns a distinguished truth-value to every theorem of **S5**. Consider the matrix $\mathscr{M}_\infty = \langle M, D, O \rangle$ defined as follows:

1. The set M of values is the power set $\wp(\mathbb{N})$ of the set $\mathbb{N}$ of all natural numbers.

2. The set D of distinguished values is the singleton $\{\mathbb{N}\}$.

3. The set of operations is $O = \{\cap, \cup, ^-, \boxempty\}$, where $\cap, \cup, ^-$ are the usual set-theoretical operations, and

$$\boxempty X = \begin{cases} \mathbb{N} & \text{if } X = \mathbb{N} \\ \varnothing & \text{otherwise.} \end{cases}$$

Let $V : Var \longrightarrow M$ be an assignment of elements of $M = \wp(\mathbb{N})$ to the propositional variables; this function can be extended to all wffs in the following way:

- $V(\bot) = \emptyset$
- $V(\alpha \supset \beta) = \overline{V(\alpha)} \cup V(\beta)$
- $V(\Box\alpha) = \boxtimes(V(\alpha))$

The last condition amounts to

$$V(\Box\alpha) = \begin{cases} \mathbb{N} & \text{if } V(\alpha) = \mathbb{N} \\ \emptyset & \text{if } V(\alpha) \neq \mathbb{N} \end{cases}$$

Consequently, for the defined connectives we have:

- $V(\alpha \wedge \beta) = V(\alpha) \cap V(\beta)$
- $V(\alpha \vee \beta) = V(\alpha) \cup V(\beta)$
- $V(\neg\alpha) = \overline{V(\alpha)}$

Proposition 3.1.4 *The infinite matrix $\mathcal{M}_\infty$ is a model for S5.*

Proof: It is easy to prove that $\mathcal{M}_\infty$ verifies the axioms of **S5** and that the rules preserve validity; as an example, consider axiom **(T)**. For any valuation V, $V(\Box p \supset p) = \overline{V(\Box p)} \cup V(p) = \overline{\boxtimes V(p)} \cup V(p)$. In fact:

- If $V(p) = \mathbb{N}$, then $\overline{\boxtimes V(p)} \cup V(p) = \overline{\boxtimes \mathbb{N}} \cup \mathbb{N} = \mathbb{N}$
- If $V(p) \neq \mathbb{N}$, then $\boxtimes V(p) = \emptyset$ and $\overline{\boxtimes V(p)} \cup V(p) = \overline{\emptyset} \cup V(p) = \mathbb{N} \cup V(p) = \mathbb{N}$

Therefore, axiom **(T)** receives a distinguished truth-value for any assignment of truth-values to p. By the same reasoning, all other axioms of **S5** can be shown to receive distinguished values, and moreover, it is easy to show that the rules **(MP)**, **(US)** and **(Nec)** preserve this property. Thus $\mathcal{M}_\infty$ is a model for **S5**. ♠

Proposition 3.1.5 *(Dugundji's Theorem) No characteristic matrix for a subsystem of S5 can have a finite number of truth-values.*

Proof: It is enough to show that no Dugundji's formula D_n can receive a distinguished value in the matrix $\mathcal{M}_\infty$, which, by Proposition 3.1.4, is a model for **S5**.

Take the following valuation V that assigns the singleton $\{k\} \subseteq \mathbb{N}$ to the propositional variable p_k. We know that for every distinct p and q, $V(p \times q) = \boxdot(\overline{P} \cup Q) \cap \boxdot(P \cup \overline{Q})$, where $V(p) = P$, $V(q) = Q$.

Note that, as P and Q are singletons, then $\overline{P} \neq \mathbb{N}$ and $\overline{Q} \neq \mathbb{N}$. Moreover, $P \subseteq \overline{Q}$ and $Q \subseteq \overline{P}$, hence $V(p \times q) = \boxdot(\overline{P}) \cap \boxdot(\overline{Q}) = \emptyset$. Consequently, every Dugundji's formula takes value $\emptyset$ in the matrix $\mathcal{M}_\infty$.

Therefore no Dugundji's formula takes a distinguished value in the infinite matrix that verifies **S5**, hence such formulas cannot be theorems of **S5**. However, for each given finite matrix there exists some Dugundji's formula which is verified in this finite matrix. Thus, no finite matrix can characterize **S5**.

To show that no finite matrix can characterize any subsystem **S** of **S5**, suppose that **S** could be characterized by a finite matrix with n truth-values. This finite matrix would verify the Dugundji's formula D_n and thus, as this matrix by hypothesis characterizes **S**, D_n would be a theorem of **S** and, consequently, it would be a theorem of **S5**, which we know to be absurd. ♠

3.2 Carnapian models and relational models

Since two-valued truth-tables are special cases of matrices, Dugundji's Theorem grants that, if by hypothesis one could characterize modal systems by tables which are n-valued extensions of usual truth-tables, n could not be a finite number. Dugundji's result suggested the incorrect idea that it is impossible to associate a rigorous semantics to modal logics and, above all, to produce a decision procedure for modal systems.

As a matter of fact, the basic idea that turns out to be the key for a semantical analysis of modal logics was already implicit in some aspects of the philosophy of Leibniz (18th century). According to a view which the tradition attributes to Leibniz, necessary propositions are those whose truth is invariant with respect to any configuration of the world, or simply those that *are true at all possible worlds* (but the reader is advised to see Section 3.6 on this topic).

In a certain sense, within truth-functional logic we already have at our disposal a notion which can be interpreted as truth at all possible worlds: the notion of a tautology. A tautology is a formula whose truth is invariant with respect to all possible assignments of truth-values to atomic variables occurring in it. Now, it is natural to view an assignment of truth-values to the atomic variables of the language as something that specifies what is true and what is false in a hypothetical state of affairs. An assignment of truth-values to all atomic variables describes, thus, what could be called, in a precise sense, a *possible world*. To say that a tautology is a formula which is true for every assignment to its atomic variables is the same as saying that this formula is true for every assignment to *all* the atomic variables of the language, that is, to say that this formula is true at all possible worlds, or *necessarily true*.

When we move from the language of standard propositional logic to the language of propositional modal logic, the notion of a possible world becomes more difficult to grasp, and, as Dugundji's Theorem shows, we cannot use any assignment with a finite number of truth-values for our purposes. Furthermore, if we see possible worlds as something depending on *possible* assignments to infinite sets of variables, the resulting circularity may be puzzling for all philosophers who are interested in reducing modal to non-modal notions.

But the basic weakness of the above outlined conception of modalities, sometimes called *extensionalist*, is that if $\Diamond p$ (for p an atomic formula) is to mean that p is true for some value assignment to atomic variables, then $\Diamond p \wedge \Diamond \neg p$ should be a logical truth, given that we always find some assignment to atomic variables which gives value 1 to p and some other assignment which gives value 0 to the same p. However, it is obvious that we cannot accept $\Diamond p \wedge \Diamond \neg p$ as an axiom or a theorem of any modal system endowed with the rule of Uniform Substitution, since a substitution instance of $\Diamond p$ would be $\Diamond \bot$ which is inconsistent with every normal modal system. On the other hand, we cannot restrict or eliminate the rule of Uniform Substitution without making unintelligible the notion of atomic variable itself, which is at the ground of the language of standard propositional logic[1].

For all the above reasons, it is convenient to take the notion of a possible world as a primitive, non-analyzed notion. As Carnap noticed in [Car47],

---

[1] The extensionalist view of modalities is grounded on the notion of necessity which is implicit in L. Wittgenstein's *Tractatus* and in R. Carnap's logicism. If our language were endowed with propositional quantifiers (for which see Chapter 9) we could accept as a thesis $\exists p(\Diamond p \wedge \Diamond \neg p)$ (i.e. $\exists p \nabla p$). This formula is actually equivalent to a postulate introduced by Lewis and Langford with the name of "Existence Postulate" (see [LL32], p.178). In lack of such linguistic resource, the semantic based on the models which in the present chapter we call "Carnapian" will not grant the logical validity of the wff $\Diamond p \wedge \Diamond \neg p$. For the coincidence between the modal logic of Wittgenstein's *Tractatus* [Wit01] and Carnap's modal logic (both equivalent to Lewis' **S5**) see Chapter VII of G.H. von Wright's [vW82].

3.2. CARNAPIAN MODELS AND RELATIONAL MODELS

a simple semantics for the system **S5** is available by interpreting, in a Leibnizian way, $\Box\alpha$ as "α is true at all possible worlds". In Carnap's theory, the notion of necessity may be rephrased in the following terms:

Definition 3.2.1 *An implicit Carnapian model is a pair* $\mathcal{M} = \langle W, V \rangle$, *where:*

1. *W is a non-empty (finite or an infinite) collection of objects called (possible) worlds and*

2. $V : Var \longrightarrow \wp(W)$ *is a function, called implicit valuation, which maps any atomic variable into a subset of W*

Intuitively, any implicit valuation V assigns to an arbitrary atomic variable the collection of exactly those worlds at which such variable is true. Still, in other words, V assigns to an arbitrary atomic variable the *proposition* that this variable expresses. Sometimes $V(p)$ is also referred to as the "set of p-worlds" (the set of worlds at which p is true), $V(q \wedge r)$ the "set of $q \wedge r$-worlds", and so on.

To help intuition, it is convenient to see that the properties of V can be defined, as much as truth-functional connectives are concerned, in the same way as the function V used to define Dugundji's matrix. For formulas of form $\Box\alpha$, the function V can be defined by the following clause:

$$V(\Box\alpha) = \begin{cases} W & \text{if } V(\alpha) = W \\ \varnothing & \text{if } V(\alpha) \neq W \end{cases}$$

Instead of mentioning sets of possible worlds, we can treat directly with truth-values 0 and 1 in place of $\varnothing$ and W; instead of writing $w \in V(\alpha)$ (w is an α-world) and $w \notin V(\alpha)$ (w is not an α-world), we can use a new symbol v such that $v(\alpha, w) = 1$ means "α is true at world w" and $v(\alpha, w) = 0$ means "α is false at world w". In this way, the notion of truth of a formula α turns out to be relativized to some possible world: we now speak of the truth of a statement α *with respect to* a possible world w. The difference between V and v is that v is a two-argument function. We may now introduce the following definition:

Definition 3.2.2 *An explicit Carnapian model is a pair* $\mathcal{M} = \langle W, v \rangle$, *where* $W \neq \varnothing$ *is a set of worlds and* $v : Var \times W \longrightarrow \{0, 1\}$ *is a map called an explicit valuation, satisfying the following properties for any world* $w \in W$:

1. $v(p, w) = 1$ *or* $v(p, w) = 0$.

2. $v(\bot, w) = 0$.

3. $v(\alpha \supset \beta, w) = 1$ iff $v(\alpha, w) = 0$ or $v(\beta, w) = 1$.

4. $v(\Box \alpha, w) = 1$ iff $v(\alpha, w') = 1$ for all $w' \in W$.

The world w, with respect to which the formula α is evaluated in an explicit model, will be said the *reference world*.

It is sometimes convenient to use specific truth conditions of a formula α with respect to a world w in a model $M = \langle W, v \rangle$. In this way, an equivalent formulation of Definition 3.2.2 is the following:

Definition 3.2.3 *We inductively define a wff α as being true at a world w in a model M (where $M, w \vDash \alpha$ has the same meaning as $v(\alpha, w) = 1$ in a model M) as follows:*

1. $M, w \nvDash \bot$.

2. $M, w \vDash p$ iff $v(p, w) = 1$.

3. $M, w \vDash \alpha \supset \beta$ iff $M, w \nvDash \alpha$ or $M, w \vDash \beta$.

4. $M, w \vDash \Box \alpha$ iff $M, w' \vDash \alpha$, for every $w' \in W$.

From such conditions, it obviously follows that:

5. $M, w \vDash \neg \alpha$ iff $M, w \nvDash \alpha$.

6. $M, w \vDash \alpha \wedge \beta$ iff $M, w \vDash \alpha$ and $M, w \vDash \beta$.

7. $M, w \vDash \Diamond \alpha$ iff there exists a world $w' \in W$ such that $M, w' \vDash \alpha$.

Clearly, the truth conditions for $(\alpha \vee \beta)$, $(\alpha \equiv \beta)$, $(\alpha \dashv \beta)$, $(\alpha \prec \beta)$ may be derived from the corresponding definitions in the language of **S5**. The choice among three ways of expressing truth-value assignments (that is, among V, v and $\vDash$) is a matter of convenience, and we will use them interchangeably depending upon context.

A formula α will be said to be *valid* in a Carnapian model M (notation: $M \vDash \alpha$) when $v(\alpha, w) = 1$ for every world $w \in W$ in M. A formula α will be said to be c-*valid* (notation: $\vDash \alpha$) when $M \vDash \alpha$ for every Carnapian model M.

To assert the truth of $\Box \alpha$ or the falsity of $\Diamond \alpha$ (that is, truth of $\Box \neg \alpha$) means to make *universal assertions*, since their sense is that *all* possible worlds of a certain class possess a certain property. Vice versa, to assert the truth of $\Diamond \alpha$ or the falsity of $\Box \alpha$ (that is, the truth of $\Diamond \neg \alpha$) means to make *existential assertions*, since their meaning is that some possible worlds of a certain class satisfy a certain property. This phenomenon suggests a parallelism between

3.2. CARNAPIAN MODELS AND RELATIONAL MODELS

quantifiers and modal operators. In **S5**, this parallelism becomes evident, due to Proposition 2.3.15, which shows that all **S5**-theorems are translatable into a fragment of first-order quantificational logic (but see also Section 3.3). In particular, one has to remark that analogues of the Reduction Theorems (cf. Proposition 2.4.1) and of the Absorption Theorems (cf. Proposition 2.4.2) are provable in first-order logic.

Remark 3.2.4 *It is actually possible to prove not only a representation theorem of **S5** into a fragment of **QL**, but the existence of a bijective correspondence between theorems of **S5** and theorems of the monadic fragment of the predicate calculus **QL**$_1$. This fragment contains only formulas of the form $P_n(x)$, for $n \in \mathbb{N}$, and their combinations of the form $\neg \alpha$, $\alpha \supset \beta$, $\forall x \alpha$ (with obvious extensions to defined connectives).*

In addition to the definition of f as in Proposition 2.3.15, we define another function $f^{-1} : \mathbf{QL}_1 \longrightarrow$ Var (which we may call converse translation, see Definition 2.3.14) , in this way:

1. $f^{-1}(P_n(x)) = p_n$

2. $f^{-1}(\bot) = \bot$

3. $f^{-1}(\alpha \supset \beta) = f^{-1}(\alpha) \supset f^{-1}(\beta)$

4. $f^{-1}(\forall x \alpha) = \Box f^{-1}(\alpha)$

It can then be proved:

Proposition 3.2.5 *(i) α is a theorem of **QL**$_1$ only if $f^{-1}(\alpha)$ is a theorem of **S5**. (ii) $\vdash_{S5} \alpha \equiv f^{-1}(f(\alpha))$ and $\vdash_{QL_1} \alpha \equiv f(f^{-1}(\alpha))$.*

Proof: (i) Easy induction on the length of the proofs in **QL**.
(ii) Induction on the length of α.   ♠

The two systems are thus inter-translatable; as **QL**$_1$ is known to be decidable, the above result indirectly grants a decision procedure for **S5**.

The first step to check the adequacy of the above formulated semantics is to show that **S5** is sound with respect to the class of Carnapian models:

Proposition 3.2.6 *If α is a thesis of **S5**, then α is c-valid (i.e., valid in all Carnapian models).*

Proof: By induction on the length of proofs. We first show that the axioms of **S5** are c-valid, and that the rules of **S5** preserve c-validity. It is useful to recall that the axioms of **S5** are the axioms of **PC** plus **(K)**, **(T)** and **(5)**. We show that each one of such axioms is c-valid reasoning by *Reductio*.

(i) The axioms of **PC** are all c-valid. Let Ax_n be any one of these axioms: suppose by *Reductio* that there exists a world w in a Carnapian model M such that $M, w \nvDash Ax_n$; then a contradiction will follow simply by applying truth-tables.

(ii) As for axiom **(K)**: suppose by *Reductio* that **(K)** is not c-valid, i.e., that there exists a world w of a Carnapian model M such that $M, w \nvDash \Box(p \supset q) \supset (\Box p \supset \Box q)$. Then we have:

1. $v(\Box(p \supset q), w) = 1$ and
2. $v(\Box p \supset \Box q, w) = 0$

- With respect to (1), it follows from clause 4 of Definition 3.2.2 that $v(p \supset q, w') = 1$ for every $w' \in M$.

- With respect to (2), it follows from standard propositional logic that $v(\Box p, w) = 1$ and $v(\Box q, w) = 0$; again, from clause 4 of Definition 3.2.2, $v(p, w') = 1$ for every $w' \in M$ and $v(q, w'') = 0$ for some world $w'' \in M$ which is incompatible with (1).

(iii) As for axiom **(T)**: suppose by *Reductio* that **(T)** is not c-valid, i.e., that there exists a world w in a Carnapian model M such that $M, w \nvDash (\Box p \supset p)$; hence $M, w \vDash \Box p$ and $M, w \nvDash p$. From clause 4 of Definition 3.2.2, $M, w \vDash \Box p$ entails that $M, w' \vDash p$ for every w' in M, which contradicts the hypothesis $M, w \nvDash p$.

(iv) For axiom **(5)**, the argument is analogous and is left to the reader (see Exercise 3.4 (a)).

(v) We now have to show that the rules of inference of **S5**, i.e. **(MP)**, **(US)** and **(Nec)** preserve c-validity. The proof concerning **(MP)** is obvious. The result for **(US)** is left as an exercise (see Exercise 3.4 (b)). For the case of **(Nec)**: suppose by *Reductio* that, for some α, $M \vDash \alpha$ and $M \nvDash \Box \alpha$ in a Carnapian model M; this leads to a contradiction, since from $M \nvDash \Box \alpha$ we conclude that there exists a world w in M where α is false, but this conflicts with the hypothesis that $v(\alpha, w) = 1$ for every w in M.

3.2. CARNAPIAN MODELS AND RELATIONAL MODELS

Therefore all theses of **S5** are c-valid. ♠

An obvious corollary of Proposition 3.2.6 is the following:

Proposition 3.2.7 *For every S such that $S \subseteq S5$, S is a consistent system.*

Proof: By *Reductio*, suppose that for some wff α, α and $\neg\alpha$ were both theorems of some **S** included in **S5**; then by Proposition 3.2.6 both would be c-valid, hence we would have that $w \vDash \alpha$ and $w \vDash \neg\alpha$ in all worlds w of all Carnapian models, which is impossible. ♠

Although the above result is enough to show that all subsystems of **S5** are consistent, the problem now is how to define a notion of truth for systems weaker than **S5**. A plausible answer to this question is the following: considering that $\square\alpha$ in **S5** intuitively means "α is true at all possible worlds", in the weaker systems $\square\alpha$ might mean that "α is true in a given subset of possible worlds".

A way to identify subsets of possible worlds appeals to the extensional notion of relation. A relation in set-theoretical terms is a collection of ordered pairs, and a relation between worlds is a collection of pairs of worlds. A relation R between worlds in a model $\mathcal{M}$ is defined as *universal* thanks to the following equivalence:

R is universal iff $\forall w \forall w'(wRw')$

Note that w and w' are not necessarily distinct worlds.

The notion of **S5**-*model* could be then defined not as a pair $\langle W, v \rangle$, but as a triple $\langle W, R, v \rangle$, where R is a universal relation. **S5** is thus sound with respect to the class of all models with universal relation R. Consequently, saying that α is true in a Carnapian model $\langle W, v \rangle$ is the same as saying that α is true in a relational model $\langle W, R, v \rangle$ in which R is universal with the obvious proviso that clause (4) of Definition 3.2.3 is replaced by:

Definition 3.2.8 (4') $v(\square\alpha, w) = 1$ *iff* $v(\alpha, w') = 1$ *for every w' such that wRw'.*

By generalizing the latter notion of **S5**-models, we obtain the abstract notion of a relational model:

Definition 3.2.9 *A relational model is a triple $\mathcal{M} = \langle W, R, v \rangle$ where $W \neq \emptyset$, R is a relation on W, i.e., a collection $R \subseteq W \times W$ of ordered pairs of elements of W, and v is defined by clauses (1)–(3) of Definition 3.2.3 and clause (4') of Definition 3.2.8.*

A conceptual advantage of dealing with relational models is that we are not obliged to regard the elements of W as possible worlds. The possible world interpretation can be suitable for certain systems, but not for all: it is sometimes more appropriate, for instance, to see the elements of W as points in space-time, individuals, or contexts. In order not to ascribe any specific interpretation to the relation R, we shall refer to it as the *accessibility relation*, while the term *world* should be intended to denote unspecified elements of W.

We also need the following definition:

Definition 3.2.10 *A wff α is valid in a relational model $\mathcal{M} = \langle W, R, v \rangle$ iff α is true at all elements of W. In this case, $\mathcal{M}$ is also said to be a model for α. When all the theorems of a system S are valid in a relational model $\mathcal{M}$, then $\mathcal{M}$ will be called a model for S.*

Note that relational models are *explicit* models. It can also be useful to define the implicit relational models $\langle W, R, V \rangle$, where V of course cannot be as in Carnapian models (Exercise 3.3).

One can plausibly conjecture that the difference between the notion of necessity axiomatized in **S5** and the one axiomatized in other systems can be mirrored by different properties of the accessibility relation R.

The concept of a model can be viewed more analytically as consisting of a part that describes the worlds and relations among them ($\langle W, R \rangle$), and another part which concerns the truth-value assignments to atomic variables. We will call *frame* (or *relational frame*) the pair $\mathcal{F} = \langle W, R \rangle$, and *model over the frame* $\mathcal{F}$ any relational model $\mathcal{M} = \langle W, R, v \rangle$ such that $\mathcal{F} = \langle W, R \rangle$ (we also say in this case that $\mathcal{F}$ is a frame *underlying* the model $\mathcal{M}$). The notions of truth and validity with respect to frames are defined in the following way:

Definition 3.2.11 *α is true at a world w of a frame $\mathcal{F}$ (notation: $\mathcal{F}, w \vDash \alpha$) iff α is true at a world w of some model $\mathcal{M}$ over $\mathcal{F}$.*

Definition 3.2.12 *A wff α is valid on a frame $\mathcal{F}$ (notation: $\mathcal{F} \vDash \alpha$) iff $\mathcal{M} \vDash \alpha$ for all relational models $\mathcal{M}$ over $\mathcal{F}$.*

These definitions are extended to sets of formulas in the obvious way (so, for example, if Γ is any set of wffs, $\mathcal{F} \vDash \Gamma$ if $\mathcal{F} \vDash \alpha$ for all wffs α in Γ).

Remark 3.2.13 *It is to be noted that a formula α is valid in a model $\mathcal{M}$ when it is globally true, that is, true at all worlds in $\mathcal{M}$. Analogously, α is valid on a frame*

3.2. CARNAPIAN MODELS AND RELATIONAL MODELS

$\mathcal{F}$ if it is globally valid, that is, valid in all models over $\mathcal{F}$. On the contrary, the notion of α being true is local, as it refers to some specific world w in a model $\mathcal{M}$. This distinction between local and global semantical properties is inherent to modal logic and helps to understand many characteristic features of modal reasoning (see Exercise 3.26).

Of course, being valid on *all* frames of a certain class C is equivalent to being valid in all models over such frames. But to be valid on *some* frame of a certain class C of frames is not the same as to be valid in some model of some frame in C: the first notion implies the second, but not vice versa (see Remark 3.2.13).

Also, the behavior of rules with respect to frames is not coincident with the behavior of rules with respect to models. It is, in fact, straightforward to prove the following:

1. (**MP**) preserves truth in an arbitrary world of a single model.
2. (**Nec**) preserves validity within a single model.
3. (**US**) does not preserve validity within a single model.

To grasp the third point, it is enough to consider a model $\mathcal{M}$ whose W consists of a single world w such that $v(p, w) = 1$. By substituting $\neg p$ for p, we do not have $v(\neg p, w) = 1$.

However, contrary to what may be the suggested by point 3, (**US**) does preserve validity within single frames (see Exercise 3.8).

We shall now examine the correspondence between the most usual modal axioms and the properties of the accessibility relation R, showing that each axiom is true on an arbitrary frame $\mathcal{F}$ if and only if the relation R of $\mathcal{F}$ satisfies some specific property of R described in quantificational language.

An illustrative table may be outlined as follows:

Table 3.2.14 *Some modal axioms and their characteristic relations*

(K)	R has arbitrary properties	
(D)	R is serial	$\forall w \exists w' (wRw')$
(T)	R is reflexive	$\forall w (wRw)$
(4)	R is transitive	$\forall w, w', w'' ((wRw' \wedge w'Rw'') \supset wRw'')$
(B)	R is symmetric	$\forall w, w' (wRw' \supset w'Rw)$
(5)	R is euclidean	$\forall w, w', w'' ((wRw' \wedge wRw'') \supset w'Rw'')$
(Ver)	R is empty	$\forall w \neg \exists w' (wRw')$

The notion of accessibility relation may be generalized in this way:

(i) $w_i R^m w_j$ means that w_j is accessible from w_i in m steps, or in other words, that there are $m - 1$ worlds $w_1 \cdots w_{m-1}$ such that $w_i R w_1 \cdots w_{m-1} R w_j$

(ii) $w_i R^0 w_j$ means that $w_i = w_j$

Using such m-step accessibility relations, generalizations of the above conditions are also frequent in the literature. So, for instance, *n-density* is defined as $\forall w, w' (w R^n w' \supset w R^{n+1} w')$. For $n=1$ this gives the usual notion of density, and for For $n=0$ this coincides with reflexivity (reading $w R^0 w'$ as $w = w'$).

It is easy to see that, for an arbitrary frame $\mathcal{F}$, $\mathcal{F}$ is n-dense iff $\mathcal{F} \vDash \Box^{n+1} p \supset \Box^n p$ (Exercise 3.15). So the above table could be appropriately extended as follows, where $(\mathbf{T}_n)$ is $\Box^{n+1} p \supset \Box^n p$:

$(\mathbf{T}_n)$	R is n-dense	$\forall w, w' (w R^n w' \supset w R^{n+1} w')$

In the case of (**Ver**), no world in $\mathcal{F}$ accesses any other world in $\mathcal{F}$, and we will say that such worlds are *terminal*.

Remark 3.2.15 *If identity is added to first-order language, other interesting properties become expressible. For instance, by considering the wff* (**F**), *which is the converse of* (**D**), *i.e.,* $\Diamond p \supset \Box p$, *we obtain the following correspondence:*

(**F**)	R is functional	$\forall w, w', w'' (w R w' \wedge w R w'') \supset w' = w''$.

We give a proof only for cases (**D**) and (**4**).

Proposition 3.2.16 *Let $\mathcal{F}$ be an arbitrary frame.*
(i) $\mathcal{F} \vDash \Box p \supset \Diamond p$ *iff R is serial.*
(ii) $\mathcal{F} \vDash \Box p \supset \Box\Box p$ *iff R is transitive.*

Proof: (i)
($\Leftarrow$) Suppose that w is a world in an arbitrary frame $\mathcal{F} = \langle W, R \rangle$, where R is serial, and let $v(\Box p \supset \Diamond p, w) = 0$, where v is an explicit valuation in some model $\mathcal{M}$ over $\mathcal{F}$ (cf. Definition 3.2.11). It follows from $v(\Box p \supset \Diamond p, w) = 0$ that $v(\Box p, w) = 1$ and $v(\Diamond p, w) = 0$. Since R is serial, there exists a w' such that $w R w'$. From $v(\Box p, w) = 1$ we have that $v(p, w') = 1$, but $v(\Diamond p, w) = 0$ implies that $v(p, w') = 0$, which (of course) is a contradiction.
($\Rightarrow$) Suppose that R is not serial. Then there exists a world w in a frame $\mathcal{F} = \langle W, R \rangle$ such that $w \in W$ and w doesnot access any other $w' \in W$,

3.2. CARNAPIAN MODELS AND RELATIONAL MODELS

that is, w is a terminal world. Hence there is no world w' such that wRw' and $v(p, w') = 0$ and also no world w' such that wRw' and $v(p, w') = 1$. So $v(\Box p, w) = 1$ and $v(\Box \neg p, w) = 1$, i.e., $v(\Diamond p, w) = 0$. Therefore $v(\Box p \supset \Diamond p, w) = 0$.

Thus $\mathcal{F}, w \models \Box p \supset \Diamond p$ for each world w of $\mathcal{F}$ iff R is serial.

(ii)
($\Rightarrow$) Suppose that R is not transitive, that is, that there are w, w', w'' such that wRw', $w'Rw''$ but $w\cancel{R}w''$. Consider the frame $\mathcal{F} = \langle\{w_1, w_2, w_3\}, \{\langle w_1, w_2\rangle, \langle w_2, w_3\rangle\}\rangle$. Let v be an explicit valuation such that $v(p, w_1) = 1$, $v(p, w_2) = 1$ and $v(p, w_3) = 0$.

1. Since $w_1 R w_2$ and $v(p, w_2) = 1$, $v(\Box p, w_1) = 1$.

2. Since $w_2 R w_3$ and $v(p, w_3) = 0$, $v(\Box p, w_2) = 0$.

3. Since $w_1 R w_2$ and $v(\Box p, w_2) = 0$, $v(\Box\Box p, w_1) = 0$.

Therefore, from $v(\Box p, w_1) = 1$ and $v(\Box\Box p, w_1) = 0$, we conclude that $v(\Box p \supset \Box\Box p, w_1) = 0$.

($\Leftarrow$) Let R be transitive and suppose by *Reductio* that in an arbitrary frame $\mathcal{F} = \langle W, R\rangle$, for some $w \in W$, $v(\Box p \supset \Box\Box p, w) = 0$. So $v(\Box p, w) = 1$ and $v(\Box\Box p, w) = 0$.

1. Since $v(\Box p, w) = 1$, $v(p, w') = 1$ for any $w' \in W$ such that wRw'.

2. Since $v(\Box\Box p, w) = 0$, there exists a $w'' \in W$ such that wRw'' and $v(\Box p, w'') = 0$; again, by the same argument, there exists a $w''' \in W$ such that $w''Rw'''$ and $v(p, w''') = 0$.

3. Since wRw'', $w''Rw'''$ and R is transitive, then wRw'''. Now from (1) $v(p, w''') = 1$, but this conflicts with (2).

We conclude that for every world w in $\mathcal{F}$, it holds $\mathcal{F}, w \models \Box p \supset \Box\Box p$ iff R is transitive. ♠

Remark 3.2.17 *Note that the correspondences exposed in Table 3.2.14 are provable with respect to frames, not to models. It can be shown, for example, that axiom* (**T**) ($\Box p \supset p$) *is valid on a frame if and only if R is reflexive, but this does not mean that* (**T**) *is valid in a model over $\mathcal{F}$ if and only if R is reflexive. To see this, it is enough to realize that there exists a non-reflexive model in which $\Box p \supset p$ holds. In fact, let us consider an implicit model* $\mathcal{M} = \langle W, R, V\rangle$, *where* $W = \{w_1, w_2\}$,

$R = \{\langle w_1, w_2 \rangle, \langle w_2, w_1 \rangle\}$ and $V(p) = \emptyset$. Then in both worlds w_1 and w_2, $\Box p$ and p are false, so we have that $\mathcal{M} \vDash \Box p \supset p$. Therefore it is false that $\Box p \supset p$ holds in a model $\mathcal{M}$ if and only if $\mathcal{M}$ is reflexive, as the given model is not reflexive. What is true is simply that, as $\Box p \supset p$ is valid on all reflexive frames, it is also valid in all reflexive models (but not only in them!).

Another possible mistake is to think that, given an arbitrary reflexive model $\mathcal{M}$, it validates all **KT**-theses and no other formulas. Actually, $\mathcal{M}$ also validates, for example, all **S4**-theses. What is true, as we shall see, is a different matter, i.e., that α is a thesis of **KT** if and only if α is true in all reflexive models. This amounts to a completeness result for the system **KT** with respect to the class of reflexive models.

Remark 3.2.18 *Terminal worlds, as we have seen, are elements in a model that do not "see" (that is, there are no arrows starting from them to) any other elements of the same model. Let us call* **semi-terminal** *a world that sees only itself. Let w be a terminal world and suppose, by Reductio, that $\Box p$ is false at w. Then there must be a world accessible from w at which p is false. But this is impossible, as w is terminal. Consequently, $\Box p$ is true at all terminal worlds for any p. This proves that the system* **Ver** *is sound with respect to the class of models whose worlds are terminal. The system* **Ban**, *on the other hand, is sound with respect to the class of all models whose worlds are semi-terminals: in fact, it is straightforward to see that $p \supset \Box p$ is true in all such models.*

3.3 Correspondence theory and bisimulations

This section presupposes that the reader has an elementary knowledge of quantificational logic. For a detailed presentation of the subject, see Chapter 9.

The examples of correspondence (mentioned in Section 3.2) between certain axioms and certain properties definable in first-order logic suggest that an algorithm may be found that associates modal formulas with properties of accessibility relations. It is not difficult, as a matter of fact, to find an algorithm that, starting from modal formulas, outputs first-order formulas expressing properties of certain simple accessibility relations in models, as we have already given rigorous truth-conditions for modal formulas.

As we have seen, an implicit model interprets each propositional variable p_n as a collection of worlds. This amounts to regarding p_n as a set variable or, also, to see it as a predicate P_n of some reference world x. The key idea is to regard the accessibility relation as a binary predicate, and to treat variables $x, y, z, \cdots$ as variables for worlds.

3.3. CORRESPONDENCE THEORY AND BISIMULATIONS

The *standard translation* s from modal wffs into first-order wffs may be defined as follows, where x is a fixed first-order variable (intuitively, the reference world):

(i) $s(p_n) = P_n(x)$

(ii) $s(\neg \alpha) = \neg s(\alpha)$

(iii) $s(\alpha \supset \beta) = s(\alpha) \supset s(\beta)$

(iv) $s(\Box \alpha) = \forall y(xRy \supset s(\alpha)[x/y])$, where y is (in some fixed enumeration of the individual variables) the first variable not occurring in α.

From (iv), it follows that $s(\Diamond \alpha) = \exists y(xRy \wedge s(\alpha)[x/y])$, where y is as stated in (iv).

As an example of the standard translation, we apply the algorithm to the axiom (**B**) (i.e., to $p_n \supset \Box \Diamond p_n$) obtaining, for a reference world x: $s(p_n \supset \Box \Diamond p_n) = P_n(x) \supset \forall y(xRy \supset \exists z(yRz \wedge P_n(z)))$, which is equivalent[2] to the simpler formula $\forall y(xRy \supset yRx)$ and expresses the symmetry of R in dependence of the reference variable x. By eliminating such reference (through quantification over x) the resulting formula is $\forall x \forall y(xRy \supset yRx)$.

We say that a relational model is an **S**-*model* if its relation R satisfies the properties exhibited by the standard translation of the axioms of **S**. For example, a transitive model is a **K4**-model, as transitivity of R results from the standard translation of axiom (**4**).

Correspondence Theory analyzes, in the first place, the connections between modal formulas and formulas of quantificational logicobtained as

---

[2]The proof of this equivalence presupposes the first-order calculus **QL** extended with the axioms of identity (see Chapter 9). The derivations in both directions are:

($\Rightarrow$)
1. $P_n(x) \supset \forall y(xRy \supset \exists z(yRz \wedge P_n(z)))$ [Hyp.]
2. $x = x \supset \forall y(xRy \supset \exists z(yRz \wedge x = z))$ $[[P_n(y)/(x = y)]$ in 1]
3. $x = x$ [QL]
4. $\forall y(xRy \supset \exists z(yRz \wedge x = z))$ [(**MP**) in 2,3]
5. $\forall y(xRy \supset (yRz_0 \wedge x = z_0))$ [instantiation of z in 4]
6. $\forall y((yRz_0 \wedge x = z_0) \supset yRx)$ [QL]
7. $\forall y(xRy \supset yRx)$ [QL in 5,6]

($\Leftarrow$)
1. $\forall y(xRy \supset yRx)$ [Hyp.]
2. $(\alpha \supset \beta) \supset (\gamma \supset (\alpha \supset (\beta \wedge \gamma)))$ [PC]
3. $P_n(x) \supset \forall y(xRy \supset (yRx \wedge P_n(x)))$ $[[\alpha/xRy, \beta/yRx, \gamma/P_n(x)]$ in 2, (**MP**) in 1, 2]
4. $P_n(x) \supset \forall y(xRy \supset \exists z(yRz \wedge P_n(z)))$ [QL in 3]

output of the standard translation. In particular, it is interesting to investigate the following equivalence:

1. A formula α is valid on a frame $\mathcal{F} = \langle W, R \rangle$ if and only if the relation R in $\mathcal{F}$ has property P.

When a property P is related to α in the above way, we say that P *corresponds* to α.

At this point, having defined an algorithm that yields a correspondence between modal formulas and first-order formulas, we have to be careful against a possible misunderstanding: to think that $s(\alpha)$, the first-order formula obtained as an output of the algorithm, is necessarily the formula which corresponds to α in the above sense. As it happens, there exist some very simple modal formulas to which no first-order formula is correspondent. The reason for this failure can be illustrated as follows: as we have already emphasized, the notion of truth on a frame $\mathcal{F} = \langle W, R \rangle$ coincides with truth in *all* models $\mathcal{M} = \langle W, R, V \rangle$ over $\mathcal{F}$. Each model specifies a value assignment to the atomic variables in the language and, as we know, the values of such assignments can be seen as subsets of W. Let $p_1, p_2, \cdots, p_n$ be the variables of α and $V(p_1), V(p_2), \cdots, V(p_n)$ be the sets of worlds associated to them. In order to be valid on a frame, the first-order formula $s(\alpha)$ has to be true for all possible assignments to the variables $p_1, p_2, \cdots, p_n$, thus for all possible sets $V(p_1), V(p_2), \cdots, V(p_n)$. The general form of the correspondence between modal formulas and quantified formulas will therefore be given as follows, noting that any relational model $\mathcal{M} = \langle W, R, V \rangle$ determines a first-order model $\mathcal{M}^* = \langle D, V^* \rangle$ in the following sense (for a rigorous definition see Chapter 9):

(i) D is a set of elements $a_1, a_2, \cdots$ biunivocally associated to $w_1, w_2, \cdots$ in W.

(ii) R is a relation over D.

(iii) V^* is defined as implying that $V^*(p_n)$ is a subset $\{a_1, a_2, \cdots\}$ of D such that the associated worlds $w_1, w_2, \cdots$ belong to $V(p_n)$ (i.e., they are p_n-worlds).

Since it is possible to identify every $\mathcal{M}$ with its associated first-order model $\mathcal{M}^*$ if $\mathcal{F}$ is a frame over $\mathcal{M}$ and if the modal formula α contains the propositional variables $p_1, p_2, \cdots, p_n$, then the following equivalence holds (where $s(\alpha)$ is the standard translation of α):

2. $\mathcal{F} \vDash \alpha$ if and only if $\mathcal{F} \vDash \forall P_1 \forall P_2, \cdots \forall P_n s(\alpha)$

3.3. CORRESPONDENCE THEORY AND BISIMULATIONS

The quantifiers in (2) bind predicate variables, which means that they are second-order formulas (see Section 9.2). Fortunately, it often happens that such formulas turn out to be equivalent to first-order formulas; in these lucky cases, second-order definability will be coincident with first-order definability. We give an example of this reduction by considering the axiom (**T**) (i.e, $\Box p_1 \supset p_1$). The second-order quantification of the standard translation of (**T**) is:

3. $\forall P_1 \forall x (\forall y (xRy \supset P_1(y)) \supset P_1(x))$

By considering xRy as an instance of the predicate $P_1(y)$, we obtain from (3) the formula $\forall x (\forall y (xRy \supset xRy) \supset xRx)$, and therefore through obvious steps also

(3') $\forall x (xRx)$

Conversely, in second-order logic with identity from (3') one derives (3) (the reason for this is that from the so-called Leibniz's law it is the same to say that $x = y$ and that two elements x and y share all properties P_1, cf. Section 9.2); so (3) and (3') are equivalent.

Unfortunately, a reduction from second-order to first-order formulas is not always at our disposal. This is particularly evident in cases where a frame is characterized not only by one property, but by a conjunction of properties. A meaningful example is the case of the system **S4.1** which results from extending **S4** with the so-called McKinsey axiom:

(**McK**) $\Box \Diamond p_1 \supset \Diamond \Box p_1$

Note that (**McK**) is easily seen to be equivalent to the wff $\Diamond (\Diamond p_1 \supset \Box p_1)$, where $\Diamond p_1 \supset \Box p_1$ is the formula (**F**) expressing the functionality of R (see Remark 3.2.15).

Thanks to the standard translation, (**McK**) corresponds to the property of *atomicity* (existence of a terminal point in every model; we say in such cases that the model or frame is *atomic*). In fact, we may prove:

4. $\forall P_1(s(\Box \Diamond p_1 \supset \Diamond \Box p_1)) = \forall P_1(\forall y(xRy \supset \exists z(yRz \wedge P_1(z))) \supset \exists y(xRy \wedge \forall z(yRz \supset P_1(z))))$

Since **S4.1** inherits the properties of **S4**, **S4.1**-frames are then reflexive, transitive and atomic. By using all such properties, it is possible to reduce this second-orderformula to a much handier first-order formula which expresses atomicity in first-order language with identity (let us call it *McKinsey property*):

5. $\forall y \exists z(yRz \wedge \forall x(zRx \supset x = z))$

It can be proven, however, that this derivation is possible relying on the transitivity of R but fails otherwise.[3] It turns out then that atomicity in arbitrary frames is not a first-order definable property.

The field of Correspondence Theory includes the discussion of a problem which is the converse of the basic problem of correspondence: given a first-order definable property P, what is the modal formula to which it corresponds (if any)?

The answer is given by a deep result in modal model theory known as Goldblatt-Thomason Theorem (see Section 3.6). A class C of relational frames is said to be *elementary* if it is definable in first-order language, i.e., if C is coincident with the class of all models of some set of sentences formulated in first-order language.

If C is defined by a single first-order sentence, then C is said to be *basic elementary*. A system S is elementary (respectively, basic elementary) if it is characterized by some elementary (respectively, basic elementary) class of frames. The Goldblatt-Thomason theorem gives a necessary and sufficient condition for an elementary class of frames C to be modally definable.

Example 3.3.1 *Take, for instance, the conjunction of the formulas expressing the following properties:*

- *Irreflexivity:* $\forall x \neg(xRx)$

- *Asymmetry:* $\forall x \forall y (xRy \supset \neg(yRx))$

- *Intransitivity:* $\forall x \forall y \forall z ((xRy \wedge yRz) \supset \neg(xRz))$

We shall see (at the end of this section) that there is no modal formula which corresponds to any of the mentioned first-order formulas. The aforementioned theorem offers a complete answer to the question, but in order to rigorously prove the present negative case, it suffices to use a simpler (but still deep) characterization. To begin, we recall here the notion of a truth-preserving map between frames known as *pseudo-epimorphism* or *p-morphism*, which describes natural homomorphisms between accessibility relations.

---

[3]For the complex proof of this fact see R. Goldblatt [Gol93], pp. 234 ff.

3.3. CORRESPONDENCE THEORY AND BISIMULATIONS

Definition 3.3.2 *A p-morphism between frames is defined as a function $f : W \longrightarrow W'$ from the worlds of a frame $\mathcal{F} = \langle W, R \rangle$ into the worlds of a frame $\mathcal{F}' = \langle W', R' \rangle$ endowed with the following properties:*

(i) *f preserves the accessibility relation, that is, wRw' implies $f(w)R'f(w')$.*

(ii) *f is semi-conservative over R, that is, $f(w)R'f(w')$ implies that, for every $w \in W$, there exists $u \in W$ such that wRu and $f(u) = f(w')$.*

This definition is specialized to:

Definition 3.3.3 *A p-morphism between relational models $\mathcal{M} = \langle W, R, v \rangle$ and $\mathcal{M}' = \langle W', R', v' \rangle$ is a p-morphism between their underlying frames which additionally satisfies:*

(iii) *$\mathcal{M}, w \vDash p$ iff $\mathcal{M}', f(w) \vDash p$, for any propositional variable p and every $w \in W$.*

A basic theorem about p-morphisms asserts that, for any wff α, if α is valid in $\mathcal{F}$, it is also valid in $\mathcal{F}'$ whenever there is a p-morphism between $\mathcal{F}$ and $\mathcal{F}'$.

A variant of p-morphism is the notion of filtration, which we shall examine in Section 5.2. Filtrations and p-morphisms are, however, special cases of the broader notion of *bisimulation*:

Definition 3.3.4 *A bisimulation between frames $\mathcal{F} = \langle W, R \rangle$ and $\mathcal{F}' = \langle W', R' \rangle$ is a relation $B \subseteq W \times W'$ which satisfies the following "back-and-forth" property:*

Forth: *If uRv, then there exists $v' \in W'$ such that vBv' and $u'R'v'$;*

Back: *If $u'R'v'$, then there exists $v \in W$ such that vBv' and uRv.*

Two worlds $u \in W$ and $v \in W'$ are said to be bisimilar if $\langle u, v \rangle \in B$.

This definition can be specialized to:

Definition 3.3.5 *A bisimulation between relational models $\mathcal{M}$ and $\mathcal{M}'$ is a bisimulation between frames which additionally preserves validity of atomic formulas:*

- *If wBw', then $\mathcal{M}, w \vDash p$ iff $\mathcal{M}', w' \vDash p$, for any propositional variable p.*

The pivotal property of bisimulations between models generalizes the case of p-morphisms in that they preserve the satisfaction of all modal formulas, or in other words, modal formulas are invariant under bisimulations. This is proved in what follows:

Proposition 3.3.6 *(Bisimulation invariance)* If B is a bisimulation between relational models $\mathcal{M} = \langle W, R \rangle$ and $\mathcal{M}' = \langle W', R' \rangle$ and wBw', then w and w' satisfy the same modal formulas (i.e, for every α, $\mathcal{M}, w \vDash \Box\alpha$ if and only if $\mathcal{M}', w' \vDash \Box\alpha$).

Proof: By induction on the complexity of modal formulas. The case of atomic wffs is immediate from Definition 3.3.4, and the inductive step for truth-functional formulas is straightforward. For formulas with $\Diamond$ (or $\Box$), the back-and-forth conditions are essential: from left to right, if $\mathcal{M}, w \vDash \Box\alpha$ and wBw', we must show that $\mathcal{M}', w' \vDash \Box\alpha$.

From Definition 3.2.8, $\mathcal{M}, w \vDash \Box\alpha$ iff $\mathcal{M}, v \vDash \alpha$ for every v such that wRv. But then, by the "Forth" condition, for each such $v \in W$ there must be a $v' \in W'$ such that vBv' and $w'R'v'$. By the induction hypothesis, $\mathcal{M}', w' \vDash \alpha$ for all such $v' \in W'$, hence $\mathcal{M}', w' \vDash \Box\alpha$ as required.

From right to left the argument is analogous, but using the "Back" condition instead of "Forth". ♦

Now consider the illustrative case of the first-order conjunctive formula in Example 3.3.1: in order to show that there is no modal formula which corresponds to this formula, it will be enough to show that irreflexivity cannot be modally characterized. We associate to the structure of natural numbers $\langle \mathbb{N}, < \rangle$ its p-morphic image formed by a single reflexive point. If, by *Reductio*, there were a modal formula expressing irreflexivity, by the theorem of p-morphism this formula would be also true on the one-point reflexive frame, and then could it not be true only on irreflexive frames. This proves that a modal formula that characterizes irreflexive frames cannot exist.

By applying the broader concept of bisimulation, an even simpler argument shows that irreflexivity is modally undefinable, as we shall see below. Bisimulations are deeply concerned with identity of models in modal logic: certain models are redundant in the sense that the informational content represented by the configuration of worlds could also be conveyed by a simplified model, but this redundancy may not be obvious. Bisimulations allow us to control redundancy, so we can make models as small as possible (by what is called *bisimulation contraction*), or also to expand models (by what is called *tree unraveling*). As an example, consider the three models above.[4]

It is easy to see that the following relation $B \subseteq W \times T$ between the sets of worlds $W_1 = \{s_1, s_2, s_3, s_4\}$ in $\mathcal{M}_1$ and $W_2 = \{u_1, u_2\}$ in $\mathcal{M}_2$ is in fact a bisimulation between $\mathcal{M}_1$ and $\mathcal{M}_2$: $B = \{\langle u_1, s_1 \rangle, \langle u_2, s_2 \rangle, \langle u_1, s_3 \rangle, \langle u_2, s_4 \rangle\}$.

---
[4]This example is adapted from P. Blackburn and J. van Benthem [BvB07].

3.3. CORRESPONDENCE THEORY AND BISIMULATIONS

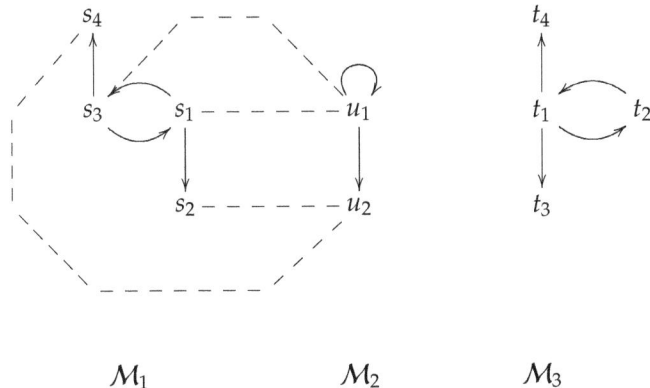

$\mathcal{M}_1$ $\qquad\qquad\mathcal{M}_2\qquad\qquad\mathcal{M}_3$

Not only this, but by considering s_1 a particular reference world (let us call it a *distinguished* world) in $\mathcal{M}_1$ and u_1 a distinguished world in $\mathcal{M}_2$, this bisimulation links the distinguished worlds.

What the bisimulation means is that $\mathcal{M}_1$ is somehow redundant: $\mathcal{M}_2$ represents the same modal information (i.e., is modally indistinguishable from $\mathcal{M}_1$) and is smaller.

However, it is impossible to find a bisimulation that links the distinguished worlds of $\mathcal{M}_2$ and $\mathcal{M}_3$: actually, t_1 sees t_2 in $\mathcal{M}_3$, but this relation has no counterpart in $\mathcal{M}_2$. In fact, u_1 sees u_2 in $\mathcal{M}_2$ but this has no representation in $\mathcal{M}_3$, as u_2 is an end point while t_2 is not (and it is not hard to realize that end points cannot bisimulate points having successors). The only other possibility would be to consider the step from u_1 to itself, but this also does not match the relation in $\mathcal{M}_3$: indeed, u_1 can see u_2, which is an endpoint, but t_1 cannot see any endpoint. This shows then that $\mathcal{M}_2$ *must be* modally distinguishable from $\mathcal{M}_3$, and there should be modal formulas that distinguish them. In fact, as one can easily see, the formula $\Box(\Box\bot \wedge \Diamond\Box\bot)$ for instance is true in $\mathcal{M}_2$ at u_1 but false in $\mathcal{M}_3$ at t_1.

Now, coming back to the issue of showing that irreflexivity is modally undefinable (Example 3.3.1), it is enough to see that the worlds s_1 in $\mathcal{M}_1$ and u_1 in $\mathcal{M}_2$ are bisimilar, although u_1 is reflexive but s_1 is not; if irreflexivity were modally definable, the modal formula which defines it, say α, would have to be invariant under bisimulations by Proposition 3.3.6. But if α were true in s_1, then it would be true in u_1, which is impossible since u_1 is a reflexive point. *A fortiori* the argument works for any wff β which might be supposed to be the modal expression of irreflexivity, asymmetry and intransitivity.

We shall discuss in Section 6.5 an application of bisimulation in establishing the non-definability of certain operators of temporal logic in terms of usual $\Box$ and $\Diamond$.

3.4 The method of relational tableaux

The first proof of completeness and decidability for normal modal systems was obtained by Saul Kripke in 1959 (limited to **KT**, **S4** and **S5**) by using relational semantics. For this historical reason, the name *Kripke semantics* is sometimes used to denote relational semantics.

The standard decision procedure for testing validity of modal formulas arises from a special codification of the method of reasoning by *Reductio ad Absurdum*. More specifically, to show that a formula α is **S**-valid (i.e., is valid in all models of a certain system **S**) we suppose, by *Reductio*, that this is not the case, i.e., suppose that α is false in some world of some **S**-model, and from this hypothesis we try to derive a contradiction. If a contradiction is reached, this amounts to a proof of the fact that the formula α is **S**-valid; otherwise, the argument shows that there exists an **S**-model that falsifies α. We will call this method of proof the *method of relational tableaux* (not to be mistaken with the method of *analytic tableaux*, which is a particular method of syntactic provability).

We will call **S**-*tableau* for a formula α (called the *input of tableau*) a relational tableau depicted by a collection of diagrams and arrows that graphically represents an attempt to build an **S**-model that falsifies α. A sequence of diagrams $w_0, w_1, \cdots$ such that $w_i \to w_j$, where the arrow represents the accessibility relation, will be called a *diagram chain*. An **S**-*tableau* for α is built by applying the following rules:

1. Enter the formula α in the first diagram w_0 (we can write this formula inside a rectangle, see example below) and assign value 0 to it. Then derive the values of all subformulas of α, applying as far as possible the truth-tables of **PC**, and write such values inside the rectangle. Rectangles intuitively describe fragments of worlds and will be labeled by the name of a world which they are supposed to represent (labels appearing on their right side). For example, if the input formula α is $p \supset q$:

$$\boxed{\begin{matrix} 1\ 0\ 0 \\ p \supset q \end{matrix}} \quad w_0$$

3.4. THE METHOD OF RELATIONAL TABLEAUX

2. In presence of universal modal assertions ($\Box$ receiving value 1 or $\Diamond$ receiving value 0), write the symbol $\forall$ over the respective truth-value (notation: $0^\forall$ or $1^\forall$), and in presence of existential modal assertions ($\Box$ receiving value 0 or $\Diamond$ receiving value 1), write the symbol $\exists$ over the respective truth-value (notation: $0^\exists$ or $1^\exists$).[5] For example, if the input formula is $\Box p \supset \Diamond q$:

$$\boxed{\begin{array}{ccc} 0^\exists & 0 & 0^\forall \\ \Box p & \vee & \Diamond q \end{array}} \; w_0$$

3. After constructing a diagram w_i, in correspondence with each one of the truth-values adorned with $\exists$, if any, (i.e, $0^\exists$ and $1^\exists$) occurring in w_i, draw arrows from w_i representing accessibility relations, and at their endings draw new diagrams, representing worlds accessible from w_0. Such new diagrams will be divided into cells; the first cell on the left will contain the immediate subformula (recall Definition 2.3.2, noting that obviously α is a subformula of $\Diamond \alpha$) of the formula which justifies the new diagram. Each of such subformulas will be marked with truth-value 1 if it is derived from the assignment $1^\exists$, or truth-value 0 if it is derived from $0^\exists$. The other cells will contain the arguments of modal operators expressing universal assertions that occur in any previous diagram w_i such that w_i sees w_j. They will receive 1 (or 0), according to the value assigned in w_i to the related universal statements. If such immediate subformula begins with a modal operator, adorn the values with $\exists$ or $\forall$ (see item (2) above). Continuing from the previous example:

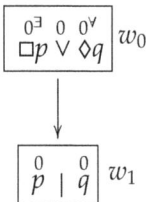

4. Arrows are introduced and drawn according to the properties of the accessibility relations of the model. When, for instance, the relation R is reflexive, every diagram is to be drawn jointly with a *loop arrow*

---
[5] Adorns $\exists$ and $\forall$ over truth-values are just used as "bookkeeping devices".

(which automatically grants the existence of an accessible world to every world in the tableau; see next figure). When R is serial, it departs at least one arrow from every diagram.

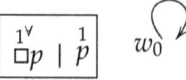

5. In case some formula receives values that do not yield univocal consequences (for example, $\alpha \wedge \beta$ with value 0, $\alpha \vee \beta$ with value 1, $\alpha \equiv \beta$ with value 1 or 0, see Section 1.5), we have to build duplicates of the relevant diagrams called *alternatives*. When, for instance, the equivocal wff belongs to w_j, the alternatives of w_j will be termed $w_j(i)$, $w_j(ii)$, $\cdots$. Each of them will be identical to w_j except for the fact that each one develops a possible consequence of the alternative assignments. Each alternative diagram will be considered to be part of a tableau that is a variant of the initial tableau.

6. The input formula α will be said **S**-*valid* when, developing all the consequences of all value assignments for all alternatives, one always reaches a contradiction (i.e., the same subformula receives both value 1 and 0). Otherwise, α is not **S**-valid. In the first case we say that the tableau for α is *closed*, while in the second case we say that the tableau for α is *open*.

It is to be noted that relational tableaux are a generalization of the semantic tableaux described in Section 1.5, in the sense that for each world there is a propositional semantic tableau to be developed, while the accessibility relation relates distinct semantic tableaux. The wffs of the degenerate modal system **PC**□ should be tested by tableaux with an empty set of worlds, so by standard semantic tableaux in which such wffs as $\Box p, \Box\Box p, \Box(p \wedge \Box p)$, etc., are treated as distinct fresh variables $p_k, p_l, p_m, \cdots$ which extend the basic language of **PC**.

Example 3.4.1 *As an illustration of the procedure, suppose that our problem is to know whether the formula $\Box p \supset \Box \Diamond p$ is **KT**-valid. Since we know that the relation R in this case is reflexive, then each diagram of the tableau will have a loop arrow. This means that in every diagram we should write the value of the arguments of the universal operators occurring in it.*

3.4. THE METHOD OF RELATIONAL TABLEAUX

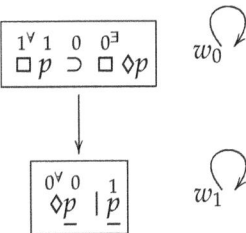

In w_1 the subformula p receives both value 0 and value 1 (the contradictory wffs being marked by an underlining), and no alternative is to be considered. Therefore there does not exist any **KT**-model that falsifies the formula above, so the formula under test is **KT**-valid.

Note that the preceding formula is neither **K**-valid nor **KD**-valid. In fact, as in these systems the relation R is not reflexive, we are not allowed to write p with value 0 inside w_1, so the procedure would end without reaching a contradiction. In this case, it would be possible to derive a **K**-model or a **KD**-model from the open diagram which is a counter-model to the input formula. The implicit **K**-counter-model, for instance, is $\langle W, R, V \rangle$, where $W = \{w_0, w_1\}$, $R = \{\langle w_0, w_1 \rangle\}$ and $V(p) = \{w_0, w_1\}$.

We have to show now that the method of validity control via tableaux is actually a decision procedure, that is, that the method is mechanical and always ends in a finite number of steps. The first condition is always granted, because at no point does the procedure open the possibility of any arbitrary choice.

The second condition, however, cannot be automatically proved for all systems. Let us see how things go for the systems **K**, **KD**, **KT** and **KT4** (i.e., **S4**). We define the *diagram degree* of a given diagram as the highest modal degree among the degrees of the formulas in the diagram. We will say that a diagram w_1 is *accessible from* w_0 if there exists an arrow from w_0 to w_1.

Tableau procedure for K

We know that the accessibility relation R in **K**-models has no specific property. No diagram w_n is accessible from another diagram (not even from itself) except in cases in which the accessibility is required by existential modal statements (i.e., formulas whose truth-value is adorned with ∃). In such cases, where w_n accesses w_{n+1}, the diagram w_{n+1} contains the immediate subformula of the existential modal statement from which it is derived and, moreover, contains the immediate subformulas of the universal modal statements in w_n. If the diagram degree of w_n is k then the diagram degree of w_{n+1} is always $k-1$. Consequently, in the worst case, either we obtain a contradiction at some step of the procedure or we obtain a diagram with no modal operators, which puts an end to the procedure.

Tableau procedure for KD

The accessibility relation R in **KD**-models is known to be serial (for each world there exists at least one world accessible from it). This establishes that each diagram w always accesses another diagram w' independently from the wffs that the diagram w contains. Since the relation R has no other property, each diagram accesses a diagram with lower diagram degree in the chain; therefore, if we do not find any contradiction in the procedure, we reach a diagram of diagram degree zero: this would end the procedure as the diagram contains only **PC**-formulas.

Tableau procedure for KT

The relation R in **KT**-models is reflexive, and thus, as already seen, each diagram has a loop arrow. This shows that each diagram is accessible from itself. For this reason, in presence of universal modal assertions, we insert in the same diagram its immediate subformulas carrying the same truth-value of the universal modal statements. Recall that reflexivity satisfies the seriality condition (as in the **KD**-models). So a distinct diagram w_{n+1} will be built only in the presence of some existential modal assertion in w_n. Since each diagram has a diagram degree which is lower than the degree of any diagrams preceding it in the chain, the procedure ends in a finite number of steps.

Tableau procedure for KT4

Given the transitivity of R in **S4**–models, we have that from each diagram an arrow is to be drawn in direction of *all* successive diagrams of the chain: so every diagram is accessible from the first one by transitivity. As a consequence, the immediate subformula of any universal modal assertion belonging to the first diagram is to be carried out to all successive diagrams accessible from this. Therefore, it is not possible to show that the procedure always ends in a finite number of steps in this way, as the diagram degree does not necessarily decrease along the chain. We cannot

3.4. THE METHOD OF RELATIONAL TABLEAUX

thus exclude the possibility that a new diagram may be identical with some diagram (or part of a diagram) which appears at an earlier point of the chain, creating a vicious circle.

In order to avoid this trouble, we need to introduce a *stop rule*, i.e, a rule that prevents the construction of infinite tableaux. A strategy that seems to be natural to follow, with regard to the problem of repeating chains, is the following: whenever all formulas that occur in a diagram w_n are already contained in a previously built diagram w_{n-k}, we say that w_n and w_{n-k} are *identifiable* and we send to w_{n-k} all arrows directed to w_n. In this way, we make sure that all the diagrams belonging to the chain contain different sets of formulas. In this way the problem of infinitely running tableaux seems to be solved, as each **S4**-tableau contains a finite number of subformulas of the formula under test, and in the worst case (given that the number of subformulas is finite), we will reach a combination of formulas that will have already occurred in some previous diagram w: at this point the stop rule asks that all arrows be sent to w, and the move should conclude the procedure.

The stop rule exposed above needs, however, a proviso and an important adjustment. First, we have to make clear that what we called "previously built" identifiable diagrams need to be part of the same chain, and not of different chains. But there is another important point, the most critical: we have to be sure that the tableau resulting from an application of the stop rule is again an **S4**-model, i.e, a model that is reflexive and transitive. The following graph illustrates that a careless use of the stop rule may be dangerous. Let us consider the following tableau:

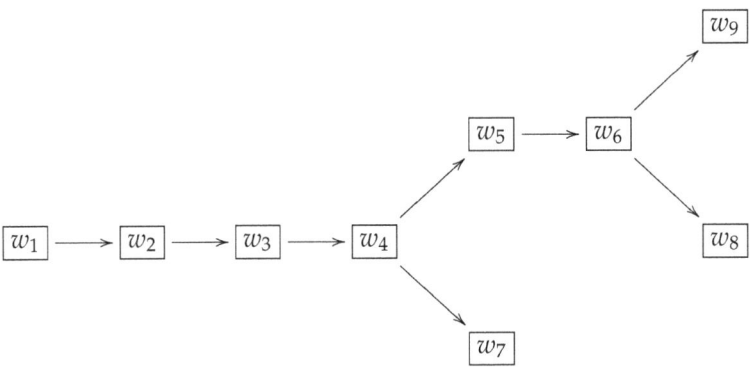

where $w_9 = w_4$, $w_8 = w_3$ and $w_7 = w_2$. By applying the stop rule, we obtain as a result a graph where $w_6 R w_4$, $w_4 R w_2$, but $w_6 R w_2$ does not hold; we also have w_6Rw_4 and w_4Rw_5, while w_6Rw_5 does not hold. Therefore the relation obtained is not transitive, and the "compressed" resulting model is not an **S4**–model. This justifies a reformulation of the stop rule with the following two additional provisos:

1. We are allowed to perform an identification of diagrams only if they belong to the same chain.

2. We should provide the resulting tableau with the transitive closure of the relation R, and consequently add to diagrams the subformulas required by the rules for tableau construction.

Tableau procedure for S5

In **S5**-models the accessibility relation is reflexive and euclidean. It is easy to see that a reflexive and euclidean relation is also transitive. In fact, let us suppose that w_0Rw_1 and w_1Rw_2. By reflexivity, w_0Rw_0 holds, and as R is euclidean, w_0Rw_1 and w_0Rw_0 imply w_1Rw_0. Again, by the same argument, w_1Rw_0 and w_1Rw_2 imply w_0Rw_2, which proves that R is transitive.

An **S5**-tableau should contain a stop rule analogous to the one introduced for **S4**, but if one wishes to avoid complications of this kind, it is possible to introduce a "mixed" procedure for **S5** by taking advantage of the fact that every higher degree formula in **S5** is reducible to a first degree formula. For this purpose, a simplified tableau decision procedure for **S5** would consist in two steps:

(a) If α is the formula to be tested, we first reduce it to the equivalent first degree formula α' by applying the procedure described in Proposition 2.4.3.

(b) We now test α' by using the rules for **S5**-tableaux described above.

It is to be noted that in this way no properties of **S5** relations (transitivity and reflexivity) will be essentially used in the tableaux construction: in fact, as the first diagram has as input a formula of first degree, the next diagrams contain consequently formulas of zero degree, so that only truth-functional rules will be employed for the wffs they contain.

We have shown that all **K**, **KD**, **KT**, **S4** and **S5** tableaux are built by applying mechanic (that is, recursive[6]) procedures that end after a finite

---

[6]For an elementary discussion on recursiveness and mechanical procedures see [EC00] specially chapter 8.

3.4. THE METHOD OF RELATIONAL TABLEAUX

number of steps. The next Proposition shows that tableaux are of help in establishing the so-called *soundness* of the basic normal systems. In Chapter 4 we will see that they are of help also to establish a property of such systems which is the converse of soundness, i.e., their completeness.

Proposition 3.4.2 *If S is K, KD, KT, S4 or S5, then every theorem of S is S-valid.*

Proof: We have to show that all axioms of each system **S** are **S**-valid and that each rule preserves **S**-validity. By applying the tableau procedures exposed above, it can be straightforwardly checked that all axioms of the systems **K, KD, KT, S4** and **S5** are, respectively, **K**-, **KD**-, **KT**-, **S4**- and **S5**-valid. It is also easy to see that validity is preserved by the rules (**US**) and (**MP**). It remains to show that (**Nec**) preserves validity. Suppose that α is a theorem of **S** and suppose, by hypothesis, that α is **S**-valid, i.e, is true at all worlds of all **S**-models. Suppose, by way of contradiction, that $\Box \alpha$ is not **S**-valid, i.e, there exists a w of some **S**-model at which $\Box \alpha$ is false. Then there must be a world w' accessible from w at which α is false. But this is absurd, as α is **S**-valid by hypothesis. This concludes the proof. ♠

Since all **S**-theses are **S**-valid, then, by contraposition, it follows that if a certain formula α is not **S**-valid, then α is not an **S**-thesis; moreover, this is verified in a finite number of steps. The following result is thus straightforward:

Corollary 3.4.3 *If an S-tableau having α as an input formula is open, then it can be proved in a finite number of steps that α is not a thesis of S.*

Let us remark that we already know that the formula (**Ban**), i.e., $p \supset \Box p$, is neither a theorem of **S5** nor of its subsystems. This result can now be proved by showing that such formula is not **S5**-valid by the tableau method (see Exercise 3.1).

The underivability of (**Ban**) (and *a fortiori* of (**Triv**)) in ordinary modal systems can also be used to mention an interesting observation about modal logics. In a certain sense, the propositional calculus **PC** can be considered a prototype of what a logical system should be, and we may wonder in which sense modal logics are similar to **PC**. This is a philosophical problem with several implications, but what we know up to this point is that in a certain way modal logics are closer to first and second-order logic than to **PC**. The key idea is to evaluate the differences in what happens when non-tautologies are added to **PC**, to modal logics and to first-order logic. What we find is that the resulting system is inconsistent in the first case, but is not so in the other cases. The reason rests upon the concept of *Post-completeness* (see Section 1.6): **PC** is Post-complete, while modal logics and predicate logic are not. Recalling Definition 1.6.3, we have:

Proposition 3.4.4 *If S is a normal system such that* $S \subseteq S5$, *then S is not Post-complete.*

Proof: We know that the formula (**Ban**) (i.e. $p \supset \Box p$) is underivable in any modal system $S \subseteq S5$. On the other hand, $S+$(**Ban**) is a consistent system for every S. In fact, there exists an **S5**-model for every $S+$(**Ban**). Such a model is $M = \langle W, R, V \rangle$, where $W = \{w_0\}$, $R = \langle w_0, w_0 \rangle$ and $V(p) = \{w_0\}$. In this model, $v(\Box p, w_0) = 1$, hence *a fortiori* $v(p \supset \Box p, w_0) = 1$. Since every thesis α of $S \subseteq S5$ is valid in every **S5**-model, then α is valid also in M. If $S+$(**Ban**) were inconsistent, then M would assign value 1 at w_0 to both p and $\neg p$, which is obviously impossible. Hence $S+$(**Ban**) is a proper consistent extension of S, so S is not Post-complete. ♠

Remark 3.4.5 *The system S5 enlightens two important facts about modal logic.*

1. *The same class of models – for instance, S5-models – can be described by the conjunction of different (but jointly equivalent) first-order formulas. It can be proven, for instance, that a relation is reflexive and euclidean if and only if it is reflexive, symmetric and transitive (i.e., an equivalence relation). In the first case, the two properties result from the standard translations of axioms* (**T**) *and* (**5**), *while in the second case the properties result from the translations of* (**T**), (**B**) *and* (**4**).

2. *The same system can be sound with respect to different classes of models. This is easy to see by examining the subsystems of S5. For example, the KT-theses are S4-theses, hence they are valid not only in all reflexive models, but in all reflexive and transitive models. Conversely, S4-models are models for S4 and for KT.*

As far as **S5** is concerned, we have already seen that the **S5**-theses are valid with respect to four classes of models:

(a) *The Carnapian models*

(b) *The relational models in which R is universal*

(c) *The relational models in which R is reflexive and euclidean*

(d) *The relational models in which R is reflexive, symmetric and transitive*

The classes of models of type (c) and (d) are coincident (they form the class of models which we called **S5**-models in Section 3.2); on the other hand, each Carnapian model is equivalent to a relational model in which R is universal, so the classes of models of type (a) and (b) are coincident. But it is not true that the classes of models of types (a)–(b) on the one side and (c)–(d) on the other side are the same class. It is enough to consider the model $\langle W, R, V \rangle$, where $W = \{w_1, w_2\}$, $R = \{\langle w_1, w_1 \rangle, \langle w_2, w_2 \rangle, \}$ and $V(p) = W$; this is an **S5**-model (why?), but R is obviously not universal. A remarkable fact is that **S5** is not only sound but also complete with respect to all the mentioned classes of models: in other words, every class of such models validates all **S5**-theses, and only them. It can thus happen that a system turns out to be not only sound but also complete with respect to different classes of models. This aspect of modal semantics will be more closely analyzed in the next chapter.

3.5 Exercises

1. Prove, by applying the tableaux method, that the formula (**Ban**) $(p \supset \Box p)$ and formula (**Triv**) $(p \equiv \Box p)$ are underivable in every system included in **S5**.

2. Prove that every theorem of **S5** takes distinguished truth–values in the infinite matrix defined in Section 3.2.

3. Give a recursive definition of the function V in an implicit relational model $\langle W, R, V \rangle$.

4. (a) Prove that axiom (**5**) is valid in all Carnapian models.
   (b) Prove that the rule of Uniform Substitution preserves validity if applied to theorems of any subsystem of **S5**.

5. Prove that the absorption laws (AT1)–(AT4) of Proposition 2.4.2 are valid in all universal models.

6. Prove that if a modal formula α is valid on a frame $\mathcal{F}$ iff $\mathcal{F}$ has a certain property P, and β is valid in $\mathcal{F}$ iff $\mathcal{F}$ has a certain property Q, then $\alpha \wedge \beta$ is valid in $\mathcal{F}$ iff $\mathcal{F}$ satisfies both P and Q.

7. Prove the correspondence between the formulas (**T**), (**B**),(**Ver**) and (**F**) and the first–order properties associated to them in the table of Section 3.2.

8. Prove that the rule of Uniform Substitution preserves validity on a frame. (Hint: Suppose that p occurs in α and, by *Reductio*, that $\mathcal{M}$ is a model over a certain frame $\mathcal{F}$ on which α is valid, such that at a world w of $\mathcal{M}$ it holds $v(\alpha[p/\beta], w) = 0$. Define a model $\mathcal{M}^*$ as $\mathcal{M}$, except that $v^*(q, w') = v(q, w')$ if $q \neq p$, and $v^*(p, w') = 1$ iff $v(\beta, w') = 1$. Prove that $\mathcal{M}^* \not\Vdash \alpha$ by induction on the complexity of α.)

9. Prove that a transitive frame $\mathcal{F}$ validates the formula (**McK**) (i.e., $\Box\Diamond p_1 \supset \Diamond\Box p_1$) if and only if $\mathcal{F}$ satisfies the McKinsey property.

10. Show that the infinite matrix $\mathcal{M}_\infty$ introduced in the proof of Dugundji's Theorem (see Proposition 3.1.4) may be converted into an implicit Carnapian model for **S5**.

11. Prove that a reflexive frame $\mathcal{F}$ is euclidean if and only if $\mathcal{F}$ is symmetric and transitive.

12. Prove that the formula $\Box p \supset \Box\Diamond p$ is not **KD**-valid.

13. Prove that the following formulas are **KT**-valid:

    (i) $\Box(p \wedge q) \supset (\Box p \wedge \Box q)$
    (ii) $\neg\Diamond(p \vee q) \supset (\neg\Diamond p \wedge \neg\Diamond q)$
    (iii) $\Diamond(p \wedge q) \supset (\Diamond p \wedge \Diamond q)$
    (iv) $((q \dashv p) \wedge (q \dashv \neg p)) \supset \neg\Diamond q$
    (v) $\Box p \supset (q \dashv p)$
    (vi) $\Box p \supset (\Diamond q \supset \Diamond(p \wedge q))$

14. Prove the following properties for an arbitrary relation R:

    (i) If R is symmetric and euclidean, then R is transitive.
    (ii) If R is reflexive and euclidean, then R is symmetric and transitive.
    (iii) If R is reflexive, then R is euclidean iff R is symmetric and transitive.
    (iv) If R is reflexive, then R is serial.
    (v) If R is symmetric and transitive, then R is euclidean.
    (vi) If R is serial, symmetric and transitive, then R is euclidean.

15. Prove that, for an arbitrary frame $\mathcal{F} = \langle W, R \rangle$, that its relation R is n-dense iff $\mathcal{F} \vDash \Box^{n+1} p \supset \Box^n p$.

3.5. EXERCISES

16. Prove that there are exactly 15 distinct systems between **K** and **S5**, obtained by combining the axioms listed in Section 2.3: **K, KD, KT= T, KB, K4, K5, KDB, KD4, KD5, K45, KTB= B, KT4= S4, KD45, KB4, KT5= S5**. Hint: from the 32 possible combinations of such axioms, use the properties of relations in the previous exercise to prove the following reductions:

    (a) **KT = KDT**, using item (iv).
    (b) **KB5 = KB4 = KB45** using items (i) and (v).
    (c) **KDTB = KTB** and **KDT4 = KT4** using item (iv).
    (d) **KT5 = KDT5 = KTB4 = KTB5 = KT45 = KDTB4 = KDTB5 = KTB45 = KDT45 = KDTB45** using items (ii), (iii), and (iv).
    (e) **KDB4 = KDB5 = KDB45** using items (i) and (v).
    (f) **KDTB4 = KDB4** using item (vi).

17. Prove that **S4** contains exactly 14 irreducible modalities, namely: $\Box$, $\Diamond$, $\Box\Diamond$, $\Diamond\Box$, $\Box\Diamond\Box$, $\Diamond\Box\Diamond$, plus the same preceded by $\neg$ and the two zero-degree modalities.

18. Prove that there exists an infinite number of irreducible modalities in **KT** and its subsystems.

19. Beyond the tableau procedure for **S5** outlined at Section 3.4, there is another practical procedure for **S5** (due to R. Carnap). Any **S5**-formula α can be reduced to an equivalent formula α' in Modal Conjunctive Normal Form, which means that α' is conjunction of sentences β_i, where each β_i is of the form $\gamma \vee \Box\delta_1 \vee \cdots \vee \Box\delta_n \vee \Diamond\epsilon$. Prove with semantical methods that any conjunct β_i is true iff any of the subformulas $\alpha \vee \epsilon, \delta_1 \vee \epsilon \cdots, \beta_n \vee \epsilon$ is **PC**–valid. (Hint: Define a Carnapian model $\langle W, V \rangle$ with $n + 1$ worlds such that, at any world in W, each one of the disjunctive subformulas above has value 0 and prove that the disjunction γ_i is invalid in this model, so it cannot be a theorem of **S5**).

20. Choose one of the modal systems introduced in this chapter–with the exception of (**Ver**)– and prove its consistency by means of the following method. Let $e(\alpha)$ be the formula obtained by eliminating every modal operator in α and show the following two points:

    (i) If α is a theorem of the chosen system then $e(\alpha)$ is a **PC**-tautology (by induction on the length of proofs).

(ii) The consistency of **PC** implies the consistency of the chosen system.

21. Prove, using Correspondence Theory, that the formula $\Box(\Box p \supset p)$ has as standard translation the property of *quasi-reflexivity*: $\forall x \forall y (xRy \supset yRy)$. Show that $\Box(\Box p \supset p)$ is a plausible deontic axiom.

22. Prove by the relational tableaux method: (i) that $\Box p \supset p$ is not a theorem of **KD**, and (ii) that $\Diamond(p \supset \Box p)$ is not a theorem of **KD**, but is a theorem of **KT**.

23. Build an **S4**-tableau whose input is the McKinsey axiom and show that without the stop rule the procedure does not end in a finite number of steps. Prove that the introduction of the stop rule allows the construction of a falsifying **S4**-model for this formula.

24. Prove that **S4** has the following property, known as Modal *Disjunction Property*: if $\vdash_{S4} \Box \alpha_1 \vee \cdots \vee \Box \alpha_n$, then $\vdash_{S4} \alpha_i$, for some $1 \leq i \leq n$.

25. Prove that the formula $\Box(\Box p \supset q) \vee \Box(\Box q \supset p)$ is a modal expression of the property called *weak connection*, that is $(\forall x)(\forall y)(\forall z)((xRy \wedge xRz) \supset (yRz \vee zRy))$.

26. At the semantical level, at least two different notions of logical consequence from a set of premises Γ can be defined: the *global* ($\vDash^G$) and the *local* ($\vDash^L$) consequence relations. Let C be a class of frames:

    - $\Gamma \vDash^G \alpha$ iff there exists a finite subset $\Gamma' = \{\alpha_1 \cdots \alpha_n\}$ of formulas of Γ such that, for every model $\mathcal{M}$ over a frame $\mathcal{F} \in C$, $\mathcal{M} \vDash \Gamma$ implies $\mathcal{M} \vDash \alpha$.
    - $\Gamma \vDash^L \alpha$ iff there exists a finite subset $\Gamma = \{\alpha_1 \cdots \alpha_n\}$ of formulas of Γ such that, for every model $\mathcal{M}$ over $\mathcal{F} \in C$, we have what follows: for each world w of $\mathcal{M}$, $\mathcal{M}, w \vDash \Gamma$ implies $\mathcal{M}, w \vDash \alpha$.

    Prove that the Semantical Deduction Theorem holds for the relation $\vDash^L$ but not for the relation $\vDash^G$. Interpret this in light of Remark 3.2.13.

3.6 Further reading

For Dugundji's theorem see J. Dugundji [Dug40], and for the standard translation of **S5** into the monadic fragment of the predicate calculus **QL**, see M. Wajsberg in [Waj33].

3.6. FURTHER READING

Dugundji's theorem in a sense blocked the interest for modal logics, as it seemed to open a wide gap between modal logics and many-valued logics and to suggest the impossibility of decision procedures for modal logics. But, on the other hand, the challenge represented by this negative result stimulated research into fundamentally new directions on modal logics, soliciting the development of non-matrix semantics.

The semantics exposed in all chapters of the present book are two-valued. This does not mean that three-valued, or generally, many-valued modelizations of modal systems are impossible or uninteresting. The early three-valued modal logics developed by Jan Łukasiewicz in 1936, in fact, lead to a "multi-valued reading" of modal notions which is highly non-standard (for recent developments in this direction, see J. M. Font and P. Hájek [FH02]). The modal logic of Łukasiewicz (of 1953) is reprinted in [Łuk70].

The first semantical analysis of modal logic was proposed in Carnap [Car47] but with reference only to **S5** and to Leibniz's conception of necessity. For the debate over the relation between **S5** and the system $\mathbf{C} = \mathbf{S5} + \Diamond p$ (in a sense, the genuine Carnap's modal logic) see G. Schurz [Sch01] where **C** receives a positive reappraisal.

S. Kripke was the first to publish a proof of completeness and decidability of some basic systems based on the relational semantics in [Kri59] and [Kri63a]. For a discussion on the genesis of possible-worlds and relational semantics, however, see J. Copeland [Cop02], where the author surveys the development of possible-worlds semantics through the work of several logicians as J. Hintikka, S. Kanger and R. Montague. Also, A.-V. Pietarinen [Pie06] defends the claim that C. S. Peirce already proposed, a century ago, a logical approach to modalities anticipating possible-worlds semantics.

The problem whether or not G. W. Leibniz had in mind the equivalence between "necessarily true" and "truth at all possible worlds" is not yet settled; it is very likely that Leibniz was convinced that necessity implied truth at all possible worlds, but it is doubtful that he would agree with the converse implication. Important references on this discussion are R. Kauppi [Kau60], B. Mates [Mat89] (chapter VI) and M. Mugnai [Mug01].

The tableau method introduced by Kripke is extensively used in G. E. Hughes and M. J. Cresswell [HC68], the first manual entirely devoted to modal logic. The problem of closure for **S4**-tableaux is discussed in B. Tapscott [Tap84]. A reference for the method of analytical tableaux in modal logic is M. C. Fitting [Fit83].

For a discussion on how propositional modal logic is better understood as a fragment of second-order logic, an important reference is S. K. Thomason [Tho72].

Correspondence Theory was introduced by J. van Benthem in [vB83b] and [vB84], but also look at R. Goldblatt in [Gol93]. For the Goldblatt-Thomason theorem, see [GT75]. Bisimulations were introduced in the context of modal logic by van Benthem in [vB76].

An interesting example of the use of bisimulations to prove non-definability results may be found in P. Blackburn, M. de Rijke and Y. Venema [BdRV01], where it is proven that the *global modalities* (sometimes called universal modalities) $\mathbb{A}$ and $\mathbb{E}$, semantically defined as:

$$\mathcal{M}, w \vDash \mathbb{A}\alpha \text{ iff } \mathcal{M}, w' \vDash \alpha, \text{ for all } w' \in M$$

$$\mathcal{M}, w \vDash \mathbb{E}\alpha \text{ iff } \mathcal{M}, w' \vDash \alpha, \text{ for some } w' \in M$$

are not definable by the usual modal operators. Though relatively simple, the argument is philosophically meaningful as it suggests that first-order logic may be treated as a special kind of modal logic (for which see also Section 9.5) endowed with such global or universal modalities.

Bisimulations were independently introduced by D. Park as a kind of equivalence between state transition systems in D. Park [Par81]. Park showed that if two deterministic automata are related by a bisimulation, then they accept the same set of inputs. This result gives evidence to the modal character of computing processes.

The notion of bisimulation is also connected to game semantics, a game-theoretical approach to formal semantics where truth and validity are grounded on concepts such as the existence of a winning strategy for a player. The idea of game semantics for logic, and of game-theoretic approach to logic in general, can be traced back to P. Lorenzen and J. Hintikka in the 1950s and it has been further developed by many authors. A survey of logic games is found in W. Hodges [Hod01]. The *p*-morphism introduced by K. Segerberg in [Seg70] are special cases of bisimulations (see also[Seg71]).

For further model-theoretical aspects of modal logic, see A. Chagrov and M. Zakharyaschev [CZ97], and for a comprehensible account of several semantical aspects of modalities, including issues of computation and complexity, see Blackburn and van Benthem [BvB07].

Chapter 4

Completeness and canonicity

4.1 The constructive completeness of K and KT

The previous chapter examined the soundness of modal systems with respect to classes of suitably defined relational frames.

In Section 1.3 we have observed that the standard Propositional Calculus **PC** may be described as complete in at least two different senses, a weak one and a strong one. Such a distinction may obviously be extended from **PC** to any logical system **S**. The strong sense of completeness is the converse of strong soundness which, as we recall, is expressed by saying that, if **S** is an arbitrary logical system, for any set of formulas Γ and for every wff α, $\Gamma \vDash_S \alpha$ implies $\Gamma \vdash_S \alpha$. The weak sense of completeness is expressed as a converse of weak soundness, i.e., by the relation expressed by saying that, for every α, $\vDash_S \alpha$ implies $\vdash_S \alpha$. The latter is the dominant sense of completeness which will be used in the present chapter. As a matter of fact, it is not difficult to prove that for most[1] normal modal systems **S**, completeness in the weak sense is equivalent to the more powerful notion of strong completeness.

Of course, strong completeness immediately implies weak completeness (indeed, weak completeness is just the particular case of strong completeness in which $\Gamma = \emptyset$), but the converse is not true: there are systems, for instance the system which will be treated under the name **KGL** (cf. Section 5.2), which are not strongly complete with respect to any class of frames, but yet are weakly complete.

As we will show in the next chapter (Section 5.2) the failure of strong completeness (see specifically page 133) goes hand in hand with the failure of a property which we call modal-semantical compactness, where a normal

---

[1] That is, at least for the the ones which will be called "canonical" (see Section 4.2).

modal system **S** is said to be *modal-semantically compact* iff for every consistent set Γ of wffs in **S**, there exists a model $M = \langle W, R, v \rangle$ for **S** and a single world $w \in W$ such that $M, w \vDash \alpha$ for every $\alpha \in \Gamma$.

Extending what has been said for **PC** in Chapter 1, we may also say that an arbitrary system **S** is Post-complete when **S** lacks proper consistent extensions. On the other hand, we have already noticed that, with the exception of such degenerated systems as **Ban**, **Triv** and **Ver**, no normal modal system **S** may be Post-complete since it can be consistently extended with (**Ban**).

Remark 4.1.1 *The underivability of (**Ban**) does not mean that its negation may be used in modal axiomatization. Indeed, the negation of (**Ban**), i.e., $\neg(p \supset \Box p)$, is inconsistent with **K** since it is equivalent to $p \wedge \neg \Box p$. If this formula were a theorem, then p would also be a theorem of **K**, which is impossible. Note that the same result holds for negations of the implicative formulas (**D**), (**T**), (**B**), (**4**) and (**5**).*

As a matter of fact, in modal logic not only Post-completeness but completeness itself should be treated more analytically than in standard logic. In fact, completeness is the converse of soundness, but we know that a modal system may be sound with respect to different classes of frames.

$\mathcal{F}$ is said to be a *frame for* **S** if every **S**-theorem is valid on $\mathcal{F}$. If we write $\mathfrak{F} \vDash \alpha$ to say that α is valid in the class of frames $\mathfrak{F}$, the form which usually receives an arbitrary completeness result for a modal system **S** is then provided by:

$$\mathfrak{F} \vDash \alpha \text{ implies } \vdash_S \alpha$$

When a system **S** satisfies the above described property, it is said to be *semantically complete* – or simply *complete* – with respect to the class of frames $\mathfrak{F}$, and the proof of this property for a system **S** is called a proof of *semantic completeness* – or simply *completeness* – of **S** with respect to $\mathfrak{F}$.

When a system **S** is both sound and complete with respect to a class of frames $\mathfrak{F}$, it is said to be *characterized by* $\mathfrak{F}$. In this case, the class of **S**-theorems coincide exactly with the class of wffs which are valid on every frame $\mathcal{F} \in \mathfrak{F}$ (of course, $\mathfrak{F}$ may be a singleton).

If α is a wff and Γ is any set of wffs, we say that α is a *valid consequence* of Γ with respect to the class of frames $\mathfrak{F}$ (notation: $\Gamma \vDash_{\mathfrak{F}} \alpha$) if, for each frame $\mathcal{F} \in \mathfrak{F}, \mathcal{F} \vDash \Gamma$ implies $\mathcal{F} \vDash \alpha$ (see Definition 3.2.12).

Soundness or completeness with respect to a class of frames $\mathfrak{F}$ is equivalent to soundness or completeness with respect to the class of models based on such frames, but soundness or completeness with respect to a class of models $\mathfrak{M}$ does not imply soundness or completeness with respect to the

4.1. THE CONSTRUCTIVE COMPLETENESS OF K AND KT

frames on which the models of $\mathfrak{M}$ are based. For a simple example, consider a class of models $\mathfrak{M}$ consisting of just one relational model $M = \langle W, R, V \rangle$; obviously, it is not the same to say that a certain formula α is valid in M and to say that α is valid on the frame $\langle W, R \rangle$ on which M is based.

A completeness proof for a system **S** is called *constructive* when it shows how to transform a proof of **S**-validity of a formula α into a syntactic proof of α within the system **S**. For the systems **K**, **T**, **S4** and **S5** examined in the previous chapter there is indeed a constructive proof of completeness. We know that a formula α is valid in **S** when the tableau that has α as input is closed, that is, ends with a contradictory assignment.

The method to prove constructive completeness in modal logics is a sophistication of the idea of "rectification under a valuation" introduced in Section 1.4. The essence of the method relies on the fact that, in a closed tableau for α, contradictory valuations to a subformula δ will yield (for a certain formula χ_k representing the content of the diagram itself) **PC**$^\square$-theses of the form $\delta \supset \chi_k$ and $\neg \delta \supset \chi_k$. From this, applications of Proof by Cases as in Proposition 1.4.2 plus (**Nec**) and (**K**) permit us to extract a syntactic proof of α in **S** from a closed tableau.

For the basic system **K**, the idea of constructive completeness can be outlined in the following steps:

1. Eliminate (using the definitions) all modal operators in the input formula α except $\square$. Then, all existential assertions consist of assignments 0 to subformulas of α with the form $\square \beta$, while universal assertions consist of assignments 1 to subformulas of α with the form $\square \gamma$. The introduction of any diagram w_j accessible from a diagram w_i will then be justified from an assignment 0 to some subformula $\square \beta$ contained in w_i. Given the normalized language, any diagram can be cast into a standard format, composed by cells in the following way:

    $$\boxed{\beta \mid \gamma_1 \mid \cdots \mid \gamma_k}$$

    where $\beta, \gamma_1, \cdots, \gamma_k$ are (immediate) subformulas of boxed formulas belonging to the preceding diagram w_i, β has value 0 and $\gamma_1, \cdots, \gamma_k$ have value 1.

2. To each diagram w_i a *characteristic formula* χ_i is associated, that is, a formula that univocally represents the contents of the diagram itself. The characteristic formula χ_i of a diagram w_i containing $\beta, \gamma_1 \cdots, \gamma_k$, as illustrated at item (1), has the form of a (possibly degenerate) disjunction $\beta \vee \neg \gamma_1 \vee \cdots \vee \neg \gamma_k$. Consequently, the characteristic formula of the first diagram w_0 is coincident with the input formula.

3. For any diagram w_k, let χ_k be its characteristic formula and δ be any subformula of χ_k.

   (a) If δ is assigned value 0 in the diagram, then $\delta \supset \chi_k$ is $\mathbf{PC}^\square$-valid.
   (b) If δ is assigned value 1 in the diagram, then $\neg\delta \supset \chi_k$ is $\mathbf{PC}^\square$-valid.

4. Let w_k be the last diagram of a closed tableau. Then its characteristic formula χ_k contains some subformula δ, such that δ receives contradictory values 0 and 1. By the argument in the preceding step, $\delta \supset \chi_k$ and $\neg\delta \supset \chi_k$ will both be $\mathbf{PC}^\square$-theses, and by Proof by Cases (see Section 1.2), χ_k is also a $\mathbf{PC}^\square$-thesis. By applying (**Nec**) it follows that $\square\chi_k$ (i.e. $\square(\beta \vee \neg\gamma_1 \vee \cdots \vee \neg\gamma_k)$) is a **K**-thesis.

5. By axiom (**K**) in the variant $\square(\neg\alpha \vee \beta) \supset (\neg\square\alpha \vee \square\beta)$, we have that $\square(\beta \vee \neg\gamma_1 \vee \cdots \vee \neg\gamma_k)$ implies $\square\beta \vee \neg\square\gamma_1 \vee \cdots \vee \neg\square\gamma_k$ or, equivalently, that $\square\chi_k$ implies $\square\beta \vee \neg\square\gamma_1 \vee \cdots \vee \neg\square\gamma_k$. Such disjunction is then also a **K**-thesis.

6. Now, as each disjunct of $\square\beta \vee \neg\square\gamma_1 \vee \cdots \vee \neg\square\gamma_k$ belongs to the previous diagram w_{k-1}, then the characteristic formula χ_{k-1} contains among its subformulas all of $\square\beta, \neg\square\gamma_1, \cdots, \neg\square\gamma_k$. By virtue of step (3), as $\square\beta$ and $\neg\square\gamma_1, \cdots, \neg\square\gamma_k$ receive value 0 in w_{k-1}, each such subformula implies χ_{k-1}. By the "Disjunction Introduction" rule (see Section 1.2), which is obviously also a derived rule of $\mathbf{PC}^\square$:

   $\vdash_{\mathbf{PC}^\square} \rho \supset \chi_{k-1},\ \vdash_{\mathbf{PC}^\square} \sigma \supset \chi_{k-1}$ implies $\vdash_{\mathbf{PC}^\square} (\rho \vee \sigma) \supset \chi_{k-1}$

   so it follows that $(\square\beta \vee \neg\square\gamma_1 \vee \cdots \vee \neg\square\gamma_k) \supset \chi_{k-1}$ is a **K**-thesis. Since $\square\chi_k$ is a theorem of **K**, by item (5) and (**MP**) one concludes that χ_{k-1} is also a theorem of **K**.

7. By reiterating the preceding argument for every diagram in the tableau until reaching the first diagram, we prove that χ_0 (that is, the input formula α) is a theorem of **K**. This concludes the proof.

Example 4.1.2 *Let the input formula α be the formula $r \vee \neg\square q \vee \square(\neg\square p \vee \square p)$. We first show that α is K-valid by the method of tableaux.*

The tableau for α consists of only two diagrams w_0 and w_1, where w_0 contains the input formula α with value 0 and w_1 contains q with value 1 and $\neg\square p \vee \square p$ with value 0, which obviously leads to a contradiction. The characteristic formula of w_1, χ_1, is $(\neg\square p \vee \square p) \vee \neg q$. Since $\square p$ receives values 0 and 1 in w_1, then by step (3) it follows that $\neg\square p \supset \chi_1$ and $\square p \supset \chi_1$ are both $\mathbf{PC}^\square$-theorems. Hence, by step (4) above, it turns out that χ_1 is a $\mathbf{PC}^\square$-theorem and hence a K-theorem. By

4.1. THE CONSTRUCTIVE COMPLETENESS OF K AND KT

*applying the rule (**Nec**) in χ_1, we have by $\mathbf{PC}^\square$ $\square(q \supset (\neg\square p \vee \square p))$, and by (**K**) and (**MP**), one obtains $\square q \supset \square(\neg\square p \vee \square p)$, an equivalent of $\neg\square q \vee \square(\neg\square p \vee \square p)$, as a **K**-theorem. But each disjunct of this formula is a subformula of α, and we know that it implies α by Step (3). So their disjunction (a **K**-theorem) also implies α and, by Modus Ponens, we obtain α as a **K**-theorem.*

The method of proof illustrated in the above example provides a very simple procedure for the system **K**, but for stronger systems some changes are mandatory. The complexity of the procedure depends on the fact that the characteristic formula χ_k of the terminal diagram may be dependent on axioms of the underlying system or on some of their instances. Such axioms express (in the object language) properties of the accessibility relations that are represented by the arrow in the tableau. In **KT**, for instance, we need to show that there is a proof of the formula $\bigwedge(\mathbf{T}) \supset \chi_k$, where $\bigwedge(\mathbf{T})$ means the conjunction of all the instances of the axiom (**T**) (that is, wffs of form $\square\alpha \supset \alpha$) such that both $\square\alpha$ and α receive a value in the diagram w_k. The strategy is then to prove χ_k with the help of the theorem $\bigwedge(\mathbf{T})$, which expresses the reflexivity of the accessibility relation R, and to follow from this point on along the same lines as in **K**. The following example illustrates the idea.

Example 4.1.3 *Let α be the formula $\square(p \supset \neg\square\neg p)$. We first show that α is **KT**-valid by the method of tableaux.*

*The tableau for α consists of two diagrams w_0 and w_1 (see next figure), where w_0 contains the input formula and w_1 contains the formula $p \supset \neg\square\neg p$ with value 0, p with value 1 and $\square\neg p$ and $\neg p$ with value 1 (see **KT** and the role of reflexivity in Section 3.4); this diagram thus contains a contradictory assignment to the variable p, as depicted in the figure below. The formulas contained in both diagrams are*

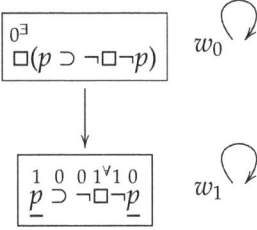

equivalent to their characteristic formulas.

Now consider the following formulas of $\mathbf{PC}^\square$, where p is the variable which receives contradictory assignments in the diagram w_1, and both $\square\neg p$ and $\neg p$ receive a value in w_1 by way of the reflexivity of R:

$$(\Box \neg p \supset \neg p) \supset (p \supset (p \supset \neg \Box \neg p))$$
$$(\Box \neg p \supset \neg p) \supset (\neg p \supset (p \supset \neg \Box \neg p))$$

*It is easy to see that both formulas are **PC**$^\Box$-valid, hence that they are theorems of **PC**$^\Box$ (by the completeness of this system); they are therefore theorems of **KT**. Since the antecedent of each formula above is an instance of the axiom (**T**), by (**MP**) the two consequents of the formulas $p \supset (p \supset \neg \Box \neg p)$ and $\neg p \supset (p \supset \neg \Box \neg p)$ are theorems of **KT**. By Proof by Cases, then, $p \supset \neg \Box \neg p$ is a **KT**-theorem, and by (**Nec**) we have that $\Box(p \supset \neg \Box \neg p)$ (that is, the characteristic formula of w_0) is a **KT**-theorem.*

Remark 4.1.4 *A rigorously mechanical application of the procedure would take into account that $\Box(p \supset \neg \Box \neg p)$ is a subformula of itself with value 0. So the required result would follow by applying (**MP**) to the thesis $\Box(p \supset \neg \Box \neg p) \supset \Box(p \supset \neg \Box \neg p)$.*

For more complex formulas of **KT** involving several diagrams, the result would follow by applying the same argument used above for the system **K**.

Note that the syntactic proof of the tested formula provided by the outlined method is usually not the quickest or the most elegant. However, it has the advantage of being mechanical, and as such it can in principle be performed by a suitable automated theorem prover. Moving from **KT** to stronger systems, the constructive proofs of completeness become more and more complex, depending upon specific properties of the respective accessibility relations described by the corresponding axioms. We have thus to rely on more general but unfortunately non-constructive methods, as we will see in next section.

4.2 Completeness by Henkin's method

An important progress in the analysis of modal systems has been attained by studying not only the properties of a single system, but the properties of an unlimited number of systems belonging to the same class. A treatment of this kind is suggested here because of the fact that, with the exception of the axioms (**K**) and (**Ver**), the most important modal axioms that we examined in the previous chapters turn out to be special cases of the following single remarkable schema $G^{k,l,m,n}$, where numerical indices stand for the number of iterated operators of the same kind:

$$G^{k,l,m,n}: \quad \Diamond^k \Box^l p \supset \Box^m \Diamond^n p$$

The axiom (**T**), for instance, is the special case provided by $G^{0,1,0,0}$, while axiom (**4**) is just $G^{0,1,2,0}$. The special case $G^{1,1,1,1}$, i.e. (**G1**) $\Diamond \Box p \supset \Box \Diamond p$,

4.2. COMPLETENESS BY HENKIN'S METHOD

however, has not been treated before (notice that it is the converse of (**McK**) introduced on page 67). (**G1**) is, in fact, the characteristic axiom of a normal modal system we call **KG1**, whose theorems turn out to be valid in the class of all convergent models, or models having the so-called *diamond property*[2]: if a world w_1 sees two worlds w_2 and w_3, then there exists a world w_4 that is accessible from w_2 and w_3. Given that $G^{k,l,m,n}$ is not an axiom, but an axiom schema, the condition on the accessibility relations R corresponding to $G^{k,l,m,n}$ is also a schema of conditions. To illustrate the diamond property, we recall notation R^m at page 62. The schema of conditions is expressed by the following first-order formula:

$$C^{k,l,m,n}: \quad \forall w_1 \forall w_2 \forall w_3 ((w_1 R^k w_2 \wedge w_1 R^m w_3) \supset \exists w_4 (w_2 R^l w_4 \wedge w_3 R^n w_4))$$

This schema can be visualized by the following convergent graph:

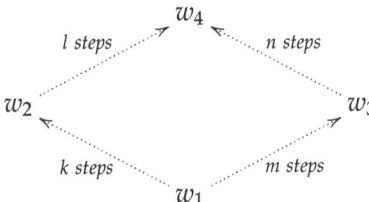

We now prove the properties of soundness and of completeness for each system that is an extension of **K** with arbitrary formulas that are instances of the schema $G^{k,l,m,n}$. Moreover, as it will be clear from what follows, such results of soundness and completeness will hold for extensions of **K** with finitely many instances of $G^{k,l,m,n}$.

Proposition 4.2.1 *If α is a wff of form $G^{k,l,m,n}$, α is valid on all frames in which R satisfies the condition $C^{k,l,m,n}$.*

Proof: Suppose by *Reductio* that some instance of the schema $G^{k,l,m,n}$ is not valid on a given frame. Then there exists a world w_1 such that:

(i) $v(\Diamond^k \Box^l \alpha, w_1) = 1$

(ii) $v(\Box^m \Diamond^n \alpha, w_1) = 0$

---
[2] Relations having this property as well as models and frames endowed with them are called *incestual* by B. F. Chellas [Che80], and also said in the literature to enjoy the *Church-Rosser property*.

– From (i), it follows that there exists a world w_2 accessible from w_1 in k steps such that $v(\Box^l \alpha, w_2) = 1$, hence that α is true in any world accessible from w_2 in l steps.
– From (ii), it follows that there exists a world w_3 accessible from w_1 in m steps in which $v(\Diamond^n \alpha, w_3) = 0$, hence that α is false in any world accessible from w_3 in n steps.

Since the relation R satisfies the condition $\mathbf{C}^{k,l,m,n}$, then, given that w_1 accesses w_2 in k steps and w_3 in m steps, condition $\mathbf{C}^{k,l,m,n}$ implies that there exists a world w_4 that is accessible from w_2 in l steps and from w_3 in n steps. So we reach a contradiction, since we have $v(\alpha, w_4) = 1$ from (i) and $v(\alpha, w_4) = 0$ from (ii). This concludes the proof. ♠

A consequence of Proposition 4.2.1 is what follows:

Corollary 4.2.2 *Each theorem of K extended with an instance of $G^{k,l,m,n}$ is valid on all frames in which R satisfies condition $\mathbf{C}^{k,l,m,n}$.*

For simplicity, from now on we will write G^∞ to denote some specific instance of $G^{k,l,m,n}$ and $\mathbf{C}^\infty$ for the corresponding condition on the accessibility relations. Different instances of $\mathbf{C}^\infty$ and G^∞ will be distinguished by subscripts.

In order to prove completeness, we now have to show:

Proposition 4.2.3 *If α is valid on all frames that satisfy a specific condition $\mathbf{C}^\infty$, then α is a theorem of the system K extended with G^∞.*

Before giving the proof of this proposition, it is useful to clarify some ideas. First of all, note that to state $\nvdash_S \neg \alpha$ means to state that α is consistent with the reference system **S**. It is recommendable at this point to recall the notion of an **S**-consistent set of wffs (Section 1.3).

Along the same lines, $\mathfrak{F} \vDash \neg \alpha$ means that $\neg \alpha$ is valid in all models based on the frames in a class $\mathfrak{F}$, and $\mathfrak{F} \nvDash \neg \alpha$ means that $\neg \alpha$ is false (hence α is true) in some model based on some frame in $\mathfrak{F}$.

Therefore, provided that α is an arbitrary formula (in particular, it can be, of course, a negation $\neg \beta$), an equivalent formulation of Proposition 4.2.3 is the following:

Proposition 4.2.4 *If α is consistent with K extended with some G^∞ ($\nvdash_{K+G^\infty} \neg \alpha$), then α is valid in at least a model $\mathcal{M}$ over some frame satisfying condition $\mathbf{C}^\infty$ ($\nvDash_{K+G^\infty} \neg \alpha$).*

4.2. COMPLETENESS BY HENKIN'S METHOD

Let us remark that Proposition 4.2.4 grants that in order to prove the completeness of a system $\mathbf{K}+G^\infty$, it is enough to find, for each formula α consistent with G^∞, a $\mathbf{K}+G^\infty$-model satisfying α. Following a standard method of proof called *Henkin's method*, the problem is solved by performing two steps:

(i) We build a special model of $\mathbf{K}+G^\infty$ (so-called *canonical model*) in which the worlds are particular sets of formulas depending on $\mathbf{K}+G^\infty$.

(ii) We show that if any formula α is consistent with $\mathbf{K}+G^\infty$, then α is valid in the canonical model.

Definition 4.2.5 *w is a maximal consistent extension of a system S iff:*

(i) $S \subseteq w$

(ii) w is consistent

(iii) For every wff α, $\alpha \in w$ or $\neg \alpha \in w$

Definition 4.2.6 *The set $\text{Den}(w) \stackrel{\text{Def}}{=} \{\alpha : \Box \alpha \in w\}$ will be called the set of the denecessitated wffs of w.*

Definition 4.2.7 *The canonical model of (or on) S is the triple $\langle W_S, R_S, v_S \rangle$ with the following properties:*

1. W_S is the set of all maximal consistent extensions of S

2. $wR_S w'$ iff $\text{Den}(w) \subseteq w'$ for every w, w' in W_S

3. For every atomic variable p,

$$v_S(p, w) = \begin{cases} 1 & \text{if } p \in w \\ 0 & \text{if } p \notin w \end{cases}$$

Definition 4.2.8 *The canonical frame of (or on) S is the pair $\langle W_S, R_S \rangle$ where $\langle W_S, R_S, v_S \rangle$ is the canonical model of S.*

The worlds in a canonical model of S are all the maximal consistent extensions of S. The accessibility relation between worlds depends then on a special correlation between formulas of a world and formulas of another: more precisely, it can be modelled on the accessibility relation among diagrams in a tableau, since the relation holds between w and w' when $\Box \alpha \in w$ and $\alpha \in w'$.

The following definition is a generalization of the definition of accessibility in a canonical model.

Definition 4.2.9 *For each $w, w' \in W$ of the canonical model of a system S, we have that $wR_S^n w'$ iff $\{\alpha : \Box^n \alpha \in w\} \subseteq w'$*

In what follows, a certain number of properties of maximal consistent sets will be presented, but proofs will only be sketched leaving the details to the reader. Lemmas 4.2.10–4.2.12 describe the well-known properties of maximal consistent sets.

Lemma 4.2.10 *If w is a maximal consistent set with respect to S, then w has the following properties:*

1. *For each formula α, exactly one element in the set $\{\alpha, \neg\alpha\}$ belongs to w;*

2. *$\alpha \wedge \beta \in w$ iff $\alpha \in w$ and $\beta \in w$*

3. *$\alpha \vee \beta \in w$ iff $\alpha \in w$ or $\beta \in w$*

4. *If $\vdash_S \alpha$, then $\alpha \in w$*

5. *If $\alpha \in w$ and $\alpha \supset \beta \in w$, then $\beta \in w$*

6. *If $\alpha \in w$ and $\vdash_S \alpha \supset \beta$, then $\beta \in w$*

Proof: We only sketch the proof of item 5 as an example, while others are left as exercises. Suppose by *Reductio* that $\alpha \in w, \alpha \supset \beta \in w$ and $\beta \notin w$. Then, since w is maximal, $\neg\beta \in w$ (since $\beta \notin w$). As $\alpha \in w$, $\alpha \supset \beta \in w$ and $\neg\beta \in w$, then $\{\alpha, \alpha \supset \beta, \neg\beta\} \subseteq w$ is consistent, given that w is S-consistent. Therefore, by definition of a consistent set, we have that $\nvdash_S \neg(\alpha \wedge (\alpha \supset \beta) \wedge \neg\beta)$, which by the maximality of S is equivalent to $\vdash_S \alpha \wedge (\neg\alpha \vee \beta) \wedge \neg\beta$, i.e., $\vdash_S \bot$, which is impossible. ♦

Lemma 4.2.11 *(Lindenbaum's Lemma) Each S-consistent set of modal formulas has at least one maximal consistent extension.*

Proof: Suppose that we enumerate all formulas of the language of some propositional modal logic S in some way

$$\alpha_1, \alpha_2, \alpha_3, \cdots, \alpha_n, \cdots$$

4.2. COMPLETENESS BY HENKIN'S METHOD

Let Γ be an **S**-consistent set and define a sequence $\Gamma_0, \Gamma_1, \Gamma_2, \cdots, \Gamma_n, \cdots$ of sets in the following way:

$$\Gamma_0 = \Gamma$$

$$\Gamma_{n+1} = \begin{cases} \Gamma_n \cup \{\alpha_{n+1}\} & \text{if } \Gamma_n \cup \{\alpha_{n+1}\} \text{ is \textbf{S}-consistent} \\ \Gamma_n \cup \{\neg \alpha_{n+1}\} & \text{if } \Gamma_n \cup \{\alpha_{n+1}\} \text{ is \textbf{S}-inconsistent} \end{cases}$$

We have to show now that if Γ_n is **S**-consistent, then Γ_{n+1} is **S**-consistent. In fact, suppose that Γ_{n+1} is **S**-inconsistent; then, by construction, we have that neither $\Gamma_n \cup \{\alpha_{n+1}\}$ is **S**-consistent, nor $\Gamma_n \cup \{\neg \alpha_{n+1}\}$ is **S**-consistent. Therefore, we have that:

(i) There are $\beta_1, \cdots, \beta_k \in \Gamma_n$ such that $\vdash_S \neg(\beta_1 \wedge \cdots \wedge \beta_k \wedge \alpha_{n+1})$

(ii) There are $\gamma_1, \cdots, \gamma_l \in \Gamma_n$ such that $\vdash_S \neg(\gamma_1 \wedge \cdots \wedge \gamma_l \wedge \neg \alpha_{n+1})$

By using **PC**$^\square$-rules we have:

$$\vdash_S \neg(\beta_1 \wedge \cdots \wedge \beta_k \wedge \gamma_1 \wedge \cdots \wedge \gamma_l)$$

Hence, Γ_n has a finite subset $\{\beta_1, \cdots, \beta_k, \gamma_1, \cdots, \gamma_l\}$ that is **S**-inconsistent, so Γ_n is **S**-inconsistent.

Now, consider the following set of formulas defined as:

$$\Gamma_{max} = \bigcup_{i=0}^{\infty} \Gamma_i$$

Clearly,

1. Γ_{max} is **S**-consistent by construction.

2. Γ_{max} is maximal. In fact, for each formula α_i, we know that $\alpha_i \in \Gamma_i$ or $\neg \alpha_i \in \Gamma_i$ and, since $\Gamma_i \subseteq \Gamma_{max}$, then $\alpha_i \in \Gamma_{max}$ or $\neg \alpha \in \Gamma_{max}$.

♠

Lemma 4.2.12 *If w is a maximal consistent extension of S containing $\neg \square \alpha$, then $Den(w) \cup \{\neg \alpha\}$ is an S-consistent set.*

Proof: The proof is by contraposition. Suppose that $Den(w) \cup \{\neg \alpha\}$ is **S**-inconsistent: then there exists a finite subset $\{\beta_1, \cdots \beta_k, \neg \alpha\}$ of $Den(w) \cup \{\neg \alpha\}$

which is inconsistent with w. This means that $(\beta_1 \wedge \cdots \wedge \beta_k) \supset \alpha$ is an **S**-thesis and by applying the rule (**DR1**) one obtains that $(\Box\beta_1 \wedge \cdots \wedge \Box\beta_k) \supset \Box\alpha$ is also such. Therefore, there exists a set $\{\Box\beta_1, \cdots \Box\beta_k, \neg\Box\alpha\}$ that is also **S**-inconsistent. ♠

A useful corollary of Lemma 4.2.10 concerning formulas with prefix ◊ is the following:

Definition 4.2.13 *The set* $Poss(w) \stackrel{Def}{=} \{\Diamond\alpha : \alpha \in w\}$ *will be called the set of the possibilitated wffs of* w.

Now, we can prove the following corollary:

Corollary 4.2.14 *Suppose that w and w' are maximal consistent extensions of an arbitrary normal modal system. Then $Den(w) \subseteq w'$ iff $Poss(w') \subseteq w$.*

Proof: Exercise 4.4. ♠

At this point we can prove the *Fundamental Theorem of Canonical Models* on modal systems.

Proposition 4.2.15 *Let $\langle W, R, v \rangle$ be a canonical model of a given normal modal system **S**. Then, for any formula α and any $w \in W$, we have:*

$$v(\alpha, w) = \begin{cases} 1 & \text{if } \alpha \in w \\ 0 & \text{if } \alpha \notin w \end{cases}$$

Proof: By induction on the length of formulas. The only non trivial case concerns $\Box\alpha$:

1. Suppose that $\Box\alpha \in w$. Then $\alpha \in w'$ for each $w' \in W$ such that wRw'; and as, by induction hypothesis, we have that $v(\alpha, w') = 1$ for each w' such that wRw', it follows $v(\Box\alpha, w) = 1$.

2. Suppose that $\Box\alpha \notin w$. Then $\neg\Box\alpha \in w$ (as w is maximal consistent). Hence, by Lemma 4.2.12 we have (for each **S**) that $Den(w) \cup \{\neg\alpha\}$ is consistent with **S**. Thus, by Lemma 4.2.11, there exists a w' such that $Den(w) \cup \{\neg\alpha\} \subseteq w'$. Therefore, $Den(w) \subseteq w'$ and $\neg\alpha \in w'$, so wRw' and $v(\alpha, w') = 0$, thus $v(\Box\alpha, w) = 0$.

♠

A corollary of the previous theorem is the equivalence between the two following properties: to be a theorem in a system **S** and to be valid in the canonical model based on **S**.

4.2. COMPLETENESS BY HENKIN'S METHOD

Corollary 4.2.16 *For any formula α, α is valid in the canonical model of S iff $\vdash_S \alpha$.*

Proof: ($\Leftarrow$) Suppose that $\vdash_S \alpha$; thus, by Lemma 4.2.10, it follows that $\alpha \in w$ for each w that extends maximally **S**, and then, by Proposition 4.2.15, $v(\alpha, w) = 1$ for each $w \in W$. Therefore α is valid in the canonical model. ($\Rightarrow$) Suppose that $\nvdash_S \alpha$; then $\{\neg\alpha\}$ is **S**-consistent. Thus, by Lemma 4.2.11, there exists a maximal consistent extension w of $\mathbf{S} \cup \{\neg\alpha\}$; hence, since w is consistent, $\alpha \notin w$. Therefore, by Proposition 4.2.15, $v(\alpha, w) = 0$, hence α is not valid in the canonical model. ♦

Proposition 4.2.15 immediately grants a proof of completeness for the system **K**.

In fact, if α is valid on all **K**-frames, it means that α is valid on every frame unexceptionably, hence in all models unexceptionably. Therefore α will also be valid in the canonical model of **K**, and then, by Corollary 4.2.16, it will be a theorem of **K**.

If the canonical frame (recall the definition of canonical frame on page 95) belongs to a certain class $\mathfrak{F}$, it follows that each formula α that is valid in such class of frames is also valid on the canonical frame, hence it is valid in the canonical model. Therefore, by Corollary 4.2.16, α is a theorem of the system whose theorems are valid on the frames of $\mathfrak{F}$. Consequently, in order to obtain a completeness proof for a system **S** with respect to $\mathfrak{F}$, it is enough to show that the canonical model based on **S** is based on a frame that belongs to $\mathfrak{F}$. We thus define:

Definition 4.2.17 *A system S is canonical if and only if the canonical frame of S is a frame for S.*

Then, to prove that a system **S** is canonical implies proving a completeness result for **S**. In order to apply this idea to some system $\mathbf{K}+G^\infty$ what we just need to prove is that this system is canonical. This will be achieved as a corollary of a stronger result which, in fact, will allow us (following E. J. Lemmon and D. Scott in [LS77], section 4) to prove at once the completeness of all the extensions of **K** with instances of the schema G^∞.

Proposition 4.2.18 *Let S be any consistent normal modal system containing $\mathbf{K}+G^\infty$ for some G^∞. Then the accessibility relation of the canonical frame of S satisfies the condition C^∞.*

Proof: Every instance of G^∞ (i.e., instance of $\Diamond^k\Box^l\alpha \supset \Box^m\Diamond^n\alpha$) belongs to every maximal consistent w which extends **S**. Suppose that $w_1 R^k w_2$ and $w_1 R^m w_3$ hold on the canonical frame of **S**. Consider the following set:

$$\Lambda = \{\alpha : \Box^l \alpha \in w_2\} \cup \{\beta : \Box^n \beta \in w_3\}$$

We have to prove that Λ is **S**-consistent.

Suppose, by *Reductio*, that Λ is not **S**-consistent; this means that there are $\alpha_1, \cdots, \alpha_r, \beta_1 \cdots \beta_s \in \Lambda$ such that $\Box^l \alpha_i \in w_2$ and $\Box^n \beta_j \in w_3$, and $\vdash_S \neg(\alpha_1 \wedge \cdots \wedge \alpha_r \wedge \beta_1 \wedge \cdots \wedge \beta_s)$. Now let α be $\alpha_1 \wedge \cdots \wedge \alpha_r$ and β be $\beta_1 \wedge \cdots \wedge \beta_s$. Then $\vdash_S \neg(\alpha_1 \wedge \cdots \wedge \alpha_r \wedge \beta_1 \wedge \cdots \wedge \beta_s)$ if and only if $\vdash_S \neg \alpha \vee \neg \beta$, if and only if $\vdash_S \alpha \supset \neg \beta$. By applying (**DR2**) n times (cf. Lemma 2.3.9), it turns out that $\vdash_S \Diamond^n \alpha \supset \Diamond^n \neg \beta$, and then, by the interchange rule, it turns out that $\vdash_S \Diamond^n \alpha \supset \neg \Box^n \beta$.

Considering that each $\Box^l \alpha_i \in w_2$ and w_2 is a maximal consistent set, then by Lemma 4.2.10, item (ii), w_2 is closed under conjunction, and by Proposition 2.3.11 (i) $\Box^l \alpha \in w_2$. By a similar argument, $\Box^n \beta \in w_3$.

Since $\Box^l \alpha \in w_2$ and $w_1 R^k w_2$, then $\Diamond^k \Box^l \alpha \in w_1$. Provided that $\Diamond^k \Box^l \alpha \supset \Box^m \Diamond^n \alpha \in w_1$, it follows by (**MP**) $\Box^m \Diamond^n \alpha$, and since $w_1 R^m w_3$, then $\Diamond^n \alpha \in w_3$. Now, from $\vdash_S \Diamond^n \alpha \supset \neg \Box^n \beta$ and (**MP**), we have that $\neg \Box^n \beta \in w_3$, which is absurd since w_3 is consistent. Therefore Λ is **S**-consistent, and by Lemma 4.2.11 there is a maximal consistent extension w_4 such that $w_2 R^l w_4$ and $w_3 R^n w_4$ (by the definition of Λ). Hence the accessibility relation of the canonical frame of **S** satisfies the condition $\mathbf{C}^\infty$. ♠

Now, Proposition 4.2.3 (or its equivalent Proposition 4.2.4) can be proven as a simple corollary of Propositions 4.2.15 and 4.2.18.

The preceding result exemplifies a general method to prove Henkin completeness for normal modal systems. To sum up, if **S** is a normal system, it is enough to show that the frame of the canonical model of **S** belongs to the class of **S**-frames, and this fact can be proved by showing that the accessibility relation of the canonical model satisfies the property of accessibility required for **S**-frames.

As already mentioned, the normal modal systems examined in the preceding chapters are special cases of $\mathbf{K}+G^{k,l,m,n}$. In a parallel way, the properties of the corresponding relation R turn out to be particular cases of the schema $\mathbf{C}^{k,l,m,n}$. To exemplify the point, it is enough to consider the cases of reflexivity, symmetry and transitivity:

$\forall w(wRw)$ iff, in $\mathbf{C}^{k,l,m,n}$ $l = 1, k = m = n = 0$

$\forall ww'(wRw' \supset w'Rw)$ iff, in $\mathbf{C}^{k,l,m,n}$ $k = l = 0, m = n = 1$

$\forall ww'w''((wRw' \wedge w'Rw'') \supset wRw'')$ iff, in $\mathbf{C}^{k,l,m,n}$
$k = l = n = 0, m = 2$

4.2. COMPLETENESS BY HENKIN'S METHOD

As a matter of fact, the axiomatic extensions of **K** obtained by combining axioms (**T**), (**4**) and (**B**) turn out to be complete thanks to a proof that is essentially the same as the one in Proposition 4.2.3, except for values of the indices. In particular, the reader can be convinced that the canonical model of **KT** is reflexive, the one of **S4** is reflexive and transitive, and the one of **S5** is reflexive and euclidean.

The completeness results proved in Proposition 4.2.4 can be shown to hold not only for singular instances of $G^{k,l,m,n}$, but also for several instances at once. We first have to be sure that the class of relations R obtained by the intersection of the classes of relations determined by $\mathbf{C}_1^\infty, \cdots, \mathbf{C}_k^\infty$ is not empty. Indeed, the following frame $\mathcal{F} = \langle \mathbb{Q}, R_\mathbb{Q} \rangle$, where $\mathbb{Q}$ is the set of rational numbers and $R_\mathbb{Q}$ is the "universal" relation on $\mathbb{Q}$ (namely, $\mathbb{Q} \times \mathbb{Q}$) satisfies *any* condition $\mathbf{C}^{k,l,m,n}$:

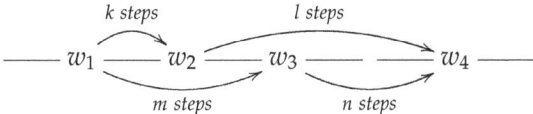

The reasons are the following: since the set $\mathbb{Q}$ of rational numbers is dense with the usual order (i.e., between any pair of rationals there exists another rational), if we understand $w R_\mathbb{Q}^n w'$ as "there exist at least $n \geq 0$ rationals ($n \geq 0$ steps) between w and w'", then from $w_1 R_\mathbb{Q}^k w_2$ and $w_1 R_\mathbb{Q}^m w_3$ we conclude that there exists w_4 such that $w_2 R_\mathbb{Q}^n w_4$ and $w_3 R_\mathbb{Q}^l w_4$. Indeed, taking any point w_4, the fact that there are infinitely many points between w_2 and w_4, as well as between w_3 and w_4, is sufficient to grant the result.

This gives us the tools to prove:

Proposition 4.2.19 *Let $G_1^\infty, \cdots, G_k^\infty$ be instances of the schema $G^{k,l,m,n}$. Then the system $\mathbf{K}+G_1^\infty + \cdots + G_k^\infty$ which results from extending $\mathbf{K}$ with $G_1^\infty, \cdots, G_k^\infty$ is sound and complete with respect to all frames in which R jointly satisfies the corresponding conditions $\mathbf{C}_1^\infty, \cdots, \mathbf{C}_k^\infty$.*

Proof: It suffices to prove the desired result for G_1^∞ and G_2^∞. Firstly, $\mathbf{K}+G_1^\infty + G_2^\infty$ is consistent, because this system is valid on the frame $\mathcal{F} = \langle \mathbb{Q}, R_\mathbb{Q} \rangle$.

Therefore, by using Proposition 4.2.18, the accessibility relation of the canonical frame of $\mathbf{K}+G_1^\infty + G_2^\infty$ satisfies the conditions $\mathbf{C}_1^\infty$ and $\mathbf{C}_2^\infty$.

On the other hand, it is clear that, if α is a thesis of $\mathbf{K}+G_1^\infty + G_2^\infty$, then α is valid on all frames in which R satisfies the conditions $\mathbf{C}_1^\infty$ and $\mathbf{C}_2^\infty$. Consequently, $\mathbf{K}+G_1^\infty + G_2^\infty$ is characterized by the class of frames in which R satisfies the conditions $\mathbf{C}_1^\infty$ and $\mathbf{C}_2^\infty$.

The same argument can be extended, reasoning two-by-two, to every system $\mathbf{K}+G_1^\infty + \cdots + G_k^\infty$. ♠

The preceding proposition immediately grants completeness results for all the subsystems of **S5** which are defined by adjoining to their axioms one or more instances of the schema $G^{k,l,m,n}$ and, in particular, for the well-known systems **KD, KT, KB, K4, K5, KDB, KD4, KD5, K45, KTB, KT4, KD45, KB4**. Of course, completeness results for **KD, KT, KB, K4** and **K5** can also be obtained as a direct consequence of Proposition 4.2.18.

For the sake of helping intuition, the reader is invited at this point to recall the semantic characterization of the best known monomodal systems (check Tables 2.3.8 and 3.2.14); this will serve as a preparation for the general form of the preceding theorem to be found at Proposition 8.6.8.

There are some axioms, however, that have received special attention by logicians but are not particular cases of the schema $G^{k,l,m,n}$. Beyond the already mentioned **Ver**, at least the following two should be recalled:

(**GL**) $\Box(\Box p \supset p) \supset \Box p$

(**D1**) $\Box(\Box p \supset q) \vee \Box(\Box q \supset p)$

This consideration shows the limits of the expressiveness of the schema $G^{k,l,m,n}$. On the other hand, there are indeed other schemas of formulas which generalize $G^{k,l,m,n}$; a particularly conspicuous one is the schema that defines the so-called Sahlqvist's monomodal systems, i.e., $\Box^n(\phi \supset \psi)$ for $n \geq 0$, where ϕ and ψ are any wffs such that:

- ϕ contains only $\Box, \Diamond, \vee, \wedge$ and $\neg$.

- $\neg$ occurs only immediately before a variable.

- no occurrences of $\Diamond, \vee$ or $\wedge$ lie within the scope of any $\Box$.

- ψ contains only $\Box, \Diamond, \vee$ and $\wedge$.

It can be seen that (**D1**) is an instance of Sahlqvist's schema, but (**GL**) (to which Section 4.4 is devoted) falls even outside them.

Multimodal versions of Sahlqvist's systems will be discussed in Section 8.4. The systems **Ver** and **S4**+(**D1**) (the later known as **S4.3**) can be axiomatized within the Sahlqvist's schema. It has been proved (see Further Reading of this chapter) that any system which results by subjoining instances of Sahlqvist's schema to **PC** is characterized by first-order definable frames. To confirm this result, let us remark that McKinsey's axiom

4.2. COMPLETENESS BY HENKIN'S METHOD

(**McK**): $\Box\Diamond p \supset \Diamond\Box p$ (which we have seen in Section 3.3 not to correspond to any first-order property) cannot be written as an instance of Sahlqvist's schema, and this imposes a limit on the expressivity of Sahlqvist's schema as well.

The system **S4.3** received special attention, since it describes frames that are qualified as *weakly connected*, i.e., reflexive, transitive and linear.

The property called *linearity* is expressible in first-order language in the following way:

(**Lin**) $\quad \forall w_i \forall w_j \forall w_k ((w_i R w_j \wedge w_i R w_k) \supset (w_j R w_k \vee w_k R w_j \vee w_j = w_k))$

In frames which are reflexive (as in the case of **S4.3** frames) since, for every w_i, $w_i R w_i$ implies $(w_i = w_k) \supset (w_i R w_k \vee w_k R w_i)$, the formula (**Lin**) is equivalent to the simpler

(**Lin'**) $\quad \forall w_i \forall w_j \forall w_k ((w_i R w_j \wedge w_i R w_k) \supset (w_j R w_k \vee w_k R w_j))$

The property of R expressed by the consequent of (**Lin**), i.e. $(w_j R w_k \vee w_k R w_j \vee w_j = w_k)$, is called *Trichotomy* or *Connectedness*.

The completeness of **S4.3** takes the following form:

Proposition 4.2.20 *S4.3 is complete with respect to the class of frames in which R is transitive, reflexive and linear.*

Proof: Let $\mathcal{M}$ be the canonical model of **S4.3**. Since $\mathcal{M}$ contains (**T**) and (**4**), it follows that $\mathcal{M}$ is reflexive and transitive. **S4.3** contains **S4**, so we already know that the mentioned canonical model is reflexive and transitive: it remains to be shown that $\mathcal{M}$ is linear. If w_i, w_j, w_k are elements of the canonical model, suppose that $w_i R w_j$ and $w_i R w_k$ (hence $Den(w_i) \subseteq w_j$ and $Den(w_i) \subseteq w_k$) and suppose, by *Reductio*, that $Den(w_j) \not\subseteq w_k$ and $Den(w_k) \not\subseteq w_j$. From the known definitions (taking into account that w_j, w_k are maximal consistent sets), we conclude that there exist p and q such that $\Box p \in w_j$, $\neg p \in w_k$, $\Box q \in w_k$ and $\neg q \in w_j$. Thus, it follows that the formula $\Diamond(\Box p \wedge \neg q) \wedge \Diamond(\Box q \wedge \neg p)$ belongs to w_i, but this is incompatible with (**D1**). Therefore the canonical model of **S4.3** satisfies the property (**Lin**), and this suffices to show the completeness of **S4.3** with respect to the mentioned class of frames. ♠

S4.3 has indeed a special position inside the spectrum of system between **S4** and **S5** (for more on this point see Section 6.3).

4.3 Completeness: models versus frames

Let us now go back to a distinction which has been outlined in Section 4.1 i.e. the distinction between completeness with respect to a class of models ($\mathcal{M}$-completeness) and completeness with respect to a class of frames ($\mathcal{F}$-completeness). We can go deeper into this analysis by introducing the following four distinct notions:

(I) Relative $\mathcal{M}$-completeness

(II) Absolute $\mathcal{M}$-completeness

(III) Relative $\mathcal{F}$-completeness

(IV) Absolute $\mathcal{F}$-completeness

We will now examine each one of these notions, keeping in mind that the notions we have used until now are (II) and (III).

(I) **Relative $\mathcal{M}$-completeness**. Relative $\mathcal{M}$-completeness is completeness with respect to some given class of models $\mathfrak{M}$. It is to be noted that a system **S** can be complete with respect to different classes of relational models $\mathfrak{M}'$, $\mathfrak{M}''$, $\mathfrak{M}'''$, $\cdots$: this fact has been already emphasized for the system **S5**, which, as seen before, can be shown to be complete with respect to the class of reflexive and euclidean frames and with respect to the class of universal frames. But even for subsystems of **S5**, there is a variety of remarkable completeness results that can be added to those already obtained. For example, the following results can also be proved:

   (a) **K** is complete with respect to the class of all irreflexive models.

   (b) **KB** is complete with respect to the class of all symmetric and irreflexive models.

   (c) **KB** is complete with respect to the class of all symmetric and intransitive models.

Normally, the procedure to establish a completeness result with respect to frames is not substantially different from the one that establishes the

4.3. COMPLETENESS: MODELS VERSUS FRAMES

same result with respect to models. However, there are some results of $\mathcal{M}$-completeness that are not derivable from parallel results of $\mathcal{F}$-completeness. In this regard, we can prove a general result that applies to all modal systems. As already mentioned, a model that validates all theses of **S** (and possibly others) is called a *model for* **S**. For example, a reflexive model consisting of just one world validates not only the theses of **KT**, but also the theses of **S4** and of **S5**: therefore, it is a model for **KT**, for **S4** and for **S5**. By using the notion of a model for **S**, we can prove the following result:

Proposition 4.3.1 *Each normal modal system **S** is complete with respect to the class of all models for **S**.*

Proof: If α is a thesis of **S**, then it is valid in all models for **S**. If α is not a thesis of **S**, then it is either inconsistent with **S** (and thus there is no model for **S** that validates α) or it is part of a maximal consistent extension of **S**; but in the latter case, its negation $\neg\alpha$ is also part of a maximal consistent extension of **S**. Hence α is false in the canonical model built on **S**, which is obviously a model for **S**. ♠

As a consequence of the above result, each system **S** is trivially complete with respect to at least a class of models, namely, the class of models for **S**.

Remark 4.3.2 *To say that a model (canonical or not) is a model for **S** is not the same as saying that it is an **S**-model (see Section 3.3).*

(II) **Absolute $\mathcal{M}$-completeness.** A system **S** is $\mathcal{M}$-complete in an absolute sense when there exists at least a class of models that validates the theses of **S** and only them. Absolute completeness is then an existential quantification over relative $\mathcal{M}$-completeness. As noted above, such a notion of $\mathcal{M}$-completeness is trivial: in fact, there is always a set of models that validates exactly the theses of **S** (namely, the class of models for **S**). Another way to understand this point is the following: Take the class $\mathfrak{M}$ consisting of exactly the canonical model of **S**. If α is a thesis of **S**, then it is valid in the canonical model; otherwise, it is not valid in the canonical model, but there exists a maximal consistent extension of **S** in which $\neg\alpha$ is true.

What we need is then to define a notion of completeness which is not trivially satisfied by all systems.

(III) **Relative $\mathcal{F}$-completeness.** Recall that to be valid on a frame $\langle W, R \rangle$ means to be valid in all models based on this frame, namely, to be valid for all assignments to the atomic variables of the given formula. Note that this implies that validity on a frame is preserved by (**US**), while this is normally not true for validity in a model.

In most interesting cases, the proof of relative $\mathcal{M}$-completeness coincides with the proof of relative $\mathcal{F}$-completeness. Normally, the proof of $\mathcal{M}$-completeness of a system **S** with respect to a certain class $\mathfrak{M}$ of models consists in showing that the frame of the canonical model (the canonical frame) enjoys the relational properties of the models in $\mathfrak{M}$. This also proves relative $\mathcal{F}$-completeness, since it implies that there exists a frame that validates **S** and belongs to $\mathfrak{M}$. Of course, this coincidence does not apply if the proof of the fact that the canonical model belongs to the class $\mathfrak{M}$ does not depends on the relational properties of the model but on some non-structural properties of it (as for example on the properties of value assignment to variables).

(IV) **Absolute $\mathcal{F}$-completeness**. The notion of absolute completeness with respect to a frame ($\mathcal{F}$-completeness) is the key to the solution of the problem formulated in (II), that is, to the problem of finding a non-trivial notion of absolute completeness: a system **S** is $\mathcal{F}$-*complete* in an absolute sense when there is some class of frames that characterizes **S**, and is $\mathcal{F}$-*incomplete* otherwise. Note that the trivial results of absolute completeness, stated above with respect to models, are not reproducible in terms of frames: in fact, validity with respect to the class of models for **S** does not coincide with validity with respect to the class of frames for **S**, and validity with respect to the canonical model of **S** does not coincide with validity with respect to the canonical frame of **S**. Indeed, it is enough to bear in mind here that validity with respect to frames is closed under Uniform Substitution, while this is not required for validity with respect to models.

To be convinced that the mentioned trivialization is impossible, we will show in the next section that one can prove some results of (absolute) incompleteness with respect to frames ($\mathcal{F}$-incompleteness).

To conclude the preceding survey, it is useful to reflect on the fact that $\mathcal{M}$-completeness and $\mathcal{F}$-completeness are compatible, but non equivalent properties. If a system is complete with respect to some class of frames $\mathfrak{F}$, then it is *a fortiori* complete with respect to the class of models based on the frames in $\mathfrak{F}$. But the converse is not generally true: a system can be complete with respect to a certain class of models, but not with respect to the class

4.4. THE LOGIC OF ARITHMETICAL PROVABILITY

of frames on which these models are based. To see this, it is convenient to reflect on the properties of the canonical model.

The Corollary 4.2.16 states that α is a theorem of **S** if and only if α is valid in the canonical model built on **S**, and this actually amounts to a completeness theorem of **S** with respect to the class of models $\mathfrak{M}$ whose only element is the canonical model only. This means that any system **S** is characterized, in the worst case, by the class of models whose only element is the canonical model only, but this does not imply that **S** is characterized by the class of frames whose only element is the frame underlying the canonical model. In general, it is not true that the frame underlying the canonical model is a frame for **S**, since the infinitely many models based on the canonical frame need not all be models for **S**. If the canonical frame is not a frame for **S**, this means that there is some thesis of **S** that does not hold in the canonical frame, hence that **S** is not $\mathcal{F}$-complete with respect to the class constituted by this frame.

Thus, it is natural to draw the following conclusion: in order to prove that a system **S** is $\mathcal{F}$-complete in absolute sense, it is enough to show that **S** is a canonical system (recall the definition of canonical system on page 99). In fact, in such a case, if **S** is consistent, there is at least a frame (that is, the canonical frame) which validates all the theorems of **S**. However, canonicity is a sufficient but not necessary condition for completeness: in fact, we will see that there are systems which are complete but not canonical.

4.4 The logic of arithmetical provability

A remarkable formula outside the schema $G^{k,l,m,n}$ is the following:

(**GL**) $\quad \Box(\Box p \supset p) \supset \Box p$

The formula (**GL**) (named after Gödel and Löb) deserves particular attention due to its history and to its peculiarity. In 1933, K. Gödel published a short article showing that, if $\Box \alpha$ were interpreted as "α is provable in Peano arithmetic (**PA**)" – and hence $\Diamond \alpha$ would read as "α is consistent with **PA**" – any system at least as strong as Lewis' **S4** becomes unacceptable. In 1955, M. Löb proved in [Löb55] that the negative result could be extended to systems which are extensions of **KT**.

The reason may be devised by looking at the following derivation:

1. $\Box p \supset p$ [(**T**)]
2. $\Box(0 \neq 0) \supset (0 \neq 0)$ [[$p/0 \neq 0$] in 1]
3. $(0 = 0) \supset \neg\Box\neg(0 = 0)$ [**PC**$^\Box$ in 2]
4. $(0 = 0)$ [**PA**]
5. $\neg\Box\neg(0 = 0)$ [(**MP**) in 3, 4]
6. $\Diamond(0 = 0)$ [Def 2.1.1 in 5]
7. $\Box\Diamond(0 = 0)$ [(**Nec**) in 7]

Since $(0 = 0)$ can be replaced by any arithmetical truth, an interpretation of this result is that the consistency of any arithmetical truth would be provable in Peano arithmetic. But this contradicts a well-known corollary of Gödel's theorem, stating that it is impossible to prove the consistency of Peano arithmetic inside arithmetic itself.

Thus (**T**) is not acceptable together with (**Nec**) in a system whose aim is to axiomatize arithmetical provability. More precisely, if $\Box p \supset p$ were a theorem, then a contradiction would be provable, hence any sentence p. Reading $\Box \alpha$ as "it is provable that α", this situation is itself described by the axiom (**GL**): just read $\Box(\Box p \supset p) \supset \Box p$ having this interpretation of $\Box$ in mind. This formula should be then adopted as an axiom instead of (**T**).

The system **KGL** presents some features that distinguish it from the more usual normal modal systems.

Let us call an *e-transform* of α ($e(\alpha)$) the formula α' that is obtained from α by erasing all modal operators (see Exercise 3.18). Then we know that any system whose axioms fall within the schema **K**+$G^{k,l,m,n}$ enjoys an easily provable property: the *e*-transforms of its axioms are **PC**-valid and the rules preserve such validity. Even axioms not covered by $G^{k,l,m,n}$ may enjoy this property; as an example, consider the so-called *Diodorean axiom* (**D1**) of **S4.3**: $\Box(\Box p \supset q) \vee \Box(\Box q \supset p)$. The *e*-transforms of (**D1**) is: $(p \supset q) \vee (q \supset p)$ which is clearly **PC**-valid.

However, the *e*-transform of (**GL**) is not **PC**-valid. In fact, $e(\Box(\Box p \supset p) \supset \Box p)$ is $(p \supset p) \supset p$, which is equivalent to p. Therefore, the consistency of this system cannot be proved by means of the reduction method to **PC**-formulas, and it is thus unavoidable to rely on other methods. The simplest alternative method, which we will use in the sequel, consists in proving soundness of this system with respect to some class of frames.

A syntactic property of the system **KGL** is that it contains the characteristic axiom of **S4**:

4.4. THE LOGIC OF ARITHMETICAL PROVABILITY

Proposition 4.4.1 $\Box p \supset \Box\Box p$ *is a theorem of* **KGL**.

Proof: An easy derivation consisting of the following steps:
1. $p \supset ((q \land r) \supset (q \land p))$ [**PC**]
2. $p \supset ((\Box p \land \Box\Box p) \supset (\Box p \land p))$ [[$q/\Box p, r/\Box\Box p$] in 1]
3. $\Box p \supset \Box((\Box p \land \Box\Box p) \supset (\Box p \land p))$ [(**DR1**) in 2]
4. $\Box(\Box p \supset p) \supset \Box p$ [(**GL**)]
5. $\Box(\Box(p \land \Box p) \supset (p \land \Box p)) \supset \Box(p \land \Box p)$ [[$p/p \land \Box p$] in 4]
6. $\Box((\Box p \land \Box\Box p) \supset (p \land \Box p)) \supset (\Box p \land \Box\Box p)$ [(**Eq**) in 5]
7. $\Box p \supset (\Box p \land \Box\Box p)$ [**PC** in 3, 6]
8. $(\Box p \land \Box\Box p) \supset \Box\Box p$ [**PC**]
9. $\Box p \supset \Box\Box p$ [**PC** in 7, 8]

♠

As a consequence of Proposition 4.4.1, any **KGL**-model must be transitive. What other properties are enjoyed by **KGL**-models? The answer is the following: we say that the relation R on W is *well-covered* if there is no infinite sequence of worlds $w_1, w_2, w_3 \cdots$ of W such that $w_1 R w_2 R w_3 \cdots$. In other words, each subset W' of W has an R-maximal element, in the sense that it possesses an element which bears the relation R to no other element of W'. This is sometimes expressed by saying that the converse of R is a well-founded relation. Note that the absence of an infinite sequence is not a notion expressible in first-order language, which implies that well-coveredness is not a first-order definable notion.

Let R be an accessibility relation that is transitive and well-covered. Through reasoning by *Reductio*, it is easy to prove the following characterization concerning the frames for **KGL**:

Proposition 4.4.2 *Let $\mathcal{F}$ be an arbitrary frame. Then $\mathcal{F} \vDash \Box(\Box p \supset p) \supset \Box p$ iff $\mathcal{F}$ is a transitive and well-covered frame.*

Proof: Exercise 4.8. ♠

At this point, the reader might think that this characterization would yield a completeness result as in Section 4.2; this is not so, as we shall make clear. Completeness for **KGL** is intimately connected to failure in finite models, as it will be proven in Section 5.2. It is instructive to see that there is another characterization for the *finite* frames for **KGL**:

Proposition 4.4.3 *If $\mathcal{F}$ is a finite frame $\langle W, R \rangle$, then R is irreflexive and transitive iff it is transitive and well-covered.*

Proof: ($\Rightarrow$) We first prove that, if W is finite and R is irreflexive and transitive, then R is well-covered.

Suppose that R is not well-covered; then there is an infinite sequence $w_1, w_2, w_3 \cdots$ of elements of W such that $w_1 R w_2 R w_3 \cdots$. Since R is a relation on a finite set W, the infinite sequence must contain some world w_k occurring at least twice (e.g. $\cdots w_k R w_{k+1} \cdots w_{k+n} R w_k \cdots$). But since R is transitive, then $w_k R w_k$ and so R is not irreflexive, which is a contradiction.

($\Leftarrow$) If W is finite and R is transitive and well-covered, then R is irreflexive. Suppose that R is not irreflexive; then there is some $w \in W$ such that $wRwRw\cdots$, but this means that R is not well-covered, a contradiction. ♦

Frames $\langle W, R \rangle$ in which W is finite and R is irreflexive and transitive are called *strict partial orders*. Hence each finite, transitive and well-covered model is based on frames which are strict partial orders.

As we will see (cf. Proposition 5.2.15), the system **KGL** turns out to be complete with respect to the class of strict partial orders frames, hence with respect to the class of finite, transitive and well-covered models. The completeness result given in what follows will use a method different from before, employing the so-called finite model property. This strategy is crucial for **KGL**: although complete, as we shall see, this property cannot be established by using canonical models. First some definitions.

Definition 4.4.4 *A model* $\mathcal{M}' = \langle W', R', v' \rangle$ *is a submodel of* $\mathcal{M} = \langle W, R, v \rangle$ *iff* $W' \subseteq W$, R' *is a restriction of* R *on* W' *and* $v(p, w) = v'(p, w)$ *for every* $w \in W'$ *and every variable* p.

Definition 4.4.5 *A frame* $\mathcal{F}' = \langle W', R' \rangle$ *is a subframe of* $\mathcal{F} = \langle W, R \rangle$ *iff any model built over* $\mathcal{F}'$ *is a submodel of some model built over* $\mathcal{F}$.

Definition 4.4.6 *A frame* $\mathcal{F} = \langle W, R \rangle$ *is generated iff there is some* $w^* \in W$ *(the generating world) such that any other world* w *is an R-descendant from* w^*, *i.e, related to* w^* *by a finite chain in* R *of the form* $w^* R^n w$ *for some* $n \geq 0$. *In this case,* $\mathcal{F}$ *is said to be generated by* w^*.

Definition 4.4.7 *A frame* $\mathcal{F} = \langle W, R \rangle$ *is strongly generated iff there is a* $w^* \in W$ *such that each* $w \in W$ *is an immediate R-descendant of* w^* *in the sense that* $w^* R w$ *(i.e., there is a world* w^* *that sees all the worlds in the frame, including itself). In this case,* $\mathcal{F}$ *is said to be strongly generated by* w^*.

From Definition 4.4.7, it follows that for transitive frames, the concepts of generated and strongly generated frames coincide.

4.4. THE LOGIC OF ARITHMETICAL PROVABILITY

We are interested in knowing which systems might be characterized by generated canonical frames. Actually, this has to do with an important property concerning disjunctions enjoyed by several modal systems which, as we will see, are not only generated, but strongly generated:

Definition 4.4.8 *A system S has the syntactical disjunction property when, if $\alpha_0, \cdots, \alpha_n$ are formulas such that $\vdash_S \Box\alpha_0 \vee \cdots \vee \Box\alpha_n$, then there exists some $i \leq n$ such that $\vdash_S \alpha_i$.*

Definition 4.4.9 *A system S has the semantic disjunction property when, if $\alpha_0, \cdots, \alpha_n$ are formulas such that $\vDash_S \Box\alpha_0 \vee \cdots \vee \Box\alpha_n$, then there exists some $i \leq n$ such that $\vDash_S \alpha_i$.*

We now prove this important relation:

Proposition 4.4.10 *The canonical frame of any system S that is consistent and has the syntactical disjunction property is strongly generated.*

Proof: Consider the set $NT = \{\neg\Box\alpha : \nvdash_S \alpha\}$. This set is consistent. In fact, supposing (by *Reductio*) that NT is not consistent, there would be formulas $\alpha_0 \cdots \alpha_n$ that are not theorems of **S** and such that $\vdash_S (\neg\Box\alpha_0 \wedge \cdots \wedge \neg\Box\alpha_n) \supset \bot$, hence, by contraposition, $\vdash_S \top \supset (\Box\alpha_0 \vee \cdots \vee \Box\alpha_n)$, and therefore $\vdash_S \Box\alpha_0 \vee \cdots \vee \Box\alpha_n$. Supposing that **S** has the syntactical disjunction property, then some α_i would be a theorem of **S** and this is incompatible with the hypothesis. Therefore NT is consistent. Thus, by Lemma 4.2.11, NT can be extended to a maximal **S**-consistent set w^*. For any formula α, if $\Box\alpha \in w^*$, then α is a theorem of **S** by construction (for, if α is not a theorem of **S**, then $\neg\Box\alpha \in NT$ and $\Box\alpha \notin w^*$, since $NT \subseteq w^*$). Hence, given that the theorems of **S** belong to every maximal consistent extension of **S**, the set $Den(w^*) = \{\alpha : \Box\alpha \in w^*\}$ is included in every consistent maximal set w. Thus, by the definition of accessibility relation between worlds of canonical frames, we have that w^*Rw for each w in the canonical frame, and by Definition 4.4.7, this means that the canonical frame is strongly generated by w^*. ♦

This result can be readily adapted to the following:

Proposition 4.4.11 *The canonical frame of any system S that is consistent and has the semantic disjunction property is strongly generated.*

Proof: By an argument analogous to the one in Proposition 4.4.10, but with the difference that the set $NV = \{\neg\Box\alpha : \nvDash_S \alpha\}$ is shown to be consistent. ♦

The next two corollaries follow by specializing the above propositions.

Corollary 4.4.12 *The canonical model of any system S that is consistent and has the syntactical disjunction property is strongly generated.*

Corollary 4.4.13 *The canonical model of any system S that is consistent and has the semantic disjunction property is strongly generated.*

An interesting observation concerning relational models is that the truth of a sentence α at a world w in a model $\mathcal{M} = \langle W, R, v \rangle$ turns out to depend only on the subformulas of α and on the R-descendants of w. We may thus concentrate (in what concerns truth in a model) on the submodels defined by the R-descendants of generators w^*. To prove the most important theorem of this section, we need the following definitions of subframes and submodels that are generated by a certain world, and then we need a lemma concerning classes of subframes and submodels:

Definition 4.4.14 *If $\mathcal{F} = \langle W, R \rangle$ is any frame and $w^* \in W$, then $\mathcal{F}' = \langle W', R' \rangle$ is a subframe generated by w^* iff $\mathcal{F}'$ is a subframe of $\mathcal{F}$ and W' is the collection of all R-descendants of w^*.*

Definition 4.4.15 *If $\mathcal{M} = \langle W, R, v \rangle$ is a model and $w^* \in W$, then $\mathcal{M}' = \langle W', R', v' \rangle$ is a submodel generated by w^* if $\mathcal{F}' = \langle W', R' \rangle$ is a subframe of $\mathcal{F} = \langle W, R \rangle$ and W' is the collection of all R-descendants of w^*.*

Remark 4.4.16 *An alternative definition for generated subframes that often appears in the literature is the following: $\mathcal{F}' = \langle W', R' \rangle$ is a subframe of $\mathcal{F} = \langle W, R \rangle$ generated by a world $w^* \in W$ iff W' satisfies the following closure condition: W' is the smallest subset of W such that $w^* \in W'$ and for any $w \in W'$, if wRw', then $w' \in W'$.*

This closure condition can be used to replace Definition 4.4.6 in the Definitions 4.4.14 and 4.4.15 for generated subframes and submodels (Exercise 4.12).

The next lemma equalizes (from the viewpoint of the truth of formulas) models and their generated submodels:

Lemma 4.4.17 *Let $\mathcal{M} = \langle W, R, v \rangle$ be any model and $\mathcal{M}' = \langle W', R', v' \rangle$ be any generated submodel of $\mathcal{M} = \langle W, R, v \rangle$. Then, for each α and $w \in W'$, $v(\alpha, w) = v'(\alpha, w)$.*

Proof: Exercise 4.13. ♠

Lemma 4.4.17 is important since it allows us, as discussed above, to restrict consideration to the generated submodels of canonical models and of ordinary models in general.

We now prove the following basic theorem.

4.4. THE LOGIC OF ARITHMETICAL PROVABILITY

Proposition 4.4.18 *The system* **KGL** *has the semantic disjunction property.*

Proof: Suppose that $\vDash \Box\alpha_0 \vee \cdots \vee \Box\alpha_n$ and that the semantic disjunction property does not hold in **KGL**. Then, it follows that, for each $i \leq n$, α_i is not valid on a **KGL**-frame. We may assume, with no loss of generality, that the domains W_i of the frames $\mathcal{F}_i = \langle W_i, R_i \rangle$ that invalidate the formulas α_i are pairwise disjoint, hence we may consider that there are worlds $w_0, \cdots, w_n$ such that each w_i ($0 \leq i \leq n$) belongs to some W_i in $\mathcal{F}_i$ and that there is an (implicit) model $\mathcal{M}_i = \langle W_i, R_i, V_i \rangle$ over the KGL-frame $\mathcal{F}_i$ such that each w_i falsifies α_i.

Let w^* be such that, for each W_i, $w^* \notin W_i$ (and consequently $w^* \neq w_i$ for every w_i). Consider the model $\mathcal{M} = \langle W, R, V \rangle$, where:

- $W = \bigcup W_i \cup \{w^*\}$
- $R = \bigcup R_i \cup \{\langle w^*, w_j \rangle : w_j \in \bigcup W_i\}$
- $V(p) = \bigcup V_i(p_i)$

Given that W consists of disjoint sets,[3] R is transitive (indeed, each R_i is transitive, and for the new element w^* we have that if $\langle w^*, w_i \rangle$ and $\langle w_i, w_j \rangle$, then $\langle w^*, w_j \rangle$). Moreover, each R_i is well-covered, therefore R is also well-covered. Now consider the submodel of $\mathcal{M}$ generated by some w_i; this submodel coincides with the submodel of $\mathcal{M}_i$ generated by w_i (in view of the fact that W consists of disjoint sets). But α_i is false at the world w_i of $\mathcal{M}_i$, hence α_i is false at the world w_i of $\mathcal{M}$ as well.

Since $w^* R w_i$, the fact that w_i falsifies α_i in $\mathcal{M}$ implies that $\Box\alpha_i$ is false at w^*. Therefore the formula $\Box\alpha_0 \vee \cdots \vee \Box\alpha_n$ is false at the world w^* of $\mathcal{M}$. This is a contradiction, given that $\langle W, R \rangle$ is a **KGL**-frame and the disjunction $\Box\alpha_0 \vee \cdots \vee \Box\alpha_n$ is valid by hypothesis on all **KGL**-frames. ♠

From this theorem two corollaries follow, the first of which is straightforward.

Corollary 4.4.19 *The canonical frame of* **KGL** *is strongly generated.*

Proof: Propositions 4.4.11 and 4.4.18. ♠

The second theorem finally establishes the fact that the method of canonical models does not apply to **KGL**.

---
[3]This method of combining models is known as *amalgamation*.

Corollary 4.4.20 *The system **KGL** is not canonical.*

Proof: Since by Corollary 4.4.19 the canonical frame of **KGL** is strongly generated, R cannot be well-covered in this frame: in fact, the world w^* which generates the canonical frame (as in the construction of Proposition 4.4.10) sees itself (i.e., it is reflexive), while we know that the accessibility relation of any finite subframe of a transitive and well-covered frame cannot be reflexive. Therefore the canonical frame of **KGL** cannot be well-covered, which means that **KGL** is not canonical. ♦

Remark 4.4.21 *The system **KGL** has other anomalous properties beyond the one stated in Corollary 4.4.20. In the next chapter, we will prove that **KGL** is complete w.r.t. to the class of transitive and irreflexive frame, but this does not imply that it is strongly complete since it lacks the property of modal semantical compactness; see Proposition 5.2.15.*

4.5 Exercises

1. In Exercise 3.19 an alternative decision procedure for the system **S5** was outlined. Show that this procedure also permits the proof of a constructive completeness result for **S5**.

2. The wff $\Diamond(p \supset \Box p)$ is **KT**-valid but not **KD**-valid. Build the validating **KT**-tableau for this wff and convert the tableau into a proof of it in **KT**.

3. Prove the properties of maximal consistent sets given at the points 1–4 and 6 in Lemma 4.2.10.

4. Prove the property of maximal consistent extensions which is stated in Corollary 4.2.14.

5. Prove the completeness of the system **KD4** with respect to the class of serial and transitive frames by applying Henkin's method. (Hint: see Proposition 4.2.19)

6. Prove the completeness of **PC**$^\Box$ by using Henkin's method.

7. By applying the proof methods of first-order logic with identity (see Correspondence Theory at Section 3.3) prove that seriality, functionality and euclideanity are special cases of the general notion of convergence expressed by the formula $\mathbf{C}^{k,l,m,n}$.

8. Prove Proposition 4.4.2. Hint: the proof from left to right is a consequence of Proposition 4.4.3, and it consists in showing that a non irreflexive frame (i.e., a frame containing a reflexive world) does not validate the formula (**GL**).

9. Prove, by induction on n, and on the ground of Definition 4.2.7, the equivalence expressed in Definition 4.2.9.

10. Prove completeness of **KB**, **S4** and **S5** as special cases of the completeness of **K**+G^∞.

11. Prove that **K** is complete with respect to the class of all irreflexive models, i.e., of all models such that, for each w, $\neg(wRw)$. (Hint: for each **K**-model $\langle W, R, v\rangle$ build a "duplicate" model $\langle W^*, R^*, v^*\rangle$ in which each world w is substituted by a couple of worlds $w+$ and $w-$ such that, if wRw holds, then $w + R^*w-$ as well as $w - R^*w+$ both hold, and, if wRw does not hold, then neither $w + R^*w-$ nor $w - R^*w+$ hold. Prove that, if a formula is true in the standard model, it is also true in the duplicate model, and that the duplicate of the canonical model is irreflexive).

12. Prove the equivalence between the notion of generated subframes given in Definition 4.4.14 and the one stated in Remark 4.4.16.

13. Prove Lemma 4.4.17.

14. Prove the soundness of **KGL** with respect to the class of transitive and well-covered frames by applying a *Reductio* argument.

15. A frame $\mathcal{F}$ is *cohesive* when any two worlds w and w' in $\mathcal{F}$ are *n*-step connected, i.e., either $wR^n w'$ or $w'R^n w$ for some n.

    (a) Prove that the frames that characterize (**Ban**) and (**Ver**) are not cohesive.

    (b) Prove that, if a frame is generated, then it is cohesive, but the converse does not hold.

4.6 Further reading

The proofs of constructive completeness for modal systems, introduced by S. Kripke in [Kri59], have been simplified by G. E. Hughes and M. J. Cresswell in [HC68]. Non-constructive methods, pioneered by

D. Makinson in [Mak66], have been generalized thanks to the elegant formulation of the schema G^∞ in the important monography of E. J. Lemmon and D. Scott [LS77] (which circulated mimeographed since 1966 and was edited by K. Segerberg in [Seg71]).

The basic ideas of Lemmon and Scott's seminal essay have been further reworked by B. F. Chellas in [Che80] and by Hughes and Cresswell in [HC84] (but see [HC86] for corrections) and [HC96]. For a more recent presentation of this approach concerning the correspondence between axioms and conditions on frames, also see A. Chagrov and M. Zakharyaschev in [CZ97] and S. Popkorn in [Pop94].

Sahlqvist's axiom schema (herein, on page 102) was proposed by H. Sahlqvist in [Sah75] generalizing a conjecture by Lemmon and Scott in [LS77], p. 78, about first-order definability of frames that characterize extensions of **PC** by a certain schema involving affirmative sentences. The conjecture was also independently proved by R. Goldblatt in [Gol75].

Classical texts about the logic of arithmetical provability logic are G. Boolos [Boo79] and especially its updated version [Boo93], as well as C. Smorynski [Smo85] and G. Japaridze and D. de Jongh [JdJ98]. An important theorem of R. Solovay, which proves that Peano arithmetic is represented in **KGL**, appears in [Sol76]. The importance of **KGL** as a tool for dealing with problems about consistency and provability of **PA** is emphasized in G. Boolos [Boo93].

A system close to **KGL** is **KGrz**, obtained by adding to **K** the axiom (**Grz**): $\Box((\Box(p \supset \Box p)) \supset p)) \supset p$. (**Grz**) is equivalent to an axiom introduced by A. Grzegorczyk in [Grz67]. The system **KGrz** characterizes provability in Intuitionistic Propositional Logic as much as **KGL** characterizes provability in Peano arithmetic (in a sense, **KGrz** can replace **S4** in the well-known inter-translation between **S4** and Intuitionistic Propositional Logic which has been proved by S. Kripke in [Kri65]). **KGL** is related to **KGrz** in the sense that any wff α is a **KGrz**-thesis iff a translation of α is a **KGL**-thesis ([Boo93], chapter 12).

Chapter 5

Incompleteness and finite models

5.1 An incompleteness result

In the last section of Chapter 4, we have met a non-canonical modal system. Up to now, our completeness results relied on canonicity, i.e., on the fact that every canonical system is automatically complete. In **KGL**, the lack of canonicity makes it problematic to prove completeness via the method applied in the preceding sections. However, it does not establish what we may call $\mathcal{F}$-incompleteness, i.e, that there is no class of frames with respect to which the given system is valid. We will be able to show that **KGL** is complete with respect to the class of irreflexive and well-covered frames. But we are going to show, nevertheless, that there are modal systems that are $\mathcal{F}$-incomplete, in the sense that no class of frames characterizes them.

There are several incompleteness results in the literature. One of them concerns the system **K+(H)**, where the axiom **(H)** is a weaker version of **(GL)**:

(**H**) $\Box(\Box p \equiv p) \supset \Box p$

The incompleteness result we are going to expose here concerns a system that we will call **KVB** (since the original proof is due to J. van Benthem). **KVB** is the system **K+(VB)**, where

(**VB**) $\Diamond \Box p \vee \Box(\Box(\Box q \supset q) \supset q)$

The proof runs in two steps:

(a) First, we show that any frame which validates (**VB**) also validates (**MV**), where:

(MV): $\Diamond\Box p \vee \Box p$

(b) Second, we show that (**MV**) is not a theorem of **KVB**.

From (a) and (b), it follows that each frame for **KVB** validates a formula that is not a theorem of **KVB**, and therefore there cannot exist any frame validating all theorems of **KVB** and only them: this means that **KVB** is $\mathcal{F}$-incomplete.

We give separate proofs of parts (a) and (b); the proof of part (a) is accomplished by the following:

Lemma 5.1.1 *If $\mathcal{F}$ is not a frame for (MV), then $\mathcal{F}$ is not a frame for (VB).*

Proof: Any frame $\mathcal{F} = \langle W, R \rangle$ which invalidates (**MV**) will have the following property: no world in $\mathcal{F}$ can be either a terminal point, nor can it see any terminal point. For, if a world w is a terminal point or sees some terminal point, the well-known fact that every wff $\Box p$ is invariably true at a terminal point (check Remark 3.2.18) entails that w validates $\Box p$ or $\Diamond\Box p$, thus validating (**MV**).

Choose, in a model $\mathcal{M} = \langle W, R, v \rangle$ over such $\mathcal{F}$ that invalidates (**MV**), worlds w and w' such that wRw'. Choose also an assignment v such that $v(p, w') = 0$, $v(q, w) = 1$ and, for every $u \neq w'$, $v(q, u) = 0$. An easy inspection shows that both disjuncts of (**VB**) are falsified in this model. The model, as defined, therefore falsifies (**VB**). ♠

Part (b) is more complicated because, differently from (a), we cannot find any frame which distinguishes (**VB**) from (**MV**). Nor would we solve the problem by finding a model in which (**VB**) were valid but (**MV**) not. For, as we remarked in Section 3.2, the rule (**US**) does not preserve validity in a single model, and this model could invalidate (**MV**) even if (**MV**) were a theorem of **KVB** (proved, for instance, by using (**US**)).

To overcome this obstacle, we define, following the original strategy worked out by J. van Benthem, a special class of frames called *general frames*, along with a restricted class of models based on such frames. In these generalized models (**VB**) and any of its substitution instances will be valid, but (**MV**) will be invalid, thus establishing part (b).

5.1. AN INCOMPLETENESS RESULT

Definition 5.1.2 *A general frame is a triple $\mathcal{G} = \langle W, R, \Pi \rangle$, where $\mathcal{F} = \langle W, R \rangle$ is a relational frame and Π is any collection of subsets of W called admissible sets closed under the following operations:*

(i) *If $X \in \Pi$, then $\overline{X} \in \Pi$.*

(ii) *If $X, Y \in \Pi$, then $X \cup Y \in \Pi$.*

(iii) *If $X \in \Pi$, then $\{w \in W : \forall w' \in W(wRw' \text{ implies } w' \in X)\} \in \Pi$.*

The closure conditions of Π grant that, for each model $\langle \mathcal{F}, \Pi, V \rangle$ and wff α, the set of worlds in which α is true (that is, $V(\alpha)$) will be a member of the collection Π.

Definition 5.1.3 *We call an admissible model any implicit model $\langle \mathcal{F}, V \rangle$ where, for each wff α, $V(\alpha)$ is an admissible set.*

Lemma 5.1.4 *A model $\mathcal{M} = \langle \mathcal{F}, V \rangle$ is admissible if $V(p)$ is admissible for any variable p.*

Proof: Exercise 5.1. ♠

Consider now the general frame $\mathcal{G}_0 = \langle W_0, R_0, \Pi_0 \rangle$ such that:

- $W_0 = \mathbb{N} \cup \{\omega, \omega + 1\}$ where $\mathbb{N}$ is the set of natural numbers and ω and $\omega + 1$ are the first transfinite ordinals.

- R_0 is defined, for each $w_i, w_j \in W_0$, as:

$$w_i R_0 w_j \quad \text{iff} \quad \begin{cases} w_i = \omega + 1 & \text{and} \quad w_j = \omega \\ w_i \neq \omega + 1 & \text{and} \quad w_j < w_i \end{cases}$$

- The collection Π_0 of admissible subsets of W_0 is specified in the following way:

    (i) $\omega \notin X$ and X is finite, or
    (ii) $\omega \in X$ and the complement of X is finite.

It is easy to check that Π_0 is indeed a collection of admissible subsets of W_0 and that the frame $\mathcal{G}_0$ can be represented by the following graph:

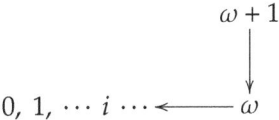

Note that $\omega+1$ can see only ω, and each number less than $\omega+1$ can only see natural numbers but not itself.

Definition 5.1.5 *A formula α is said to be Π_0-valid if α is valid in every admissible model over $\mathcal{G}_0 = \langle \mathcal{F}_0, \Pi_0 \rangle$, where $\mathcal{G}_0$ is the above specified general frame.*

Our purpose is to show that the property of being Π_0-valid is satisfied by **KVB**, but not by (**MV**). The desired result is obtained from the conjunction of the two following propositions.

Proposition 5.1.6 *(**MV**) is not a Π_0-valid wff.*

Proof: Let $\mathcal{M} = \langle \mathcal{G}_0, V \rangle$ be a model based on the frame $\langle \mathcal{G}_0, \Pi_0 \rangle$ at which every p is false for every $w \in W$. This is an admissible model after Lemma 5.1.4, as the empty set, by construction, is an element of Π_0. Then $\Box p$ is false at the world ω, hence $\Diamond \Box p$ is false at $\omega + 1$. Given that p is false at ω, then $\Box p$ is false at $\omega + 1$. Since each disjunct of (**MV**) is false at $\omega + 1$, then (**MV**) is also false at $\omega + 1$. ♠

Proposition 5.1.7 *Every theorem of **KVB** is Π_0-valid.*

Proof: It is enough to show that every substitution instance of (**VB**) is Π_0-valid, as (**K**) is obviously valid. First, as 0 is a terminal point, each formula of the form $\Box \gamma$ is vacuously true at 0 (inasmuch as no accessible world exists where γ is false). Thus, in particular, $\Box(\Box(\Box \beta \supset \beta) \supset \beta)$ is true at 0, and, consequently, so is (**VB**). Any other world except 0 and $\omega+1$ sees a terminal point where $\Box \alpha$ is true, therefore $\Diamond \Box \alpha$ is true at this world, and so is (**VB**).

The only remaining world to be analyzed is $\omega + 1$. Suppose, by way of contradiction, that (**VB**) is false at $\omega + 1$. Then $\Box(\Box(\Box \beta \supset \beta) \supset \beta)$ is false at $\omega + 1$, but as $\omega + 1$ sees only ω, then $\Box(\Box \beta \supset \beta) \supset \beta$ must be false at ω, which means that:

(i) $\Box(\Box \beta \supset \beta)$ is true at ω, and

(ii) β is false at ω. From (i), as ω sees every natural number n, it follows that:

(iii) $\Box \beta \supset \beta$ is true at every n.

Now it can be proven by induction on natural numbers that β is true at any natural number n: in fact, in view of (iii), $\Box \beta \supset \beta$ is true at 0, and as we have

5.1. AN INCOMPLETENESS RESULT

seen above, 0 is a terminal point, therefore $\Box\beta$ is true at 0. So β is also true at 0. Now, suppose by induction hypothesis that β is true at every $m < n$. This means that $\Box\beta$ is true at n, and because (iii) β is true at n. Consider now the set X constituted by all the worlds at which β is true; X is infinite, as β is true at every natural number, but $\omega \notin X$ since β is false at ω (in view of (ii)). Therefore X is not an admissible set. However, this is a contradiction (see Definition 5.1.2). ♠

From Propositions 5.1.6 and 5.1.7, it thus follows that (**MV**) is not a theorem of **KVB**, and this concludes the argument.

It is to be noted that the incompleteness proof given above relies on the divergence between frames and general frames. In fact, although (**MV**) is valid on any frame in which (**VB**) is valid, it is not so for the *general* frames in which (**VB**) is valid. General frames are essential to understand the phenomenon of modal incompleteness. On the one hand, they intuitively play the role of the kind of models for quantificational logic which are qualified as "non-standard",[1] while, on the other hand, they can be seen as kinds of relational frames equivalent to algebraic semantic frames (see next section).

To appraise such points, we can, as a matter of fact, state the following result:

Proposition 5.1.8 *Each normal modal system S is characterized by the class of all general frames for S.*

Proof: Exercise 5.2. ♠

A consequence of Proposition 5.1.8 is the following: if to the notions of completeness listed in Section 4.3 we add the notion of *completeness with respect to general frames* (call it $\mathcal{G}$-completeness), then there is no normal modal system which is $\mathcal{G}$-incomplete.

Any treatment of incompleteness in modal logic cannot omit mentioning a special kind of incompleteness which has been called Halldén-incompleteness:

---

[1] As we have seen in Section 3.3, the notion of frame validity is essentially second-order. The general frames, in a sense, "domesticate" the second-order character of frame validity by restricting the universe of valuations, and in this sense behave similarly to the non-standard models of a first-order theory, i.e. to models that are not isomorphic to the intended (or standard) models.

Definition 5.1.9 *A system S is Halldén-incomplete iff, for some wffs α and β which have no common variables, $\alpha \vee \beta$ is an S-thesis, but neither α nor β are S-theses. A system which is not Halldén-incomplete is said to be Halldén-complete.*

A simple observation in this connection is that **K** is Halldén-incomplete. It suffices to remark that the wff $\Box(p \wedge \neg p) \vee \Diamond(q \vee \neg q)$ is clearly a **K**-theorem whose disjuncts have no common variables, but $\Box(p \wedge \neg p)$ and $\Diamond(q \vee \neg q)$ are not **K**-theorems.

This feature is slightly paradoxical for a complete modal system. In fact, if some disjunction $\alpha \vee \beta$ is a thesis of a system **S**, soundness implies that all assignments to atomic variables satisfy $\alpha \vee \beta$. Given that α and β lack common variables, it seems that one of the two wffs should turn out to be valid for all assignments to its atomic variables; so, if both α and β turn out to be non-theorems of **S** and **S** is frame-complete, this appears to be quite anomalous. As a matter of fact, a system **S** is Halldén-incomplete if it is the intersection of two systems **S'** and **S''**, neither of which contains the other: the reason is that if $\alpha \vee \beta$ is an **S**-thesis and **S** is Halldén-incomplete, then α and β are both theorems of two distinct extensions of **S**. Since $\Box(p \wedge \neg p) \vee \Diamond(q \vee \neg q)$ is valid in all normal modal systems, all such systems which are Halldén-complete should contain as a theorem either $\Box(p \wedge \neg p)$ or $\Diamond(q \vee \neg q)$. Using the rule (**Eq**) which holds in all normal modal systems (cf. Proposition 2.3.10), it is then clear that all normal systems which are Halldén-complete should contain either $\Box\bot$ or $\Diamond\top$ as a theorem. It is easy to see that $\Diamond\top$ is equivalent to axiom (**D**) (to see this, just think of $\Diamond\top$ as equivalent to $\Diamond(p \vee \neg p)$ and use the distributivity of $\Diamond$ over disjunctions, cf. Proposition 2.3.11(ii), to derive (**D**); the converse direction is obvious). Therefore any Halldén-complete normal modal system is either an extension of **K**+ $\Box\bot$ (i.e., **Ver**) or of **K**+ $\Diamond\top$ (i.e., **KD**). Several extensions of **KD**, including **KD** itself plus **KT**, **KD4**, **S4** and **S5**, are known to be Halldén-complete.

Halldén-incompleteness has some interesting connections with Post-completeness (recall Definition 1.6.3) and with the so-called *Craig interpolation property*. The standard way to define such property with respect to an arbitrary modal system **S** is as follows:

Whenever $\vdash_S \varphi \supset \psi$ and φ and ψ have at least one common variable, there exists a formula ρ such that $\vdash_S \varphi \supset \rho$ and $\vdash_S \rho \supset \psi$, where all the propositional variables occurring in ρ occur simultaneously in φ and in ψ.

This property holds for standard propositional logic (cf. Exercise 1.15) and for first-order logic but is not generally valid for modal logic. In

5.2 Finite model property and filtrations

$\mathcal{F}$-incompleteness results such as the ones reported in Section 5.1 are remarkable, since they show that relational semantics provides a meaningful modelization of modal logic. Relational frames would be, in a sense, unimportant or trivial if we could show that any normal modal system is complete with respect to some class of relational frames. There are other semantical approaches, however, that may be legitimately suspect of providing a trivial semantics. An example is given by the so-called *algebraic semantics*, whose basic notions can be defined in the following way:

Definition 5.2.1 *A modal algebra or a Boolean algebra with operators is a quadruple $\mathcal{A} = \langle \mathcal{B}, -, \cap, \tau \rangle$, where $\langle \mathcal{B}, -, \cap \rangle$ is a Boolean algebra and τ is a monadic operation defined on $\mathcal{B}$.*

Two distinguished elements of modal algebras are:

- $0 \stackrel{\text{Def}}{=} a \cap \bar{a}$
- $1 \stackrel{\text{Def}}{=} \bar{0}$

Algebras basically manipulate terms and equations, and there are very precise rules for deriving equations from other basic equations (the so-called *equational logic*), so algebras can be seen as part of logic. But, on the other hand, many logics have equation systems as an algebraic counterpart. This is the case of all normal modal logics, where the basic equations are given as follows:

Definition 5.2.2 *A modal algebra is regular iff, for any $a, b \in \mathcal{B}$:*

(i) $\tau(a \cap b) = \tau(a) \cap \tau(b)$

(ii) $\tau(1) = 1$

Of course, it is possible to define a dual operator of τ, the operator μ, in the following way: $\mu(a) = \overline{\tau(\bar{a})}$, besides other operations and relations such as:

- $a \cup b \stackrel{\text{Def}}{=} \overline{\bar{a} \cap \bar{b}}$
- $a \leq b \stackrel{\text{Def}}{=} (a \cap b) = a$
- $a \to b \stackrel{\text{Def}}{=} \overline{a \cap \bar{b}}$

Now, let us introduce a function V associating to the atomic variables $p, q, r \cdots$ elements of the power set of $\mathcal{B}$. This function is defined as in the implicit relational models (see Section 3.2) with the obvious difference that $V(\Box \alpha) = \tau(V(\alpha))$.

A formula α is said to be *valid* on a given class $\mathcal{Z}$ of algebras when $V(\alpha) = 1$ in every algebra in $\mathcal{Z}$.

Definition 5.2.3 *For any system S of normal modal logic, a modal algebra is an S-algebra if, for any wffs of S α and β, the identity $V(\alpha) = V(\beta)$ is true if and only if $\vdash_S \alpha \equiv \beta$.*

The conditions which define the regular modal algebras correspond to the following equivalences valid in all normal modal systems:

(i) $\Box(p \wedge q) \equiv \Box p \wedge \Box q$

(ii) $\Box \top \equiv \top$

It can be proven that **K** is complete with respect to the class of all regular algebras, so the class of **K**-algebras[2] coincides with the class of regular algebras. The proof is analogous to the Henkin style completeness proof and may be sketched as follows:

1. First, it can be shown that there exists a specific algebra (the so-called *Lindenbaum algebra*)[3] in which the elements are the equivalence classes with respect to the relation $\vdash_K \alpha \equiv \beta$. More exactly, a Lindenbaum algebra is a quadruple $\langle \mathcal{B}^*, -^*, \cap^*, \tau^* \rangle$, where $\mathcal{B}^*$ is a collection of equivalence classes of formulas $\alpha, \beta, \cdots$ such that $\vdash_K \alpha \equiv \beta$. Operations on equivalence classes are defined as follows:

   - $[\alpha] \cap^* [\beta]$ is $[\alpha \wedge \beta]$
   - $\overline{[\alpha]}^*$ is $[\neg \alpha]$
   - $\tau^*([\alpha])$ is $[\Box \alpha]$

2. On the ground of the definition of modal algebras given above, it is not difficult to show that the Lindenbaum algebra is a modal algebra. The distinguished element **1** is $[p] \cup^* \overline{[p]}^*$, i.e., $[p \vee \neg p]$ in this case (where $\cup^*$ is defined as expected).

---

[2] A class of algebras determined by a collection of equations is called a *variety*.
[3] Also known in the literature as *Lindenbaum-Tarski algebra*.

5.2. FINITE MODEL PROPERTY AND FILTRATIONS

It can be shown that, if α is not a thesis of **K**, then there is some assignment V such that $V(\alpha) \neq 1$ in some regular modal algebra. In fact, suppose by *Reductio* that α is not a thesis of **K**. Then $\alpha \equiv (p \vee \neg p)$ is not a **K**-thesis, so the formulas α and $p \vee \neg p)$ do not belong to the same equivalence class. Since $[p \vee \neg p] = 1$, it follows that $V(\alpha) \neq 1$. This immediately grants the completeness of **K**, that is, if α is **K**-valid, then α is a theorem of **K**.

It is now straightforward to see that each modal axiom (**A**) is mirrored by a property of modal algebra represented by a translation of (**A**) into the algebraic language. For example, a **KT**-algebra will be a **K**-algebra that also satisfies the additional condition $\tau(\alpha) \leq \alpha$, and, by virtue of this property, the completeness proof for **KT** with respect to the class of **KT**-algebras (also called *transitive algebras*) is obtained with appropriate modifications. For **S4**, the relevant property will be $\tau(\alpha) = \tau(\tau(\alpha))$, giving rise to the so-called *topological Boolean algebras*. The possibility of translating each modal axiom into a specular property of a modal algebra explains the general method to prove algebraic completeness. Furthermore, it is not difficult to realize that every general frame $\langle W, R, \Pi \rangle$ may be converted into a modal algebra $\langle \mathcal{B}, -, \cap, \tau \rangle$, where $\mathcal{B}$ is a set of elements $a_1, a_2, \cdots$ biunivocally associated to the worlds $w_1, w_2, \cdots \in W$ and $\tau(a_i)$ is defined as the set of all a_j which are associated to some w_j such that $w_i R w_j$. Any system which is complete with respect to some class of modal algebras will be called *algebraically complete*. Now we can prove the following result:

Proposition 5.2.4 *Every normal modal system is algebraically complete.*

Proof: Exercise 5.3. ♠

This result encouraged the impression that algebraic semantics is only "syntax in disguise" and that it is uninformative, in the sense that it adds nothing to the syntactical objects that it aims to modelize. As a matter of fact, relational semantics is prominent not only because it is conceptually informative, but also because it provides in many interesting cases practical decision procedures, as for instance the already seen tableau procedures.

On the other hand, algebraic semantics suggests that the decidability problem can be treated in abstract terms for wide classes of normal modal systems. Suppose, in fact, that it is possible to show that a certain system **S** is characterized by a certain class $\mathfrak{M}$ of models of finite cardinality. By $\mathcal{M}$-completeness (see Section 4.3) each formula α that is valid in the models of this class is a theorem of **S**. A different way to express the same concept is to say that each non-theorem of **S** is invalid at some world of some finite model for **S**. A precise definition of this concept is the following:

Definition 5.2.5 *If **S** is a normal modal system, then it has the finite model property (fmp) iff, for each wff α that is not a theorem of **S**, there exists an **S**-model $\langle W, R, v \rangle$, where W is finite and:*

(i) *There is some $w \in W$ such that $v(\alpha, w) = 0$.*

(ii) *If β is a theorem of **S**, then $v(\beta, w) = 1$ for each $w \in W$.*

The decidability of a modal system is related to the finite model property, given that in presence of the (fmp) we have a procedure to check, for every α, whether α is a theorem or a non-theorem of the given system.

But before giving a rigorous proof of this fact it is useful to answer to the following question: how do we know if **S** enjoys the finite model property or not? To answer to this question, we expose the method of *filtrations*, which has the purpose of establishing this property for an arbitrary system **S**. The idea of this method can be summarized in this construction:

- We know that every system **S** is characterized by some class of models, in particular, by the class of models for **S**. This means that, if α is not a theorem of **S**, then α is invalid in some model for **S**. If the falsifying model is infinite, then it is possible to construct an equivalent finite model with a particular technique that makes use of the finiteness of the formulas under test. In fact, given a wff α, the set of subformulas of α is finite. Hence it is possible to define a finite model whose elements, analogously to the case of Lindenbaum algebra, consist of equivalence classes, with the difference that now we will use classes of worlds instead of classes of formulas.

Let $\mathcal{M} = \langle W, R, v \rangle$ be a model and Σ be any set of wffs closed under subformulas. Consider the equivalence relation $\cong_\Sigma$ defined as follows:

$$w \cong_\Sigma w' \text{ iff, for each } \beta \in \Sigma, \ v(\beta, w) = 1 \text{ iff } v(\beta, w') = 1$$

Definition 5.2.6 *A filtration of $\mathcal{M}$ through Σ is a model $\mathcal{M}^* = \langle W^*, R^*, v^* \rangle$ such that:*

(F1) $W^* = W/ \cong_\Sigma$ *(that is, W^* is the collection of equivalence classes $w/ \cong_\Sigma$ with respect to $\cong_\Sigma$. The subscript Σ will be omitted when such reference is contextually clear).*

5.2. FINITE MODEL PROPERTY AND FILTRATIONS

(F2) R^* is a suitable relation in the sense that, for each $w, w' \in W$:

(a) If wRw', then $[w]R^*[w']$.

(b) If $[w]R^*[w']$ then, for each sentence $\Box\alpha \in \Sigma$, it holds: if $v(\Box\alpha, w) = 1$, then $v(\alpha, w') = 1$.

(F3) For each atomic sentence p and each $w \in W$:
$v^*(p, [w]) = 1$ iff $v(p, w) = 1$.

It is important to have in mind that there may be several distinct suitable relations R^* that satisfy the conditions (F1) to (F3) required in Definition 5.2.6. There is at least one suitable relation R^*_Σ with respect to Σ defined by requiring "if and only if" in place of "if ... then "in clause (F2b):

1. $[w]R^*_\Sigma[w']$ iff, for each sentence $\Box\alpha \in \Sigma$, it holds: if $v(\Box\alpha, w) = 1$, then $v(\alpha, w') = 1$

R^*_Σ is the suitable relation with the greatest number of elements (sometimes also called the "largest filtration"), but there is another relation (not necessarily distinct) having the least number of elements (coincident with the intersection of all suitable relations, and also called the "least filtration"):

1. $[w]R^*_{\Sigma'}[w']$ iff there are $w_1 \in [w]$ and $w_2 \in [w']$ such that $w_1 R w_2$

It is not difficult to check that R^*_Σ and $R^*_{\Sigma'}$ satisfy conditions (F2a) and (F2b) and are well-defined, in the sense of being independent of any particular class representatives w_1 and w_2 (Exercise 5.6).

The so-called *Filtration Theorem* carries over the generalization of clause (F3) of Definition 5.2.6 to arbitrary formulas in Σ:

Proposition 5.2.7 *(Filtration Theorem) For every $\alpha \in \Sigma$,*

$$v^*(\alpha, [w]) = 1 \quad \text{iff} \quad v(\alpha, w) = 1$$

Proof: By induction on the length of formulas. The non-trivial step is to prove $v^*(\Box\beta, [w]) = 1$ iff $v(\Box\beta, w) = 1$.
($\Rightarrow$) Suppose that $v^*(\Box\beta, [w]) = 1$ and let w' be an arbitrary world such that wRw'. From Definition 5.2.6, item (a), it follows that $[w]R^*[w']$, thus $v^*(\beta, [w']) = 1$. By the induction hypothesis, $v(\beta, w') = 1$. As w' is arbitrary and wRw', $v(\Box\beta, w) = 1$.
($\Leftarrow$) By the induction hypothesis, the property holds for β. Suppose that $v(\Box\beta, w) = 1$, and let $[w']$ be an element of W^* such that $[w]R^*[w']$. Since

$\Box\beta \in \Sigma$, it follows from Definition 5.2.6, item (b), that $v(\beta, w') = 1$. Again, by the induction hypothesis $v^*(\beta, [w']) = 1$. Since $[w]R^*[w']$, then $v^*(\Box\beta, [w]) = 1$.

♠

Two facts among others about filtrations deserve attention. First, the most interesting filtrations are the ones where the set of wffs Σ is *finite*, because the set $W^* = W/\cong_\Sigma$ is also finite in such cases, and we can construct finite models.

The second remark is that the properties of R in the original model are not necessarily inherited by *all* the accessibility relations R^* in the filtrations: it can be shown that seriality and reflexivity, for example, are directly transmitted to any R^*, while symmetry, euclideanity and transitivity are not (cf. Exercise 5.9). In some cases, supplementary conditions can be introduced in the definition of R^* in order to preserve the properties of the original model even if we lack a universal strategy to be applied to all cases. For symmetry and transitivity, we can prove what follows:

Proposition 5.2.8 *For every symmetric (respectively, for every transitive, and for every symmetric and transitive) model $\mathcal{M} = \langle W, R, v\rangle$ and for any Σ closed under subformulas, there is a symmetric (respectively, a transitive, and a symmetric and transitive) filtration of $\mathcal{M}$ through Σ.*

Proof: The proof consists in defining a suitable relation R_Σ^S preserving symmetry, a suitable relation R_Σ^T preserving transitivity, and combining them to define a third relation R_Σ^{ST} preserving symmetry and transitivity.

(i) Given a symmetric model $\mathcal{M}$ and a set of sentence Σ closed under subformulas, define R_Σ^S over $w/\cong_\Sigma$ as follows: $[w]R_\Sigma^S[w']$ iff for each sentence α such that $\Box\alpha \in \Sigma$ it holds: if $v(\Box\alpha, w) = 1$, then $v(\alpha, w') = 1$, and, if $v(\Box\alpha, w') = 1$, then $v(\alpha, w) = 1$.

It is easy to see that R_Σ^S is well defined (i.e., the definition is independent of any particular class representatives). Also, it follows from the definition that R_Σ^S is symmetric and that it fulfills (F2b). As for (F2a), suppose that wRw'. If $\Box\alpha \in \Sigma$ and $v(\Box\alpha, w) = 1$, then $v(\alpha, w') = 1$. Since R is symmetric, $w'Rw$; consequently, if $v(\Box\alpha, w') = 1$ then $v(\alpha, w) = 1$. Therefore $[w]R_\Sigma^S[w']$.

(ii) Given a transitive model $\mathcal{M}$ and a set of sentences Σ closed under subformulas, define R_Σ^T over $w/\cong_\Sigma$ by:
$[w]R_\Sigma^T[w']$ iff for each sentence α such that $\Box\alpha \in \Sigma$ it holds: if $v(\Box\alpha, w) = 1$, then $v(\alpha, w') = 1$ and $v(\Box\alpha, w') = 1$.

It is again easy to see that R_Σ^S is well defined (i.e., the definition is independent of any particular class representatives). It can also be easily verified that R_Σ^T is transitive (Exercise 5.10). Also, condition (F2b) follows immediately by definition. To check (F2a), suppose that wRw'. If $\Box\alpha \in \Sigma$ and $v(\Box\alpha, w) = 1$, then $v(\alpha, w') = 1$. Since R is transitive, then $\mathcal{M}$ validates $\Box\alpha \supset \Box\Box\alpha$ (cf. Proposition 3.2.16), and consequently, $v(\Box\Box\alpha, w) = 1$ and thus $v(\Box\alpha, w') = 1$. Hence $[w]R_\Sigma^T[w']$.

(iii) Given a symmetric and transitive model $\mathcal{M}$ and a set of sentence Σ closed under subformulas, define R_Σ^{ST} over $w/\cong_\Sigma$ as follows: $[w]R_\Sigma^{ST}[w']$ iff for each sentence α such that $\Box\alpha \in \Sigma$ it holds: if $v(\Box\alpha, w) = 1$, then $v(\alpha, w') = 1$ and $v(\Box\alpha, w') = 1$, and, if $v(\Box\alpha, w') = 1$, then $v(\alpha, w) = 1$. It is easy to check that R_Σ^{ST} is well defined and that (as we are putting together the stipulations for R_Σ^S and R_Σ^T) it is symmetric and transitive while fulfilling condition (F2).

♠

It is important to note that a filtration through a finite set Σ has a finite number of elements, bounded by $2^{|\Sigma|}$, where $|\Sigma|$ is the cardinality of Σ. Therefore, if such a filtration is an **S**-model, then it is a finite **S**-model.

In order to prove the finite model property for a system **S**, in case **S** is finitely axiomatized and known to be sound with respect to the class of **S**-models, the procedure is this: let α be a non-theorem of **S** and consider its negation $\neg\alpha$, which is surely consistent with **S**; consider an **S**-consistent maximal set containing $\neg\alpha$. If the canonical model of **S** belongs to the class of **S**-models,[4] then we have a counter-model to α. Now, let Σ be the set of all subformulas of $\neg\alpha$ (including $\neg\alpha$); then any filtration of the canonical model through Σ which has the properties of an **S**-model will be a finite counter-model to α. If this can be achieved, we can be sure that **S** has the finite model property.

The last step is to introduce the following definition:

Definition 5.2.9 *A normal modal system **S** has the finite frame property iff, for each wff α which is not an **S**-theorem, there is a frame $\mathcal{F}$ in which W is finite and*

(i) *There is some model $\mathcal{M}$ over $\mathcal{F}$ in which, for some $w \in W$ of $\mathcal{M}$, $v(\alpha, w) = 0$.*

(ii) *Every theorem of **S** is valid in $\mathcal{F}$.*

---

[4]Note that referring to the canonical model of **S** does not mean that **S** is canonical; recall Definition 4.2.17.

Luckily the finite model property and the finite frame property go hand in hand. It can be shown[5] that a system satisfying the finite model property also satisfies the finite frame property and vice versa. From this point of view, talking about models or frames makes no difference. We now have the tools to prove what follows:

Proposition 5.2.10 *If a modal system S has the finite model property and is finitely axiomatizable, then S is decidable.*

Proof: Suppose that **S** has the (fmp) and is finitely axiomatizable. Let α be a wff to be tested. The idea is to run two procedures in parallel:

1. Enumerate all proofs in some order of increasing complexity. This procedure is possible because **S** may be formulated, by hypothesis, by using a finite number of axioms.

2. Arrange all finite frames in some order of increasing cardinality, disregarding isomorphic duplicates.

3. Since **S** has the (fmp), then it also has the finite frame property (explained above).

4. Check in parallel the two lists: if α is a theorem, there is a proof that has α in exactly the last line, and thus α appears in the first list. If α is a non-theorem, we will eventually find a (finite) frame containing a model that invalidates α in the second list, i.e., a frame which has a world w such that $v(\alpha, w) = 0$.

♠

As a consequence of Proposition 5.2.8, Proposition 5.2.10 and the fact that filtrations always preserve reflexivity and seriality, recalling the properties of the accessibility relations described in Table 3.2.14, we immediately obtain that all the systems treated in Section 2.3, namely **KD**, **KT**, **KB**, **K4**, **KDB**, **KD4**, **KTB**, **KT4=S4**, **K45**, **KD45**, **KB4** and **KTB4= KT5=S5**, are all decidable since they are finitely axiomatizable.

Although we can find a suitable relation preserving transitivity, filtrations are not so well behaved with respect to euclideanity and, consequently, the decidability of **K5** and **KD5** (the remaining ones in the list of systems between **K** and **S5**) has to rely on different methods which we do not treat here.

---

[5]See, for instance, P. Blackburn, M. de Rijke and Y. Venema [BdRV01], Section 3.4.

5.2. FINITE MODEL PROPERTY AND FILTRATIONS

It is clear that decidability via the (fmp), besides its great theoretical interest, is not necessarily efficient. For example, a simple formula such as $p \vee \Box q$ has 4 subformulas, so we have to check whether this formula receives value 1 in all frames with at most $2^4 = 16$ worlds. There are difficult questions of algorithmic complexity involved here and, even with the help of fast computers, only formulas of limited size can be checked. But this does not mean that decidability via the (fmp) could not be made efficient; the search for more efficient decision procedures belongs to an independent line of research which will not be treated here.

To conclude, two observations are in order. First, there are systems that enjoy the finite model property, but are not decidable due to the fact that they are not finitely axiomatizable. Secondly, completeness and decidability of a system are not interdependent properties. For example, it can be shown that the incomplete system **KVB**, analyzed in Section 5.1, has the finite model property and is finitely axiomatized, so it is decidable, even if it is provably incomplete.

Remark 5.2.11 *It is not easy to identify a system which lacks the (fmp). D. Makinson has been able to prove that the system* **Mk**= **KT**+(**Mk**), *where* (**Mk**) *is* $\Box(\Box\Box p \supset \Box\Box\Box p) \supset (\Box p \supset \Box\Box p)$ *is characterized by reflexive frames which are infinite and non-transitive. However, one may also show that every finite, reflexive frame for* **Mk** *is transitive, and that* **(4)** *(i.e,* $\Box p \supset \Box\Box p$*) is not an* **Mk***-theorem. But we know from Proposition 3.2.16 that* **(4)** *holds on an arbitrary frame* $\mathcal{F}$ *iff* $\mathcal{F}$ *is transitive. Conclusion: if* **Mk** *had the (fmp),* **(4)** *would be invalid in a finite transitive model, which is impossible.* **Mk** *is then a system lacking the (fmp).*

After these digressions on decidability, let us go back to the system **KGL**. As a matter of fact, as we have seen, this system is not canonical. In particular, its canonical model is not irreflexive (cf. Corollary 4.4.20), hence it is not a model over a well-covered frame, but we know from Proposition 4.4.2 that **KGL**-frames must be transitive and well-covered. Hence the analysis of **KGL** in terms of canonical models is not sufficient to provide a completeness result.

Nonetheless, it is interesting to see that the completeness of **KGL** *can* be proved by using the concept of filtration. In fact, to obtain the expected completeness result, it is enough to show that, if α is not a theorem of **KGL**, then α is not valid in some finite irreflexive and transitive model. Now, let W be the collection of all the maximal consistent extensions of **KGL** (i.e., W is the set of worlds of the "candidate" canonical model of

KGL, which we know not to be a **KGL**-model, since **KGL** is not canonical). However, we can define a sort of filtration over W which will turn out to be a **KGL**-model.

Let $\langle W^+, R^+, v^+ \rangle$ be a model in which W^+ and v^+ are defined as in the filtration of the canonical model $\langle W, R, v \rangle$ built on **KGL** (such canonical model obviously exists) through the set of subformulas Σ_α of a wff α, with the difference that R^+ is defined for arbitrary elements of W^+, as follows:

$[w]R^+[w']$ iff both following conditions are satisfied :

(i) For each $\Box\beta \in \Sigma_\alpha$, if $\Box\beta \in w$, then $\Box\beta \in w'$ and $\beta \in w'$

(ii) For some $\Box\gamma \in \Sigma_\alpha$, $\neg\Box\gamma \in w$ and $\Box\gamma \in w'$

It is easy to see that the relation R^+ is well-defined (in the sense that its definition does not depend upon any particular class representative), i. e., if $w \cong_{\Sigma_\alpha} v$ and $w' \cong_{\Sigma_\alpha} v'$, then $[w]R^+[w']$ iff $[u]R^+[u']$ (Exercise 3.26).

Now we are able to prove the following properties of this model:

Lemma 5.2.12 *For R^+ above defined, it holds, for every $\Box\beta \in \Sigma_\alpha$ and every $w \in W$, that $\Box\beta \in w$ iff for every w' such that $[w]R^+[w']$, $\beta \in w'$.*

Proof: ($\Leftarrow$) If $\Box\beta \in w$ and $[w]R^+[w']$, then clearly $\beta \in w'$ by the definition of R^+.

($\Rightarrow$) We have to show, for $\Box\beta \in \Sigma_\alpha$, that, if $\Box\beta \notin w$, then there exists a w' such that $[w]R^+[w']$ and $\beta \notin w'$.

Let Δ be the following collection of wffs: $\Delta = \{\neg\beta, \Box\beta\} \cup \{\gamma, \Box\gamma : \gamma \in w\}$.

Now suppose that Δ is **KGL**-inconsistent. Then there are $\gamma_1, \cdots, \gamma_n$ such that that:

$\vdash_{KGL} \neg(\neg\beta \wedge \Box\beta \wedge \gamma_1 \wedge \Box\gamma_1 \cdots \gamma_n \wedge \Box\gamma_n)$ or equivalently,

$\vdash_{KGL} (\gamma_1 \wedge \Box\gamma_1 \cdots \gamma_n \wedge \Box\gamma_n) \supset (\Box\beta \supset \beta)$ and thus, by (**Nec**) and **K**:

$\vdash_{KGL} (\Box\gamma_1 \wedge \Box\Box\gamma_1 \cdots \Box\gamma_n \wedge \Box\Box\gamma_n) \supset \Box(\Box\beta \supset \beta)$.

But since $\vdash_{KGL} \Box\gamma_i \supset \Box\Box\gamma_i$ (Proposition 4.4.1) and $\Box(\Box\beta \supset \beta) \supset \Box\beta$ is an instance of (**GL**), it follows that $\vdash_{KGL} (\Box\gamma_1 \cdots \Box\gamma_n) \supset \Box\beta$.

Then, since $\Box\gamma_1, \cdots, \Box\gamma_n$ are all in w, $\Box\beta \in w$. Thus, under the supposition that Δ is **KGL**-inconsistent, $\Box\beta \in w$; therefore, if $\Box\beta \notin w$, then Δ is **KGL**-consistent, and there exists a maximal consistent set w' such that $\Delta \subseteq w'$, $\neg\beta \in \Delta \subseteq w'$ and thus $\beta \notin w'$.

It is now easy to see that $[w]R^+[w']$: indeed,

(i) For each $\Box\gamma \in \Sigma_\alpha$, if $\Box\gamma \in w$, then γ and $\Box\gamma$ are in $\Delta \subseteq w'$.

(ii) By hypothesis, for $\Box\beta \in \Sigma_\alpha$, $\Box\beta \notin w$, and then $\neg\Box\beta \in w$ (since w is maximal consistent) and also $\Box\beta \in \Delta \subseteq w'$. ♠

5.2. FINITE MODEL PROPERTY AND FILTRATIONS

Now, we have to prove a variant of Proposition 5.2.7, more specifically for the model $\langle W^+, R^+, v^+ \rangle$:

Lemma 5.2.13 *For each $\beta \in \Sigma_\alpha$ and each $[w] \in W^+$, $v^+(\beta, [w]) = 1$ iff $\beta \in [w]$*

Proof: By induction on the length of formulas. Since each $[w] \in W^+$ is an equivalence class in the collection W of the maximal consistent extensions of **KGL**, the result holds for atomic variables and is clearly preserved by connectives. The only non trivial case concerns formulas of the form $\Box \gamma$, but this is granted by Lemma 5.2.12. ♠

Lemma 5.2.14 $\langle \mathcal{M}^+, R^+, v^+ \rangle$ *is an irreflexive, transitive and finite model.*

Proof: Since Σ_α is finite, W^+ is also finite. In order to prove transitivity, suppose that $[w_1]R^+[w_2]$ and $[w_2]R^+[w_3]$. As $\vdash_{KGL} \Box p \supset \Box\Box p$ (cf. Proposition 4.4.1), if $\Box\beta \in [w_1]$, then $\Box\Box\beta \in [w_1]$. This clearly implies $\beta \in [w_3]$, and thus $[w_1]$ and $[w_3]$ fulfill clause (i) of the definition of R^+.

For clause (ii), since $[w_1]R^+[w_2]$, we have, for some $\Box\gamma \in \Sigma_\alpha$, $\neg\Box\gamma \in w_1$ and $\Box\gamma \in w_2$. But as $[w_2]R^+[w_3]$ then $\Box\gamma \in w_3$. Therefore $[w]R^+[w_3]$.

For irreflexivity, suppose, by *Reductio*, that $[w]R^+[w]$. From clause (ii) of the definition of R^+, an immediate contradiction follows. ♠

Proposition 5.2.15 *If $\nvdash_{KGL} \alpha$, then α is false in some irreflexive and transitive finite model.*

Proof: If $\nvdash_{KGL} \alpha$, then $\alpha \notin w$ for some w in the collection W of the maximal consistent extensions of **KGL**. Let $\langle W^+, R^+, v^+ \rangle$ be defined as above. Since $\alpha \in \Sigma_\alpha$, $v^+(\alpha, [w]) = 0$ from Lemma 5.2.13, because $\alpha \notin [w]$. Thus α is falsified in the model $\langle W^+, R^+, v^+ \rangle$, but by Lemma 5.2.14, this model is finite, irreflexive and transitive. ♠

This last result simultaneously establishes the completeness and decidability of **KGL**, which in itself is a surprising result. In fact, we are aware of the so-called "arithmetical completeness" of **KGL** (see Section 4.6) which means that α is provable in **KGL** if and only if all arithmetical translations of α (that assign a sentence of Peano arithmetic **PA** to any propositional variables and interpret $\Box$ in**KGL** as *Bew*, the standard "provability" predicate of **PA**[6]) are provable in **PA**: it turns out then that **KGL** is decidable, even if **PA** is known not to be decidable.

---

[6] The predicate *Bew(x)*, meaning *beweisbar* – German for "provable"– was constructed by K. Gödel to express that x is the value of the so-called Gödel number of a sentence provable in **PA**.

Failure of canonicity is not the only anomalous property of **KGL**. Indeed, it can be proven that **KGL** is not modal-semantically compact in the sense (see page 88) that there are consistent sets of **KGL**-wffs that cannot be satisfied at a single world of a **KGL**-frame; this can be proved by exhibiting an appropriate consistent set Γ of **KGL**-wffs such that every finite subset of Γ is simultaneously satisfied in some **KGL**-frame $\mathcal{F} = \langle W, R \rangle$, but Γ cannot be satisfied in any **KGL**-frame.

Note that any canonical system is modal-semantically compact: it is enough to observe (cf. Lemma 4.2.11) that any consistent set Δ of wffs can be extended to a maximal consistent set, hence all the elements of Δ are simultaneously satisfied at some world of the canonical model (see Proposition 4.2.15). Unfortunately, the converse of this proposition is not known to be true – this an open problem in modal logic.

To see that **KGL** is not modal-semantically compact, consider the set of **KGL**-wffs $\Gamma = \{\Diamond p_0\} \cup \{\Box(p_i \supset \Diamond p_{i+1})\}$ for $i \geq 1$. It can be proven that (i) Γ is consistent, but (ii) Γ cannot be satisfiable on any **KGL**-frame (Exercise 5.14).

Such a failure of modal-semantic compactness, incidentally, gives another proof of the non-canonicity of **KGL**.

The failure also proves that **KGL** cannot be strongly complete. In fact, suppose by contradiction that **KGL** is strongly complete with respect to some class of **KGL**-frames $\mathfrak{F}$. It can be proved that any system **S** is strongly complete with respect to a class of **S**-frames $\mathfrak{F}$ iff every **S**-consistent set of sentences is satisfiable on some $\mathcal{F} \in \mathfrak{F}$ (Exercise 5.13). But Γ above is not satisfiable on any **KGL**-frame.

The analysis of finite models has led to two results (not proven here; for details, the reader is referred to Section 5.4) which give a special prominence to two modal systems, **S4.3** and **S5**.

The first theorem, known as *Bull's Theorem*, states what follows:

Proposition 5.2.16 *S4.3 and all of its consistent extensions have the (fmp).*

The second theorem, known as *Scrogg's Theorem*, states what follows:

Proposition 5.2.17 *Every proper consistent extension of S5 is characterized by a single finite frame.*

When concerning Scrogg's Theorem some remarks are in order:

1. Let us recall the sequence of Dugundji's formulas D_n, for $1 \leq i \leq n+1$ and $1 \leq j \leq n+1$ (also recall, from Definition 2.2.2, that $p_i \asymp p_j$ means $\Box(p_i \supset p_j) \wedge \Box(p_j \supset p_i)$):

5.2. FINITE MODEL PROPERTY AND FILTRATIONS

$$D_n \stackrel{\text{Def}}{=} \bigvee_{i \neq j}(p_i \asymp p_j)$$

and let us see what happens by subjoining them as axioms to **S5**.

2. As suggested in Chapter 3, every Carnapian implicit model may be converted into a finite or infinite matrix having the form which has been outlined to prove Dugundji's theorem (see Proposition 3.1.3) and vice versa.

3. Every Dugundji's formula is satisfied by a finite matrix for all assignments to the variables, hence also by a finite model over some Carnapian frame. **S5**+ D_1 is an inconsistent system (since by (**US**) we reach $\top \asymp \bot$). But **S5**+ D_2 is equivalent to **S5** + $\Box p_1 \vee \Box \neg p_1$. For the proof of the equivalence, it is enough to reason as follows. In one direction, starting from D_2, take $\bot$ for p_2 and $\top$ for p_3. In the other direction, consider that, for every p_i, $\Box p_i \vee \Box \neg p_i$ equals $\Diamond p_i \supset \Box p_i$. But $(\Diamond p_1 \supset \Box p_1) \wedge (\Diamond p_2 \supset \Box p_2)$ implies $(\Diamond p_1 \vee \Diamond p_2) \supset (\Box p_1 \wedge \Box p_2)$ or equivalently $(\Box \neg p_1 \wedge \Box \neg p_2) \vee (\Box p_1 \wedge \Box p_2)$; this wff implies $\Box(p_1 \equiv p_2)$, so a fortiori $p_1 \asymp p_2 \vee p_2 \asymp p_3 \vee p_1 \asymp p_3$.

Also note that $\Box p_1 \vee \Box \neg p_1$, when subjoined to every system containing **T**, is equivalent to (**Triv**) (see Section 2.2).

4. It is therefore easy to check that all Dugundji's formulas of the preceding list from D_2 on, when added as axioms, may be replaced by the corresponding wffs which appear in the following sequence:

   (a) Alt_1: $\Box \neg p_1 \vee \Box p_1$ (for D_2)

   (b) Alt_2: $\Box \neg p_1 \vee \Box(p_1 \vee \neg p_2) \vee \Box(p_1 \vee p_2)$ (for D_3)

   (c) Alt_3: $\Box \neg p_1 \vee \Box(p_1 \vee \neg p_2) \vee \Box(p_1 \vee p_2 \vee \neg p_3) \vee \Box(p_1 \vee p_2 \vee p_3)$ (for D_4)

   (d) $\vdots$ and so on

We leave to the reader the proof of this conclusive theorem:

Proposition 5.2.18 *If L_n is **S5**+ Alt_n, the canonical model of L_n has m elements, for some $m < n$.*

The last proposition implies a completeness theorem for all members of the class of systems **S5**+ Alt_n, in addition to giving a proof of their decidability.

5.3 Exercises

1. Prove Lemma 5.1.4. (Hint: induction on the length of formulas. Note that $V(\Box \alpha)$ is admissible independently of the fact that $V(\alpha)$ be admissible or not.)

2. Prove Proposition 5.1.8. (Hint: take the frame of the canonical model $\langle W, R \rangle$ of any normal modal system **S** and define the collection Π of admissible sets by stipulating that $X \in \Pi$ iff there exists some formula α that is true in each world of X but not true in any other world. Show that $\langle W, R, \Pi \rangle$ is a general frame that characterizes **S**).

3. Prove Proposition 5.2.4. (Hint: show that every modal algebra may be converted into a general frame.)

4. Give direct proofs for the facts that **KT** and **S4** have the (fmp). (Hint: prove that the relation R^* of all **KT**-filtrations is reflexive, and that there is an **S4**-filtration which is reflexive and transitive. For the latter define R^* in the following way: for any $w, w' \in W$, $[w]R^*[w']$ iff for each $\Box\beta \in \Phi$, if $v(\Box\beta, w) = 1$, then $v(\Box\beta, w') = 1$).

5. Prove the completeness of **KT** with respect to the class of regular algebras such that $\tau(\alpha) \leq \alpha$.

6. Prove that R_Σ^* and $R_{\Sigma'}^*$, as defined on page 126, are well-defined and satisfy conditions (a) and (b). Prove also that R_Σ^* is maximal and $R_{\Sigma'}^*$ is minimal, in the sense that, for any suitable R^*, $R_{\Sigma'}^* \subseteq R^* \subseteq R_\Sigma^*$.

7. Prove that seriality and reflexivity of R are preserved in passing from models to filtrations.

8. Show that the definition of R_Σ^* in the largest filtration is equivalent to $[w]R_\Sigma^*[w']$ iff for each sentence $\Diamond\alpha \in \Sigma$ it holds what follows: if $v(\alpha, w') = 1$, then $v(\Diamond\alpha, w) = 1$.

9. Show that not all filtrations preserve transitivity, symmetry or euclideanity. (Hint: let $w_1, \cdots, w_5$ be worlds;

    (a) For transitivity, consider the following model where the relation is transitive.

    Then filter through $\Sigma = \{p, \Diamond p\}$ using the relation R_Σ^* as characterized in Exercise 5.8 and show that the filtration is not transitive.

5.3. EXERCISES

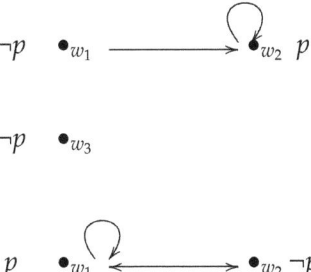

(b) For symmetry, consider the following model where the relation is symmetric, and again filter through $\Sigma = \{p, \Diamond p\}$ using the relation R^*_Σ as characterized in Exercise 5.8. Show that the filtration is not symmetric.

(c) For euclideanity, consider the following model where the relation is euclidean and filter through $\Sigma = \{p, \Box p\}$ using the relation R^*_Σ.

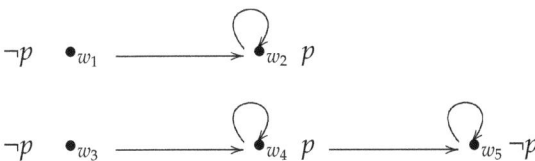

Show that the filtration is not euclidean.[7])

10. Prove that R^T_Σ in Proposition 5.2.8 is transitive.

11. Prove that, if a system **S** has only one Post-complete extension and is Halldén-incomplete, then interpolation fails in **S**. (Hint: Suppose **S** has only one Post-complete extension **C**, that $\alpha \vee \beta$ is an **S**-theorem and **S** is Halldén-incomplete. If p is any variable foreign to α and β, since $\alpha \vee \beta$ is an **S**-theorem, it follows that $(\neg \alpha \wedge (p \supset p)) \supset (\beta \wedge (p \supset p))$ is also an **S**-theorem having only p as a common variable between the two clauses ($\neg \alpha \wedge (p \supset p)$ and $\beta \wedge (p \supset p)$). Suppose that interpolation holds in **S** for some γ. Show that neither γ nor $\neg \gamma$ can be **S**-theorems, but that both are **S**-consistent. Then derive a contradiction from the supposition that **S** would have two maximal consistent extensions, one containing γ and another containing $\neg \gamma$.)

---

[7]Counter-examples suggested by V. Goranko (personal communication, 2007).

12. Prove that, for any α, if $w \cong_{\Sigma_\alpha} v$ and $w' \cong_{\Sigma_\alpha} v'$, then $[w]R^+[w']$ iff $[u]R^+[u']$, where R^+ is defined as on page 132. (Hint: just note that $w \cong_{\Sigma_\alpha} v$ iff w and v satisfy the same subformulas of α, and that $\neg\Box\gamma \in [w]$ iff $\Box\gamma \notin [w]$ since w is maximal consistent).

13. Prove that any system **S** is strongly complete with respect to the class of **S**-frames $\mathfrak{F}$ iff every **S**-consistent set of sentences is satisfiable on some $\mathcal{F} \in \mathfrak{F}$. (Hint: the left-to-right direction is obvious. To prove the right-to-left direction, suppose that **S** is not strongly complete, i.e., suppose that there is a $\Delta \cup \{\alpha\}$ such that $\Delta \vDash_\mathfrak{F} \alpha$ but $\Delta \nvdash \alpha$. Then $\Delta \cup \{\neg\alpha\}$ is consistent but cannot be satisfiable on any $\mathcal{F} \in \mathfrak{F}$).

14. Prove that the set of **KGL**-wffs $\Gamma = \{\Diamond p_0\} \cup \{\Box(p_i \supset \Diamond p_{i+1})\}$ for $i \geq 1$ is consistent but lacks modal-semantic compactness. Use this fact to prove the semantic non-compactness of **KGL**. (Hint: show that for Γ to be satisfiable on an irreflexive and transitive frame, the frame would need to have an infinite chain, so it cannot be a frame for **KGL** in view of Proposition 4.4.2).

5.4 Further reading

The algebraic analysis of arithmetical provability has been developed by the so-called "Siena-school" (C. Bernardi, F. Montagna, G. Sambin, A. Ursini and S. Valentini among others) founded by R. Magari in the Sixties. A complete survey of all material about provability logic may now be found in S. Artemov and L. Beklemishev [AB05].

As incompleteness is concerned, the first work is due to S. K. Thomason [Tho74], but the result exposed in Section 5.1 for **KVB**, originally due to J. van Benthem ([vB78]), has been reproduced following the lines of G. E. Hughes and M. J. Cresswell [HC84]. The incompleteness of **KH** was first proved by R. Magari in [Mag82], and a simplified proof appears in G. Boolos and G. Sambin [BS85]. A detailed proof can be found in [HC96], chapter 9. The latter also proves (by elementary means) an interesting connection between **KH** and **KGL**: a wff α is valid on all frames for **KH** iff α is a **KGL**-theorem.

There is a different sense of completeness, the so-called *structural completeness* for a logic, which has to do with the question of which rules can be added to a system **S** without increasing the set of derivations of **S**; rules with such a property are called *admissible rules* (a concept introduced by P. Lorenzen in [Lor55]). In less informal terms, if Δ/ϕ is a rule (for $\Delta \cup \{\phi\}$

5.4. FURTHER READING

a finite collection of **S**-wffs), Δ/ϕ is admissible if, adding substitution instances of Δ to the theorems of **S**, then the substitution instance of ϕ is also a theorem of **S**. A rule Δ/ϕ is *derivable* in **S** when $\Delta \vdash_S \phi$. Of course, any derivable rule is admissible in any system **S**, but a rather interesting point is that the converse is in many cases not true, and is true precisely when **S** is structurally complete. For instance, the rule $p/\Box p$ is an admissible rule of **K**, but it is not derivable in **K**. A good source on the interesting question of admissibility of inference rules is V. V. Rybakov [Ryb97].

For the system **Mk** see D. Makinson in [Mak69]; a class of systems satisfying (fmp) although undecidable (as not finitely axiomatized) was defined by A. Urquhart in [Urq81].

For Halldén incompleteness, Bull's Theorem and Scroggs's Theorem, see E. J. Lemmon [Lem66c], R. A. Bull [Bul66] and S. J. Scroggs [Scr51].

For details on the Filtration Theorem, see K. Segerberg in [Seg71]. A good exposition about elementary properties of filtrations (in Spanish) is R. Jansana [Jan90]. For a proof of Proposition 5.2.18, see P. Gärdenfors [Gär73].

We use here the term "modal-semantic compactness" to distinguish a sense of compactness which is different from the usual "classical" sense, as defined on page 11, by which a system **S** is (semantically) compact if and only if a set Γ of **S**-wffs has a model in an **S**-frame when each finite subset of Γ is satisfied on some **S**-frame. As shown in G. F. Schumm [Sch89], the two senses of compactness coincide for canonical systems, but differ for non-canonical ones (the former implies the latter, but not vice-versa).

Historically important references for modal algebras are J. Lemmon [Lem66a, b] where McKinsey-Tarski algebraic methods of [MT48] are generalized to several modal calculi. For a more recent treatment see W. Rautenberg [Rau79], especially pages 144–161. The interconnection between relational frames and modal algebras examined in the present chapter might suggest that these two kinds of semantics exhaust the panorama of the semantical treatment of modal logic. As a matter of fact, even if relational semantics is still today the core of modal semantics, other approaches have been proposed in the last decades as alternatives or generalizations of relational semantics. In 1970 D. Makinson proposed a generalization of relational semantics in which frames are seen as triples $\langle W, N, R \rangle$ where N is a non-empty subset of W (see [Mak70]). In the so-called *neighborhood semantics*, essentially due to D. Scott and R. Montague (see [Sco70] [Mon70]) a modal frame is a couple $\langle U, N \rangle$ where U is a set and $N : U \mapsto \wp(\wp(U))$ is a function assigning to every $w \in U$ a class of subsets of U. $\Box \alpha$ is true at a point w in U iff the set of the α-worlds (i.e. the set of worlds in U at which α is true) belongs to $N(w)$. Every relational frame may be converted into a neighborhood frame, but the converse does not hold. The strength of this and other generalizations of relational semantics has been studied in a comparative way by B. Hansson and P. Gärdenfors in [HG73].

Another generalization of relational semantics has been offered by the so-called *bidimensional semantics* , i.e. by the idea that accessibility should be seen not as a relation between worlds but between ordered couples of worlds $\langle w, w' \rangle$ endowed with certain definable interconnections (see L. Humberstone [Hum81] and S. Kuhn [Kuh89]). For an investigation along similar lines see K. Segerberg [Seg73] and the rich trend of inquiries derived from Y. Venema's Ph.D. thesis [Ven93]. In this direction of inquiry multidimensional semantics has a syntactic counterpart in systems whose language has more than one primitive modal operator, so it may be considered as belonging to the field of multimodal logic.

Chapter 6

Temporal logics

6.1 Logics with two primitive modal operators

The logics that have been examined in previous chapters were defined in terms of languages containing a unique primitive monadic operator, usually chosen in the set $\{\Box, \Diamond\}$, or also (for logics containing the axiom (**T**)), in the set $\{\Delta, \nabla\}$. But, in principle, it is not difficult to introduce, as primitives in the language, two or more distinct modal operators or even an infinite number of them.

Logics having modal languages extended in this way are called *multimodal logics*. In Chapter 8 we shall study, in an abstract way, the properties of multimodal logics containing an arbitrary number of primitive modal operators. The logics studied up to now turn out to be just special cases of such multimodal logics and should be properly called *monomodal logics.*

To appraise the philosophical interest of multimodal logics and their prospective applications, it is enlightening to consider some simple examples of *bimodal* logics, that is, of logics whose languages have two primitive modal operators.

The logic of arithmetical provability examined in the previous chapter can be easily include a bimodal extension: in fact, one could be interested in giving a modal representation not only of provability in Peano arithmetic, but also of provability in one of its subsystems. In this perspective, two distinct modal operators would have to be introduced as primitives.

A simple case of bimodal logic is the logic of physical modalities, i.e, a logic whose language allows us to express two different notions of necessity: logical necessity, symbolized by $\Box$, and physical necessity, symbolized by $\boxdot$. The formation rules for this calculus are the same as for the monomodal ones described in Definition 2.3.1, extended with the following obvious clause:

(iii') If $\alpha \in \mathit{WFF}$, then $\boxdot \alpha \in \mathit{WFF}$

From this clause, it follows that $\Box \boxdot \alpha$ and $\boxdot \Box \alpha$ are also wffs of this language. By analogy with the standard possibility operator, the operator of physical possibility is defined as:

$$\Diamondblack \alpha \stackrel{\text{Def}}{=} \neg \boxdot \neg \alpha$$

When axioms for two or more primitive modal operators are introduced, it is convenient to describe the relationship between them, so as to avoid that the resulting systems be simply the juxtaposition of different modal systems. In the case of physical modalities, we have to introduce an axiom which governs the relationship between logical necessity and physical necessity. The simplest connection between them is provided by the following *logical-physical bridge axiom*:

(**LP**) $\Box p \supset \boxdot p$

The meaning of this mixed axiom is that logical necessity is stronger than physical necessity: anything that is logically necessary is physically necessary. To understand this point, we have to mention that what is physically necessary is that which is implied by logical or physical laws, while what is logically necessary is that which is simply implied by logical laws, or that which is self-contradictory to deny.

As an axiomatic basis for $\Box$, we can select the one for **KT**, while axioms for $\boxdot$ can be obtained by duplicating those of **KT** for physical necessity, that is:

(**K$^\boxdot$**) $\boxdot(p \supset q) \supset (\boxdot p \supset \boxdot q)$

(**T$^\boxdot$**) $\boxdot p \supset p$

The logic axiomatized in this way (that is, **KT**+(**K$^\boxdot$**) +(**T$^\boxdot$**)+(**LP**)) will be called **KT$^\boxdot$**.

Note that in view of (**LP**) the (**Nec**) rule of **KT** holds also for $\boxdot$, since we easily derive the rule: if $\vdash_{KT^\boxdot} \alpha$, then $\vdash_{KT^\boxdot} \boxdot \alpha$.

The semantical analysis of **KT$^\boxdot$** is particularly simple. In the class of logically possible worlds that are accessible from a certain world, we need to identify a subclass of worlds that are "physically accessible" (that is, that are physically possible with respect to the reference world). Thus we define

6.1. LOGICS WITH TWO PRIMITIVE MODAL OPERATORS   143

a **KT$^\square$**-frame as a triple $\langle W, R, R^\square \rangle$, where W and R are are the same as in **KT** and $R^\square$ satisfies:

(a) for each $w, w' \in W$, if $wR^\square w'$, then wRw' $\hspace{2em}(R^\square \subseteq R)$

(b) for each $w \in W$, $wR^\square w$ $\hspace{2em}(R^\square$ is reflexive)

If we wish to define assignments in explicit **KT$^\square$**-models, we will have to add the following clause to Definition 3.2.8:

(5) $\quad v(\square \alpha, w) = 1 \quad$ iff $\quad v(\alpha, w') = 1 \quad$ for every w' such that $wR^\square w'$.

The definition of **KT$^\square$**-validity is an obvious extension of the definition of **KT**-validity. We can also formulate rules on how to build tableaux for this logic. Besides the usual arrows that represent the logical accessibility relations, we also need to introduce arrows indexed by $R^\square$ to indicate the physical accessibility depending upon existential statements occurring in the diagrams. Moreover, as worlds that are physically accessible are also logically accessible (that is, $R^\square \subseteq R$), we stipulate that when an indexed arrow is introduced, it also stands for the usual arrow. We introduce then the following additional rule:

(**RR$^\square$**) Let w' be a diagram such that there is an arrow indexed by $R^\square$ which goes from w to w'. Then reproduce in w' the wff α with value 1 if $\square \alpha$ or $\square \alpha$ has value 1 in w and α with value 0 if $\lozenge \alpha$ or $\lozenge \alpha$ has value 0 in w:

$$v(\alpha, w') = \begin{cases} 1 & \text{if } v(\square \alpha, w) = 1 \text{ or } v(\square \alpha, w) = 1 \\ 0 & \text{if } v(\lozenge \alpha, w) = 0 \text{ or } v(\lozenge \alpha, w) = 0 \end{cases}$$

Example 6.1.1 *Consider, for instance, the formula $\lozenge p \supset \lozenge \lozenge p$. The **KT$^\square$**-tableau having this formula as input is the following:*

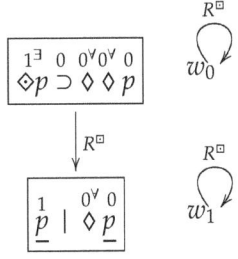

Since $v(\Diamond p \supset \Diamond\Diamond p, w_0) = 0$, then $v(\Diamond p, w_0) = 1$ and $v(\Diamond\Diamond p, w_0) = 0$. By the reflexivity of R, we have that $v(\Diamond p, w_0) = 0$ and, consequently, $v(p, w_0) = 0$. From $v(\Diamond p, w_0) = 1$ and by (5), we have $v(p, w_1) = 1$. On the other hand, from $v(\Diamond\Diamond p, w_0) = 0$ and (**RR**$^\square$), we have that $v(\Diamond p, w_1) = 0$ and, by the reflexivity of R, $v(p, w_1) = 0$: contradiction.

Since each diagram of the tableau has a lower modal degree[1] with respect to all previous diagrams in the same chain, the method of tableau with two accessibility relations ends in a finite number of steps, hence it provides a decision procedure for this logic. It is also possible to work out a constructive completeness proof for **KT**$^\square$ modeled on the one given for **KT** in Section 4.2 (Exercise 4.1).

The logic of physical and logical necessity **KT**$^\square$ thus receives a semantical characterization expressed by saying that one of the accessibility relations is included in the other. But there are other equally simple connections that could hold between accessibility relations: an intuitive one consists in requiring that the relations in the frames be *reciprocally converse*. As an example, consider pairs of accessibility relations as "to the left of" and "to the right of", or "before" and "after". With regard to the temporal relation "before-after", it is easy to find an interpretation for the necessity operators: indeed, such an interpretation is "it has always been the case that" and "it will always be the case that", symbolized by ▣ and ▢, respectively. The formation rules are extended in a obvious way so that ▣α and ▢α are wffs if α is. The dual operators are defined by:

- ⧫α $\stackrel{\text{Def}}{=}$ ¬▣¬α

- ◆α $\stackrel{\text{Def}}{=}$ ¬▢¬α

⧫α can be read as "it has been the case that α" and ◆α as "it will be the case that α".[2] Logics sharing this language are called *PF-logics* or *tense logics*. The minimal system of PF-logic is called **K**$_t$ and is axiomatized by adding the following axioms to the **PC**-axioms:

(**K**▢)   ▢$(p \supset q) \supset (▢p \supset ▢q)$ (axiom (**K**) for ▢)

(**K**▣)   ▣$(p \supset q) \supset (▣p \supset ▣q)$ (axiom (**K**) for ▣)

---

[1] The notion of modal degree defined in Section 2.1 can be adapted to the bimodal tableaux by treating ▣ as ▢.

[2] Elsewhere in the literature the symbols H and G are used, respectively, instead of ▣ and ▢. Also, F and P are used, respectively, instead of ◆ and ⧫.

6.1. LOGICS WITH TWO PRIMITIVE MODAL OPERATORS

(**BA1**) $p \supset \boxdot \diamondsuit p$ (past-future bridge axiom)

(**BA2**) $p \supset \Box \diamondsuit p$ (future-past bridge axiom)

The rules are the following:

(**Nec$_\boxdot$**) $\vdash_{K_t} \alpha$ implies $\vdash_{K_t} \boxdot \alpha$

(**Nec$_\Box$**) $\vdash_{K_t} \alpha$ implies $\vdash_{K_t} \Box \alpha$

Just as we called **PC**$^\Box$ the **PC** system within modal language, we will call **PC**$^{\boxdot\Box}$ the system that is obtained by dropping from K_t the axioms ($K_\Box$), ($K_\boxdot$), (**BA1**) and (**BA2**), as well as the rules (**Nec$_\boxdot$**) and (**Nec$_\Box$**). We will call $K^{\boxdot\Box}$ the system obtained by dropping the two bridge axioms (**BA1**) and (**BA2**) from K_t.

Note that the two bridge axioms allow us to derive two variants of the necessitation rule in K_t (which are not derivable in $K^{\boxdot\Box}$):

(**Nec$_{\boxdot\diamondsuit}$**) $\vdash_{K_t} \alpha$ implies $\vdash_{K_t} \boxdot \diamondsuit \alpha$

(**Nec$_{\Box\diamondsuit}$**) $\vdash_{K_t} \alpha$ implies $\vdash_{K_t} \Box \diamondsuit \alpha$

To help the intuitive interpretation of temporal frames, instead of referring to sets of worlds we will refer to sets of *instants* of time $T = \{t_0, t_1, t_2, \cdots\}$. For heuristic reasons, we will read $t \overleftarrow{R} t'$ as "t follows t'" and $t \overrightarrow{R} t'$ as "t precedes t'".

A *temporal frame* is a triple $\mathcal{F} = \langle T, \overleftarrow{R}, \overrightarrow{R} \rangle$. The notion of *temporal model* is defined in the obvious way. The correspondence theory for temporal logic is the same as the one for monomodal systems except for obvious changes. In particular, we replace the clause for $s(\Box \alpha)$ in the definition of the standard translation, given in Section 3.3, with the following:

- $s(\boxdot \alpha) = \forall t'(t \overleftarrow{R} t' \supset s(\alpha)[t/t'])$, where $t \neq t'$

- $s(\Box \alpha) = \forall t'(t \overrightarrow{R} t' \supset s(\alpha)[t/t'])$, where $t \neq t'$

By applying this translation rule to the bridge axioms (**BA1**) and (**BA2**) and by making use of the laws of first-order logic, we obtain (see Exercise 6.1):

- $s(p \supset \boxdot \diamondsuit p) = \forall t \forall t'(t \overleftarrow{R} t' \supset t' \overrightarrow{R} t)$ (that is[3] $\overleftarrow{R} \subseteq (\overrightarrow{R})^{-1}$)

- $s(p \supset \Box \diamondsuit p) = \forall t \forall t'(t \overrightarrow{R} t' \supset t' \overleftarrow{R} t)$ (that is $\overrightarrow{R} \subseteq (\overleftarrow{R})^{-1}$)

---

[3] R^{-1} means the inverse relation of R, i.e, $xR^{-1}y$ iff yRx. In logico-temporal contexts all inverse relations are usually called *converse*.

Thus, while in $\mathbf{K}^{\boxdot\boxdot}$-frames the two relations $\overrightarrow{R}$ and $\overleftarrow{R}$ have no specific properties, in $\mathbf{K}_t$-frames $\overleftarrow{R}$ and $\overrightarrow{R}$ are converse relations. The definitions of truth and validity for both systems are parallel to the ones given for monomodal systems:

- $\mathcal{M}, t \vDash \boxdot \alpha$ iff $\mathcal{M}, t' \vDash \alpha$ for every t' such that $t \overrightarrow{R} t'$.

- $\mathcal{M}, t \vDash \boxdot \alpha$ iff $\mathcal{M}, t' \vDash \alpha$ for every t' such that $t \overleftarrow{R} t'$.

It is obvious that for $\diamondsuit$ and $\diamondsuit$ we have:

- $\mathcal{M}, t \vDash \diamondsuit \alpha$ iff $\mathcal{M}, t' \vDash \alpha$ for some t' such that $t \overrightarrow{R} t'$.

- $\mathcal{M}, t \vDash \diamondsuit \alpha$ iff $\mathcal{M}, t' \vDash \alpha$ for some t' such that $t \overleftarrow{R} t'$.

Since $\mathbf{K}^{\boxdot\boxdot}$ is nothing else than a duplicate version of $\mathbf{K}$, $\mathbf{K}^{\boxdot\boxdot}$-tableaux are duplications of $\mathbf{K}$-tableaux, in the sense that their diagrams are ordered by the two distinct relations $\overrightarrow{R}$ and $\overleftarrow{R}$. To adopt a graphic device, the arrows owing to the operators for the future ($\boxdot$ and $\diamondsuit$) will be labeled by $\overrightarrow{R}$, while the arrows owing to the operators for the past ($\boxdot$ and $\diamondsuit$) will be labeled by $\overleftarrow{R}$. The initial diagram (that contains the formula under test) is indicated by $t_{(0)}$. Any diagram placed in the immediate future (respectively, past) of $t_{(0,u,\cdots,v)}$ will be labeled by $t_{(0,u,\cdots,v,z)}$ (respectively $t_{(0,u,\cdots,v,-z)}$), where $t_{(0,u,\cdots,v,1)} \cdots t_{(0,u,\cdots,v,z-1)}$ are all the alternatives of $t_{(0,u,\cdots,v)}$.[4] In the following examples, the names of the diagrams will be simplified by omitting parentheses and numerals when the resulting symbol is unequivocal.

Example 6.1.2 *Consider the formula $\boxdot q \vee \boxdot \alpha$, where α is $(\boxdot p \wedge q) \supset (\boxdot \diamondsuit q \vee \boxdot (p \vee q))$. The $\mathbf{K}^{\boxdot\boxdot}$-tableau for this formula is the following:*
Since $v(\boxdot q \vee \boxdot \alpha, t_{(0)}) = 0$, we have $v(\boxdot q, t_{(0)}) = 0$ and $v(\boxdot \alpha, t_{(0)}) = 0$. Hence there exist a $t_{(0,-1)}$ and a $t_{(0,-2)}$ such that $t_{(0)} \overleftarrow{R} t_{(0,-1)}$ and $t_{(0)} \overleftarrow{R} t_{(0,-2)}$, and such that $v(q, t_{(0,-1)}) = 0$ and $v(\alpha, t_{(0,-2)}) = 0$, but α is $(\boxdot p \wedge q) \supset (\boxdot \diamondsuit q \vee \boxdot (p \vee q))$, thus $v((\boxdot p \wedge q) \supset (\boxdot \diamondsuit q \vee \boxdot (p \vee q)), t_{(0,-2)}) = 0$. Consequently,

(a) $v((\boxdot p \wedge q), t_{(0,-2)}) = 1$

(b) $v((\boxdot \diamondsuit q \vee \boxdot (p \vee q)), t_{(0,-2)}) = 0$

---

[4]We stipulate that the bottom up arrows represent the direction from past to future, while the top down arrows represent the direction from future to past.

6.1. LOGICS WITH TWO PRIMITIVE MODAL OPERATORS

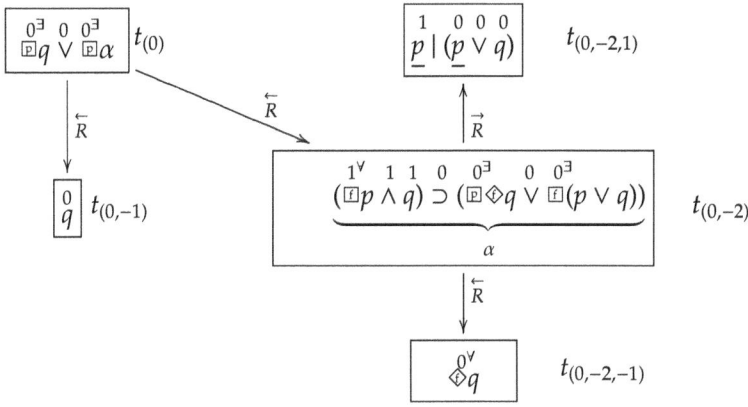

- From (b) $v(\boxdot \Diamond q, t_{(0,-2)}) = 0$ and $v(\Box(p \vee q), t_{(0,-2)}) = 0$. From the latter, there exists $t_{(0,-2,1)}$ such that $t_{(0,-2)} \vec{R} \, t_{(0,-2,1)}$ and $v((p \vee q), t_{(0,-2,1)}) = 0$, therefore $v(q, t_{(0,-2,1)}) = 0$ and $v(p, t_{(0,-2,1)}) = 0$.
- From (a), $v(\Box p, t_{(0,-2)}) = 1$ and $v(q, t_{(0,-2)}) = 1$. Since $t_{(0)} \vec{R} \, t_{(0,-2,1)}$, from $v(\Box p, t_{(0,-2)}) = 1$ we have that $v(p, t_{(0,-2,1)}) = 1$.

Note that in diagram $t_{(0,-2,1)}$ the tableau is closed since the atomic variable p receives contradictory assignments, and that such result does not depend on specific properties of the accessibility relation.

The completeness proof for $\mathbf{K}^{\boxdot\Box}$ is parallel to the one for $\mathbf{K}$, as the definition of a characteristic formula is, *mutatis mutandis*, the same given in Section 4.1. The difference is that the transition from the characteristic formula of a diagram t_i to the one of a diagram t_j, which is accessible to t_i, is now possible due to two distinct necessitation rules: if the arrow from t_i to t_j is $\vec{R}$, we apply (Nec$_\Box$); otherwise we apply (Nec$_\boxdot$).

An instance of how to convert a closed $\mathbf{K}^{\boxdot\Box}$-tableau into a proof of its input formula is the following (provided by Example 6.1.2 after eliminating definitionally all modal symbols from the tableau with the exception of $\Box$ and $\boxdot$); note that the characteristic formula $\chi_{t_{0,-2,1}}$ of the inconsistent diagram is a **PC**-tautology.

1. $(p \vee q) \vee \neg p$ $[\chi_{t_{0,-2,1}}]$
2. $\Box((p \vee q) \vee \neg p)$ [(Nec$_\Box$) in 1]

3. $\boxdot(p \supset (p \vee q))$ [(Eq) in 2]
4. $\boxdot(p \supset (p \vee q)) \supset (\boxdot p \supset \boxdot(p \vee q))$ [($\mathbf{K}^{\boxdot\boxdot}$)]
5. $\boxdot p \supset \boxdot(p \vee q)$ [(MP), 3, 4]
6. $\neg \boxdot p \vee \boxdot(p \vee q)$ [$\mathbf{PC}^{\boxdot\boxdot}$, 5]
7. $\neg \boxdot p \supset ((\boxdot p \wedge q) \supset (\boxdot\neg\boxdot\neg q \vee \boxdot(p \vee q)))$ [$\mathbf{PC}^{\boxdot\boxdot}$]
8. $\boxdot(p \vee q) \supset ((\boxdot p \wedge q) \supset (\boxdot\neg\boxdot\neg q \vee \boxdot(p \vee q)))$ [$\mathbf{PC}^{\boxdot\boxdot}$]
9. $(\neg \boxdot p \vee \boxdot(p \vee q)) \supset ((\boxdot p \wedge q) \supset (\boxdot\neg\boxdot\neg q \vee \boxdot(p \vee q)))$ [$\mathbf{PC}^{\boxdot\boxdot}$, 7, 8]
10. $(\boxdot p \wedge q) \supset (\boxdot\neg\boxdot\neg q \vee \boxdot(p \vee q))$ [(MP), 6, 9]
11. $\boxdot((\boxdot p \wedge q) \supset (\boxdot\neg\boxdot\neg q \vee \boxdot(p \vee q)))$ [($\mathbf{Nec}_{\boxdot}$) in 10]
12. $\boxdot q \vee \boxdot((\boxdot p \wedge q) \supset (\boxdot\neg\boxdot\neg q \vee \boxdot(p \vee q)))$ [$\mathbf{PC}^{\boxdot\boxdot}$, 11]

Note that $\neg\boxdot(\boxdot p \wedge q)$ and $\boxdot(\boxdot\neg\boxdot\neg p \vee \boxdot(p \vee q))$ both imply (12) by the methods of $\mathbf{K}^{\boxdot\boxdot}$.

Since every closed $\mathbf{K}^{\boxdot\boxdot}$-tableau may be converted into a proof for the input formula, the following completeness result for $\mathbf{K}^{\boxdot\boxdot}$ straightforwardly follows by a suitable extension of the completeness proof of $\mathbf{K}$ discussed in Section 4.1.

Proposition 6.1.3 α *is a* $\mathbf{K}^{\boxdot\boxdot}$-*valid wff iff* α *is a theorem of* $\mathbf{K}^{\boxdot\boxdot}$.

Proof: The argument is a suitable adaptation of the constructive completeness for $\mathbf{K}$ given in Section 4.1 by using the two necessitation rules.   ♠

The construction of tableaux for $\mathbf{K}_t$ is like the one for $\mathbf{K}^{\boxdot\boxdot}$ with the difference that we apply two *circuit rules*:

(**CR1**) Suppose that a $\mathbf{K}_t$-tableau has a bottom-up arrow (represented by $\overrightarrow{R}$) from a diagram t_i to t_j.[5] If there are universal past operators in t_j (i.e, $\boxdot\alpha$ with value 1 or $\diamondsuit\alpha$ with value 0), then draw an arrow from t_j to t_i (represented by $\overleftarrow{R}$) and introduce the values of formulas required by the new accessibility relation into t_i.

(**CR2**) Analogous to the preceding (**CR1**) with arrows in the opposite direction, but instead with $\boxdot$ and $\diamondsuit$.

---
[5]The indices i and j are here sequences of numbers that label diagrams, as defined before.

6.1. LOGICS WITH TWO PRIMITIVE MODAL OPERATORS 149

With these two new rules, it is easy to check that the bridge axioms (**BA1**) and (**BA2**) are $\mathbf{K}_t$-valid; as an example, for (**BA1**):

Example 6.1.4

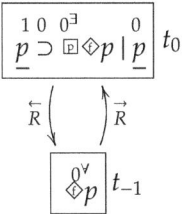

Note that if the arrows were not converse (that is, if (**CR2**) had not been applied) we would not know that t_0 is in the accessibility sphere of t_{-1}, hence we would not be allowed to introduce p with value 0 in diagram t_0 in order to obtain a contradiction. A simple argument of the kind exposed in Section 3.2 shows that the formula $p \supset \boxdot \Diamond p$ holds on a temporal frame $\mathcal{F}$ if and only if $\overrightarrow{R}$ and $\overleftarrow{R}$ are converse relations in $\mathcal{F}$.

It is easy, in any case, to prove the following soundness and decidability results:

Proposition 6.1.5 *The system $\mathbf{K}_t$ is sound with respect to the class of frames $\langle T, \overleftarrow{R}, \overrightarrow{R} \rangle$, in which $\overleftarrow{R}$ and $\overrightarrow{R}$ are converse relations.*

Proof: It is enough to observe that the axioms are $\mathbf{K}_t$-valid with respect to all models based on such frames, and that the rules preserve $\mathbf{K}_t$-validity. ♠

Proposition 6.1.6 *The tableau method for $\mathbf{K}^{\boxdot\boxdot}$ extended with rules (**CR1**) and (**CR2**) is a decision procedure for $\mathbf{K}_t$.*

Proof: It is enough to observe that rules (**CR1**) and (**CR2**) codify the information that $\overleftarrow{R}$ and $\overrightarrow{R}$ are converse relations and that no application of such rules modify the modal degree of any diagram. ♠

We keep examining the $\mathbf{K}_t$-diagrams.

Definition 6.1.7 *We call upper circuit (respectively, lower circuit) a pair of diagrams connected by a double arrow and built by means of one or more applications of (**CR1**) (respectively (**CR2**)).*

We shall use the notation $t_j \to t_i \to t_j$ to indicate any circuit that starts from t_j and turns back to t_j but only after passing through t_i. Considering that we have at our disposal a completeness proof for $\mathbf{K}^{\square\square}$ obtained by transforming a closed tableau for a formula α into a proof of α, we expect that an analogous proof can be given for $\mathbf{K}_t$. In fact, to transform a closed tableau into a proof is not problematic when the central diagram t_0 containing the input formula is the origin of a chain of upper or lower diagrams with no circuits: tableaux with no circuits, in fact, exhibit the same pattern of $\mathbf{K}^{\square\square}$-tableaux. The difficulty arises when tableaux contain lower or upper circuits.

A way to avoid this problem consists in showing that, by a simple modification, each tableau with circuits can be transformed into another tableau without circuits that validates the same formulas. Furthermore, we may show that it is possible to transform a closed $\mathbf{K}_t$-tableau into a closed $\mathbf{K}^{\square\square}$-tableau with some special features. We call an *upper circuit formula* (respectively, *lower circuit formula*) any formula that is introduced in a diagram by one or more applications of (**CR1**) rule (respectively (**CR2**)). In Example 6.1.4, for instance, the atomic formula p in the diagram t_0 with value 0 is a lower circuit formula.

The last example gives evidence of this fact: the occurrence of a lower circuit depends on the validity of the bridge axiom $p \supset \square\diamondsuit p$ or of some of its instances, while the occurrence of an upper circuit depends on the validity of the future-past bridge axiom $p \supset \square\diamondsuit p$ or of some of its instances. The following theorem shows that introducing instances of such axioms into a diagram is equivalent to eliminating circuits.

Proposition 6.1.8 *Let $\mathcal{T}$ be a closed tableau that has α as an input formula and let t_i be a diagram that belongs to an upper circuit $t_i \to t_{i+1} \to t_i$ which is terminal, that is, belongs to the last circuit introduced in $\mathcal{T}$ before reaching the closure. Let β be an upper circuit formula introduced in t_i. Let $\mathcal{T}'$ be the tableau that is obtained by modifying $\mathcal{T}$ in the following way:*

1. *The arrow $t_{i+1} \to t_i$ is suppressed jointly with the cell inside t_i that contains β.*

2. *The reduced diagram t'_i is extended with the formula $\beta \supset \square\diamondsuit\beta$ with value 1 if the eliminated β had previously value 0, and with $\neg\beta \supset \square\diamondsuit\neg\beta$ with value 1 if β previously had value 1.*

Then $\mathcal{T}'$ is also a closed tableau for α.

6.1. LOGICS WITH TWO PRIMITIVE MODAL OPERATORS

Proof: Let β be a circuit formula in t_i such that $v(\beta, t_i) = 0$ (the argument with $v(\beta, t_i) = 1$ is analogous and will be omitted). Since **(CR1)** has been used to introduce β in t_i, the formula $\Diamond\beta$ appears in t_{i+1} with value 0. We obtain a new diagram t'_i by eliminating the circuit formula β from t_i (remember that β is a subformula of some formula occurring in t_i) and by deleting the arrow from t_{i+1} to t_i, while introducing into t_i the formula $\beta \supset \Box\Diamond\beta$ with value 1.

Distribute truth-values in t'_i by taking into account the subformulas already evaluated in t_i. It may happen that this new distribution of values will yield a contradiction in the diagram t'_i or in any other diagram, and we are done. In the case of no such contradiction, we have to analyze two possibilities: after having assigned value 1 to $\beta \supset \Box\Diamond\beta$, the antecedent β can receive value 1 or 0. Thus, it may happen that we need to build two variants of the same tableau $\mathcal{T}'$, which we will call $\mathcal{T}'_1$ and $\mathcal{T}'_2$ (see Figure 6.1). We have to show in this case that $\mathcal{T}'_1$ and $\mathcal{T}'_2$ are both closed tableaux. In fact:

(i) Let $\mathcal{T}'_1$ be the tableau variant in which β receives value 0. It is clear that this assignment yields a contradiction in the terminal branch of $\mathcal{T}$ (possibly formed by t_i alone) free of circuits, since we are

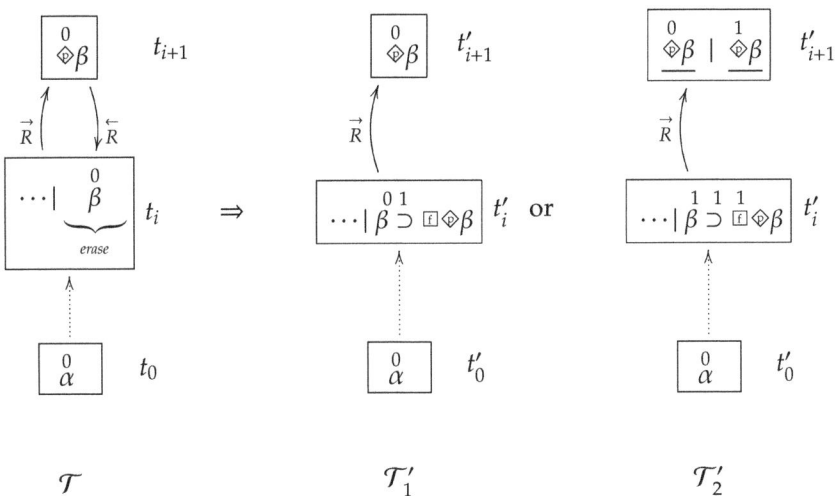

Figure 6.1: Tableau variants

re-introducing the formula β in t'_i with the same value that granted the closure in $\mathcal{T}$.

(ii) Let $\mathcal{T}'_2$ be the tableau variant where β receives value 1. Then $\Box \Diamond \beta$ must receive value 1 by the truth-table for material implication. But this means that in every diagram $\overrightarrow{R}$-accessible from t_i, $\Diamond \beta$ has value 1. This however yields a contradictory assignment in t_{i+1}, since $v(\Diamond \beta, t_{i+1}) = 0$ and $v(\Diamond \beta, t_{i+1}) = 1$.

Therefore both tableaux $\mathcal{T}'_1$ and $\mathcal{T}'_2$ are closed, and thus $\mathcal{T}'$ is a closed tableau. ♠

Proposition 6.1.9 *The same as in Proposition 6.1.8, replacing "upper circuit" by "lower circuit", $\beta \supset \Box \Diamond \beta$ by $\beta \supset \Box \Diamond \beta$ and $\neg \beta \supset \Box \Diamond \neg \beta$ by $\neg \beta \supset \Box \Diamond \neg \beta$.*

Proof: An obvious modification of the proof of Proposition 6.1.8. ♠

By iterated applications of Propositions 6.1.8 or 6.1.9, we transform any $\mathbf{K}_t$-tableau $\mathcal{T}$ into a tableau $\mathcal{T}^*$ free of circuits, called a *corrected tableau*, that validates the same input formula of $\mathcal{T}$. Since the procedure ends in a finite time, there is a proof what follows:

Proposition 6.1.10 *If $\mathcal{T}$ is a closed $\mathbf{K}_t$-tableau for a wff α, there exists a corrected tableau $\mathcal{T}^*$ which is also a closed tableau for α.*

Proof: Immediate application of Propositions 6.1.8 and 6.1.9. ♠

The transformation of $\mathcal{T}$ into $\mathcal{T}^*$ can be graphically visualized by drawing a sequence of tableaux and indicating which circuit formulas are to be erased.

Example 6.1.11 *Consider the formula $(p \wedge q) \supset \Box(p \supset \Diamond(p \vee q))$.*

Here the lower circuit formula of t_0 is $(p \vee q)$ and has value 0; this formula is erased and $(p \vee q) \supset \Box \Diamond (p \vee q)$ is introduced into diagram t'_0 with value 1. As $(p \wedge q)$ has value 1, then it is possible to assign value 1 to both p and q. Hence, in t'_0, $\Box \Diamond (p \vee q)$ receives value 1 by the truth-table for material implication, thus in all previous diagrams of t'_0 such that $t'_{-1} \overleftarrow{R} t'_0$ the wff $\Diamond(p \vee q)$ receives value 1, which yields a contradiction in t'_{-1}. The closed tableau obtained this way is free of circuits.

6.1. LOGICS WITH TWO PRIMITIVE MODAL OPERATORS

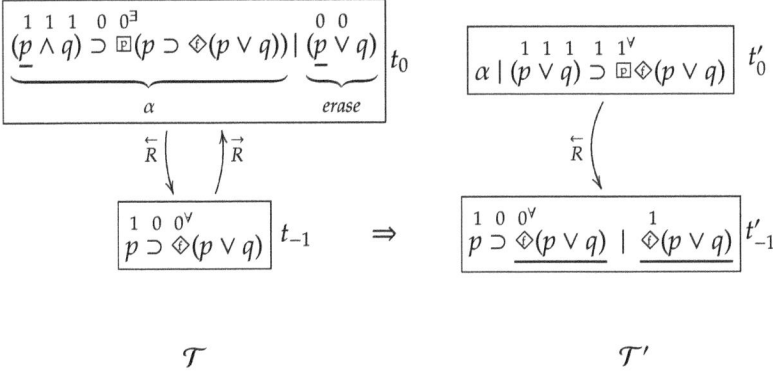

Example 6.1.12 As an example, consider the formula $\Diamond\Box p \supset \Box\Diamond p$.

This example is interesting because the diagram t_0 contain two cells where the circuit formula p itself receives contradictory truth-values. Supposing that the last inserted circuit was the lower one, the lower circuit arrow yields p in t_0 with value 0, so let it be the first to be discharged. Therefore we insert the wff β, i.e., $p \supset \Box\Diamond p$ (an instance of axiom (**BA1**)) in the diagram t'_0, obviously with value 1, since it is an axiom. But $\Box\Diamond p$ has value 1 in t'_0 because p has value 1 in t'_0 and β has value 1; this yields a contradiction in t'_0 as $\Box\Diamond p$ already had value 0 in t_0 (see figure above).

Now, the last step is to remove the second circuit formula p (which has value 1 in t'_0). Together with p, we also erase the arrow from t'_{+1} to t'_0, introducing into t''_0

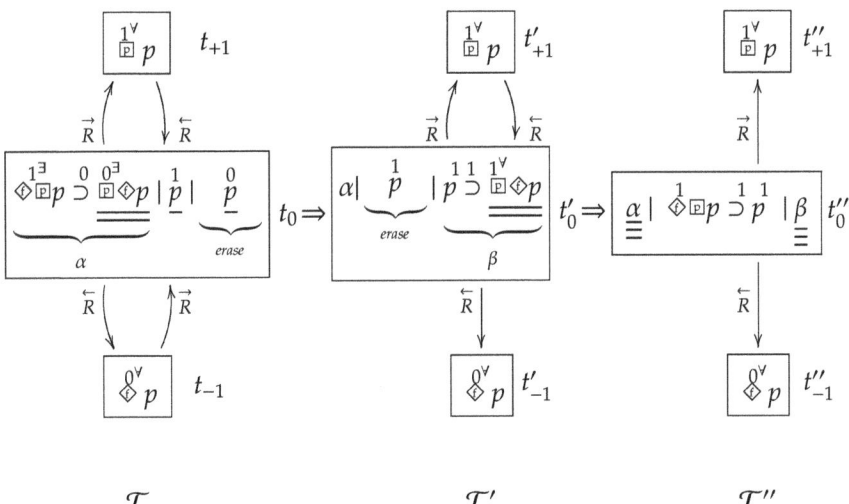

the wff $\neg p \supset \Box \Diamond \neg p$ (an instance of axiom **(BA2)**, again with value 1, since it is an axiom). For sake of simplicity, we may insert into t''_0 the equivalent wff $\Diamond \Box p \supset p$. Thus, as $\Diamond \Box p$ has value 1 in t_0, due to the value of α, then p has value 1 in t''_0. Therefore the wff $\Box \Diamond p$ receives value 0 in α and 1 in β, a contradiction.

Now we have the tools to prove the completeness of $\mathbf{K}_t$ by showing that any corrected tableau may be expanded to a tableau which has the structural properties of a $\mathbf{K}^{\Box\Box}$-tableau.

Proposition 6.1.13 *Let α be a $\mathbf{K}_t$-valid wff. Then there exists a proof of α in $\mathbf{K}_t$.*

Proof: Let us suppose that α contains no modal symbol except $\Box$ and $\Box$, or that it is reduced to this form by suitable transformations. If α is $\mathbf{K}_t$-valid, α is the input of a closed $\mathbf{K}_t$-tableau $\mathcal{T}$. From Proposition 6.1.10, $\mathcal{T}$ may be transformed into a corrected tableau $\mathcal{T}^*$, i.e., a closed tableau free of circuits with the same input formula α.

All the wffs belonging to the cells of any diagram of $\mathcal{T}^*$ fall into two distinct classes: either they are subformulas of the tested wff α introduced by some application of the rules for the tableaux construction, or they are substitution instances of one of the two bridge axioms of $\mathbf{K}_t$ ((**BA1**) or (**BA2**)) with value 1, introduced by the elimination of circuits. We now extend $\mathcal{T}^*$ to a tableau $\mathcal{T}^{*+}$, which we will call the *fully corrected* tableau in the following way:

- Whenever a diagram t_j of $\mathcal{T}^*$ contains a substitution instance of axioms (**BA1**) or (**BA2**) prefixed with an n number of $\Box$ and/or an m number of $\Box$ (for $n, m > 0$) —let us conventionally call it $K^{n,m}$ — with value 1, proceed as follows:

    (i) If there is a diagram t_i such that $t_i \overrightarrow{R} t_j$, put the wff $\Box K^{n,m}$ into t_i with value 1, and

    (ii) If there is a diagram t_k such that $t_j \overleftarrow{R} t_k$, put the wff $\Box K^{n,m}$ into t_k with value 1.

Beginning from the terminal diagrams, the outlined procedure of completion may be reiterated and ends when all the relevant formulas are introduced in the first diagram t^*_0. This extended variant of t^*_0 will be called t^{*+}_0. So at the end t^{*+}_0 will have the input formula α along with a certain number of substitution instances of axioms (**BA1**) or (**BA2**) prefixed by a certain number of $\Box$ and $\Box$, all with value 1. It is easy to prove that the fully corrected tableau $\mathcal{T}^{*+}$ is a closed $\mathbf{K}_t$-tableau for a formula which is

6.1. LOGICS WITH TWO PRIMITIVE MODAL OPERATORS

not α but a wff χ_0^{*+} of form $\bigwedge K^{n,m} \supset \alpha$ where $\bigwedge K^{n,m}$ is the conjunction of the mentioned instances of (**BA1**) or (**BA2**) prefixed by a certain number of ⊡ and ⊡. Suppose, in fact, that we assign value 0 to the implicative wff χ_0^{*+}: the antecedent will receive value 1 and the consequent, α, will receive value 0. Now all the subformulas of α and $\bigwedge K^{n,m}$ occurring in the diagrams of $\mathcal{T}^{*+}$ may be seen as consequences of the application of the known rules for the tableau construction established for the system $\mathbf{K}^{⊡⊡}$. Furthermore, the modal degree of every diagram t_j is lower than the modal degree of any diagram t_i preceding it in the chain of diagrams.

Being $\mathcal{T}^{*+}$ a closed $\mathbf{K}^{⊡⊡}$-tableau, the characteristic wff of any diagram can be proved in $\mathbf{K}^{⊡⊡}$ with the known tableau method, which is a trivial extension of the method formulated for the monomodal **K**. A proof may then be given of the characteristic wff of the first diagram t_0^{*+}, which is equivalent to $\bigwedge K^{n,m} \supset \alpha$. This wff is then a theorem of $\mathbf{K}^{⊡⊡}$, and so it is *a fortiori* a theorem of $\mathbf{K}_t$. But the antecedent $\bigwedge K^{n,m}$ is obviously a $\mathbf{K}_t$-theorem. Thus by applying Modus Ponens to $\bigwedge K^{n,m} \supset \alpha$, we reach α as a theorem of $\mathbf{K}_t$. ♠

Let us see an example illustrating the procedure to convert tableaux into proofs.

Example 6.1.14 *Consider the wff ⊡α where α is $(p \wedge \Diamond q) \supset \Diamond(\Diamond p \wedge q)$ (see figure below).*

By the described procedure, the tableau $\mathcal{T}$ is transformed into the fully corrected tableau $\mathcal{T}^{*+}$ consisting of the diagrams t_a^*, t_b^*, t_c^* where $a = 0$, $b = (0,-1)$, $c = (0,-1,1)$ (for sake of simplicity, $\Diamond$ and $\Diamond$ appear here, not eliminated by definition).

- $\chi_a^* = ⊡\alpha \vee \neg⊡(p \supset ⊡\Diamond p)$
- $\chi_b^* = \alpha \vee \neg(p \supset ⊡\Diamond p)$
- $\chi_c^* = \neg q \vee (\Diamond p \wedge q) \vee \neg\Diamond p$

The preceding wffs are the characteristic formulas of the diagrams of $\mathcal{T}^{*+}$. Since χ_c^* is obviously a $\mathbf{PC}^{⊡⊡}$-thesis, it is also a $\mathbf{K}^{⊡⊡}$-thesis.

*By applying (**Nec**$_⊡$) and axiom (**K**$_⊡$) to χ_c^*, it follows that*

(i) $\neg⊡q \vee ⊡(\Diamond p \wedge q) \vee \neg⊡\Diamond p$

is also a $\mathbf{K}^{⊡⊡}$-thesis. Each disjunct of the previous formula implies χ_b^*, as the reader can see by employing $\mathbf{K}^{⊡⊡}$-tableaux (see Exercise 6.6). Therefore, the disjunction (i) implies χ_b^*, and, consequently, χ_b^* is a $\mathbf{K}^{⊡⊡}$-thesis.

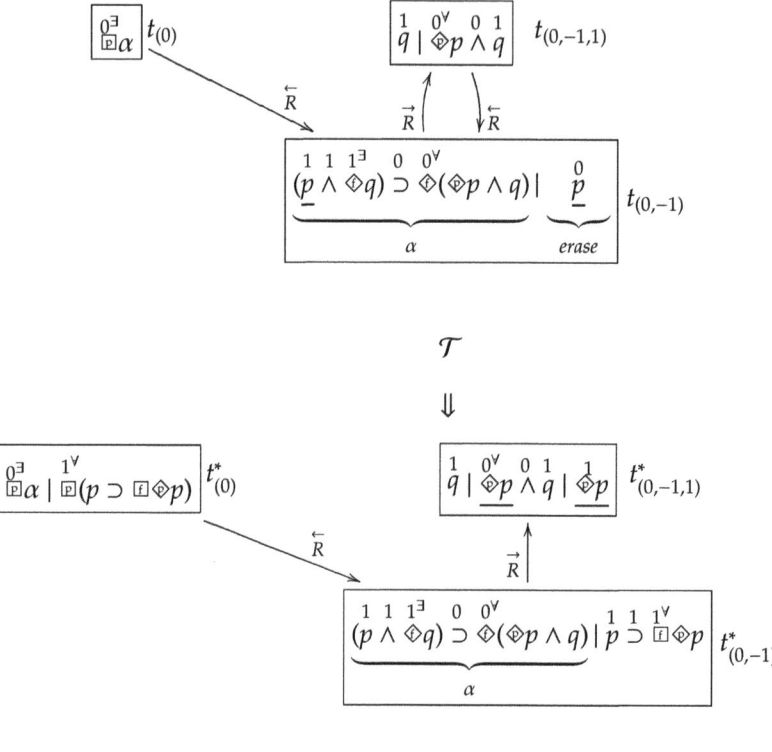

Now, by applying (**Nec**$_{\boxdot}$) and axiom (**K**$_{\boxdot}$) to $\chi_{b'}^*$, it follows by definitional transformations that

(ii) $\boxdot(p \supset \boxdot \diamondsuit p) \supset \boxdot((p \wedge \diamondsuit q) \supset \diamondsuit(\diamondsuit p \wedge q))$

is also a $\mathbf{K}^{\boxdot\boxdot}$-thesis. But this is precisely χ_a^* rewritten in implicational form. Since $\boxdot(p \supset \boxdot \diamondsuit p)$ is clearly also a $\mathbf{K}_t$-thesis, just by applying (**Nec**$_{\boxdot}$) to axiom (**BA2**), it follows by (**MP**) that α is a $\mathbf{K}_t$-thesis, and α is the formula under test.

6.2 Completeness and incompleteness of *PF*-logics

We have seen in Section 6.1 that temporal frames for systems which are extensions of $\mathbf{K}_t$-system are triples $\langle T, \overleftarrow{R}, \overrightarrow{R} \rangle$, where $\overleftarrow{R}$ and $\overrightarrow{R}$ are converse relations. After the basic semantical analysis, it is convenient to distinguish

6.2. COMPLETENESS AND INCOMPLETENESS OF PF-LOGICS

the properties of temporal frames concerning temporal order (ramification, convergence, linearity) from the properties concerning the features of temporal medium (density, discreteness, etc.).

From a taxonomic viewpoint, the most important distinction is drawn between systems characterized by branching as opposed to linear frames, even if systems exhibiting mixed properties also exist. Here are some remarkable systems treated in the literature, all sharing the basic presupposition that the temporal order is transitive:

Branching time

- **CR** (N. B. Cocchiarella): $\mathbf{K}_t + (\Diamond\Diamond p \supset \Diamond p) + (\lozenge\lozenge p \supset \lozenge p)$ (branching in the past and in the future).

Note that axiom $\Diamond\Diamond p \supset \Diamond p$ describes transitive frames (see Proposition 3.2.16) which are branching in the sense that they may be convergent or non-convergent towards some point in the future (as we know – see page 93– convergence towards the future is expressed by $\Diamond\Box p \supset \Box\Diamond p$, which is independent from $\Diamond\Diamond p \supset \Diamond p$.)

- $\mathbf{K}_b$ (N. Rescher): $\mathbf{CR} + (\lozenge p \wedge \lozenge q) \supset (\lozenge(p \wedge q) \vee \lozenge(p \wedge \lozenge q) \vee \lozenge(\lozenge p \wedge q))$ (branching in the future and linearity in the past).

The intuitive reading of axiom $(\lozenge p \wedge \lozenge q) \supset (\lozenge(p \wedge q) \vee \lozenge(p \wedge \lozenge q) \vee \lozenge(\lozenge p \wedge q))$ is made clear by observing that $\lozenge(p \wedge \lozenge q)$ means that, in some past moment, p was present and q was antecedent to p. The sense of the axiom is then that, provided that p and q were true in some moment of the past, either p and q occurred jointly or they occurred in some chronological order, which could not be said if they had taken place in different possible past histories.

Linear time

- **CL** (N. B. Cocchiarella): $\mathbf{K}_b + (\Diamond p \wedge \Diamond q) \supset (\Diamond(p \wedge q) \vee \Diamond(\Diamond p \wedge q) \vee \Diamond(p \wedge \Diamond q))$ (linearity in the past and in the future).

This is just a duplication of the previous case: see the argument for $\mathbf{K}_b$ applied to both past and future tense.

- **SL** (D. Scott): $\mathbf{CL} + (\Box p \supset \Diamond p) + (\boxdot p \supset \lozenge p)$ (linearity and seriality in the past and in the future).

Axiom $\Box p \supset \Diamond p$ is a temporal version of (**D**), which we know to express seriality (Section 3.2, page 61). The second axiom describes seriality in the past (note that, in the absence of reflexivity, a serial sequence of points forms an infinite sequence).

- **PL** (A. N. Prior) : **SL**+ ($\Diamond p \supset \Diamond\Diamond p$)+ ($\blacklozenge p \supset \blacklozenge\blacklozenge p$) (linearity, seriality and density in the past and in the future).

Axiom $\Diamond p \supset \Diamond\Diamond p$ describes density, as shown in the following Proposition 6.2.1.

Circular time

- **PCr**: $\mathbf{K}_t$+ ($\Diamond\Diamond p \supset \Diamond p$) + ($\Box p \supset p$) + ($\Box p \supset \blacksquare p$) (circularity and transitivity).

  Since accessibility here turns out to be an equivalence relation, the properties of such a system recall the properties of Lewis' **S5**. This opens the difficult problem of giving meaning to the distinction between past and future operators, which is often solved by going beyond the resources of bimodal language. Notice also that having $p \supset \Diamond p$ as a theorem implies having density as a property of the temporal series.

Axiom $\Box p \supset \blacksquare p$ states that what happened is going to happen again (see Nietzsche's "eternal return"), or semantically that if $t \overrightarrow{R_S} t'$ then $t \overleftarrow{R_S} t'$.

Other systems can be obtained by combining some of the mentioned axioms (once the independence of such axioms is granted) or by consistently introducing new axioms. In this way, it is possible to axiomatize principles that are characteristic of certain religious or cosmological views: the system $\mathbf{K}_b$ may be seen, for instance, as viewing the future as an "open tree of possibilities" (which implies the refusal of determinism), while the view that "time had a beginning, but will have no end" can be axiomatized by adding $\blacksquare\bot \vee \Diamond\blacksquare\bot$ and $\Box p \supset \Diamond p$ to $\mathbf{K}_t$ (for the first formula simply consider that one of the two disjuncts must be vacuously true at a first instant of time or at some posterior instant).

From our treatment in Chapter 3, it should be clear that results of $\mathcal{F}$-completeness for all these systems can be reached in a non constructive way by analyzing canonical frames, that is, by showing case by case that the accessibility relation of the canonical frame for the system under investigation

6.2. COMPLETENESS AND INCOMPLETENESS OF PF-LOGICS

satisfies the required properties for that system. It should be clear that the temporal canonical frame $\langle T_S, \overleftarrow{R_S}, \overrightarrow{R_S} \rangle$ for a system **S** consists of a set T_S of maximal consistent extensions of **S** and of two relations defined as:

(a) $t \overrightarrow{R_S} t'$ if only if $Den_\square(t) \subseteq t'$, where $Den_\square(t) \overset{\text{Def}}{=} \{\alpha : \square\alpha \in t\}$.

(b) $t \overleftarrow{R_S} t'$ if only if $Den_\boxminus(t) \subseteq t'$, where $Den_\boxminus(t) \overset{\text{Def}}{=} \{\alpha : \boxminus\alpha \in t\}$.

Analogously to what was established in Corollary 4.2.14, it also holds that $Den_\square(t) \subseteq t'$ iff $Poss_\diamond(t') \subseteq t$, where $Poss_\diamond(t') = \{\diamond\alpha : \alpha \in t'\}$ (*mutatis mutandis* for $\boxminus$ and $\diamondsuit$). Here is an example on how to prove a completeness result with respect to a class of temporal frames:

Proposition 6.2.1 *If **S** is the system $K_t + \diamond\alpha \supset \diamond\diamond\alpha$, **S** is sound and complete with respect to the class of temporal frames $\langle T, \overleftarrow{R}, \overrightarrow{R} \rangle$ with the following properties:*

(i) $\overleftarrow{R}$ *and* $\overrightarrow{R}$ *are converse relations.*

(ii) $\overrightarrow{R}$ *is dense.*

Proof: Checking soundness is routine and is left to the reader. For completeness, the procedure is standard, and it is enough to verify that the canonical model enjoys the above properties. Let t, t', $t''\cdots$ denote the maximal consistent extensions of the system $\mathbf{S}=K_t+\diamond\alpha \supset \diamond\diamond\alpha$. The proof is divided in two parts.

(i) First, we need to show that the accessibility relations of the canonical frame for **S** are converse.[6]

Let p be an arbitrary wff in t. As the past-future bridge axiom (**BA1**) (i.e, $p \supset \boxminus\diamond p$) belongs to t, then t contains $\boxminus\diamond p$ and hence $\diamond p$ will belong to all maximal consistent extensions t' such that $t \overleftarrow{R_S} t'$. Suppose by *Reductio* that, for some t', $t \overleftarrow{R_S} t'$ holds but not its converse $t' \overrightarrow{R_S} t$. This means that in at least one t, without loss of generality, $\square\neg q \in t'$ and $\neg q \notin t$. But t is maximal, hence $q \in t$. In this case, therefore, $\diamond q \in t'$; but $\square\neg q$ is equivalent to $\neg\diamond q$, and thus both $\diamond q$ and $\neg\diamond q$ belong to t', which is absurd in view of the consistency of t'. Then $t \overleftarrow{R_S} t'$ implies $t' \overrightarrow{R_S} t$. An analogous argument shows that $t \overrightarrow{R_S} t'$ implies $t' \overleftarrow{R_S} t$ by using bridge axiom (**BA2**).

---

[6]Note that this amounts to giving another completeness proof for K_t. As a matter of fact, the completeness of K_t with respect to the same class of frames has been already proved in Proposition 6.1.13 with a constructive method.

(ii) Second, it remains to be shown that $\vec{R}_S$ is a dense relation. We need to show that, for each t and t' of the canonical model such that $t\ \vec{R}_S\ t'$, there exists a t'' such that $t\ \vec{R}_S\ t''$ and $t''\ \vec{R}_S\ t'$. It follows from the hypothesis that $t\ \vec{R}_S\ t'$ and, from the fact that $\vec{R}_S$ and $\overleftarrow{R}_S$ are converse, that $t'\ \overleftarrow{R}_S\ t$. By supposing that $p \in t$ and $q \in t'$, from the past-future bridge axiom $q \supset \boxdot \Diamond q \in t'$ it follows that $\Diamond q \in t$. By applying the tableaux method, it is not difficult to check that $(p \wedge \Diamond q) \supset \Diamond (\Diamond p \wedge q)$ is a $\mathbf{K}_t$-thesis (cf. Example 6.1.14), hence it belongs to any maximal consistent extension. By substituting q by $\Diamond q$ in the previous theorem, we also have that $(p \wedge \Diamond \Diamond q) \supset \Diamond (\Diamond p \wedge \Diamond q)$ is an element of any maximal consistent extension. But p and $\Diamond q$ are in t, so $\Diamond \Diamond q$ (by $\Diamond q$ and $\Diamond q \supset \Diamond \Diamond q$) and therefore $p \wedge \Diamond \Diamond q$. Then, by (**MP**), $\Diamond (\Diamond p \wedge \Diamond q) \in t$. This implies that there exists a maximal consistent set t'' to which belong both $\Diamond p$ and $\Diamond q$ and such that $t\ \vec{R}_S\ t''$. But $q \in t'$. Hence, from $\Diamond q \in t''$ and $q \in t'$, by definition of the accessibility relation it follows $t''\ \vec{R}_S\ t'$ (since $Poss_{\Diamond}(t') \subseteq t''$).

♠

As one can expect, it may be possible to have a multiplicity of results of $\mathcal{F}$-completeness at our disposal for one and the same system. For example, it is possible to show that $\mathbf{K}_t$ is characterized by all temporal frames in which $\overleftarrow{R}$ and $\vec{R}$ are irreflexive and converse and also by all frames in which $\overleftarrow{R}$ and $\vec{R}$ are intransitive and converse (see Exercise 6.8). Analogous results have to do with the properties of the temporal medium. Let $\mathbb{Q}$ be the set of rational numbers and let $\mathbb{R}$ be the set of real numbers. It can be shown – but we will not do it here – that:

1. **SL** is characterized by $\langle \mathbb{Q}, <, > \rangle$

2. **SL** is characterized by $\langle \mathbb{R}, <, > \rangle$

As far as incompleteness results are concerned, it may be surprising to find that the presence of two accessibility relations in the frames simplifies the search for negative results. A very simple result is the following, which uses first-order undefinability of some axioms. As we may suspect by analogy with monomodal logic, the most remarkable of such axioms are temporal versions of McKinsey and Gödel-Löb axioms (cf. Sections 3.3 and 4.3), i.e:

(**McK**$_f$) $\boxdot \Diamond p \supset \Diamond \boxdot p$

(**GL**$_p$) $\boxdot (\boxdot p \supset p) \supset \boxdot p$

6.2. COMPLETENESS AND INCOMPLETENESS OF PF-LOGICS

Let us recall that (**McK**$_f$) corresponds to atomicity (existence of terminal points, see Section 3.3), while (**GL**$_p$) characterizes frames whose accessibility relation is transitive and well-covered (see Proposition 4.4.2). Here, however, such properties pertain to different accessibility relations, even if mutually converse. The formula (**McK**$_f$) is valid on a frame $\mathcal{F}$ iff in $\mathcal{F}$ there is an instant, posterior to the instant of reference, after which all propositions stop changing truth-values. We know that this property is only first-order definable in a transitive frame and is expressed by

$$\forall t \exists t'(t \overrightarrow{R} t' \wedge \forall t''(t' \overrightarrow{R} t'' \supset (t'' = t')))$$

On the other hand, (**GL**$_p$) defines the conjunction of the transitivity of $\overleftarrow{R}$ with the well-coveredness of $\overleftarrow{R}$: this amounts to the non-existence of any infinite sequence of the form $t \overleftarrow{R} t' \overleftarrow{R} t'' \overleftarrow{R} \cdots$ (intuitively, the non-existence of an infinite regress towards the past), a property known not to be first-order definable.

We can now prove the following incompleteness result.

Proposition 6.2.2 *Let $K_t^{\circ\circ}$ be the system K_t + (**McK**$_f$)+(**GL**$_p$). $K_t^{\circ\circ}$ is consistent but not characterizable by any class of temporal frames.*

Proof: The consistency of $K_t^{\circ\circ}$ is shown as a corollary of soundness, i.e., by showing that each theorem of $K_t^{\circ\circ}$ holds in all atomic, transitive and well-covered frames. For the rest of the proof, the remaining part of the argument is as follows. First of all, let $\mathcal{F}$ be any frame that validates the conjunction of (**McK**$_f$) and (**GL**$_p$). Each frame which validates (**McK**$_f$) is a frame in which the formula $\forall t \exists t'(t \overrightarrow{R} t' \wedge \forall t''(t' \overrightarrow{R} t'' \supset (t'' = t')))$ holds. As (by Definition 3.2.9) the set of worlds W is non-empty, for $t \in W$ there exists at least one t' such that $t \overrightarrow{R} t'$ and $t'' = t'$ for any t'' such that $t' \overrightarrow{R} t''$. By applying substitutivity of identities from $t'' = t'$ we have that $\forall t \exists t'(t \overrightarrow{R} t' \wedge t' \overrightarrow{R} t')$. Consequently, the temporal order must contain infinite sequences of instantaneous loops $t' \overrightarrow{R} t' \overrightarrow{R} t' \overrightarrow{R} t' \cdots$, where $t' \overrightarrow{R} t'$ iff $t' \overleftarrow{R} t'$. But this is incompatible with the well-coveredness of $\overleftarrow{R}$, which is granted by frames that satisfy (**GL**$_p$). Therefore, there cannot be any frame which validates both formulas. ♠

6.3 Monomodal fragments of *PF*-logics

The problem of how to define alethic necessity and possibility within temporal logic turns out to be interesting from the philosophical viewpoint inasmuch as one of the first conceptions of possibility historically known, due to Diodorus Cronus (4th century BC), is based on the notion of time.

The interest in isolating the (mono)modal fragments of various temporal logics played a significant role in the historical development of the subject, and we shall see that such fragments of *PF*-logics coincide with the familiar alethic systems.

While disputing with Aristotle, in fact, Diodorus defined the possible as "what is or will be true" (the necessary, hence, is "what is and will always be true"). It is intuitive that in referring to the future he was not referring to some possible future (for, in this case, the definition would be circular), but to the actual future, that is to the only future that will actually take place. Diodorus' idea was that if something is never realized, neither now nor in the future, it cannot be said to be possible: it makes no sense, for Diodorus, to speak of non-realizable possibilities. This principle was later called *Principle of Plenitude*. Contrarily, the Aristotelian notion of potentiality was compatible with the idea that non realizable possibilities exist. For example, according to Aristotle, it is correct to say that a shell at the bottom of the sea may be seen by anyone even if nobody will ever see it, while for Diodorus this is simply a contradiction. An idea of possibility inspired by Diodorus Cronus is expressed by the following definition:

$$\bar{\Diamond}\alpha \stackrel{\text{Def}}{=} \alpha \vee \Diamond\alpha$$

Another less limitative definition of possibility (or potentiality), called Megaric or Aristotelian-Megaric, is the following:

$$\bar{\bar{\Diamond}}\alpha \stackrel{\text{Def}}{=} \alpha \vee \diamondsuit\alpha \vee \Diamond\alpha$$

This couple of definitions allows us to establish a relation between different definable modal notions: for example, in $\mathbf{K}_t$ and in all its extensions $\bar{\Diamond}\alpha$ implies $\bar{\bar{\Diamond}}\alpha$ while the converse does not hold. Based on the definition of $\bar{\Diamond}$, a Diodorean definition of necessity will be given by:

$$\bar{\Box}\alpha \stackrel{\text{Def}}{=} \alpha \wedge \Box\alpha$$

while the Megaric one will be:

$$\bar{\bar{\Box}}\alpha \stackrel{\text{Def}}{=} \alpha \wedge \boxdot\alpha \wedge \Box\alpha$$

6.3. MONOMODAL FRAGMENTS OF PF-LOGICS

It is to be noted that there are also different ways to define modalities in temporal terms. Think, for example, of the following pair of plausible dual definitions:

- $\widehat{\Diamond}\alpha \stackrel{\text{Def}}{=} \diamondsuit\!\!\!\diamond\,\alpha$

- $\widehat{\Box}\alpha \stackrel{\text{Def}}{=} \boxed{p}\boxed{f}\alpha$

or of the two following:

- $\widehat{\widehat{\Diamond}}\alpha \stackrel{\text{Def}}{=} \diamondsuit\!\!\!\diamond\,\alpha \vee \alpha \vee \diamond\alpha$

- $\widehat{\widehat{\Box}}\alpha \stackrel{\text{Def}}{=} \boxed{p}\boxed{f}\alpha \wedge \alpha \wedge \Box\alpha$

We call the *Diodorean fragment* (respectively, *Megaric fragment*) of a system **S** a subsystem of **S** where the alethic notions of possibility and necessity are defined according to the Diodorean (respectively, Megaric) conception.

Which one is the Diodorean fragment of the minimal system $\mathbf{K}_t$? The answer to this question is given by Proposition 6.3.1.

In the case of the definition of the Diodorean modalities, we translate the language of **KT** into the language of $\mathbf{K}_t$ (recall Definition 2.3.14) by defining a function f as follows:

(i) $f(p) = p$

(ii) $f(\neg\alpha) = \neg f(\alpha)$

(iii) $f(\alpha\#\beta) = f(\alpha)\#f(\beta)$ where # is any arbitrary two place truth-functional operator

(iv) $f(\Diamond\alpha) = f(\alpha) \vee \diamond f(\alpha)$

What we have to prove is the following metatheorem, which shows that f is a strong translation (by way of the preceding definition) between **KT** and $\mathbf{K}_t$.

Proposition 6.3.1 *For every α, α is a **KT**-thesis if and only if $f(\alpha)$ is a $\mathbf{K}_t$-thesis.*

Proof: The proof runs in two parts.

1. We first prove by induction on the length of proofs that, if α is a **KT**-thesis, then $f(\alpha)$ is a $\mathbf{K}_t$-thesis. It is straightforward to see that the axiom $\Box p \supset p$ has an f-image which is the $\mathbf{K}_t$-theorem $(p \wedge \boxed{\cdot}p) \supset p$, while $f(\Box(p \supset q) \supset (\Box p \supset \Box q))$ is $(\boxed{\cdot}(p \supset q) \wedge (p \supset q)) \supset ((\boxed{\cdot}p \wedge p) \supset (\boxed{\cdot}q \wedge q))$,

which turns out to be a $\mathbf{K}_t$-theorem via $\mathbf{K}_t$-tableaux. It is easily shown that the inference rules preserve the described property; in particular, if α is a $\mathbf{K}_t$-thesis, then both α and $\Box\alpha$ are such: so $\vdash_{KT} \alpha$ implies $\vdash_{K_t} f(\alpha)$.

2. For the converse, we show that, if α is not a **KT**-thesis, $f(\alpha)$ is not a $\mathbf{K}_t$-thesis. If α is not a **KT**-thesis, by the completeness of **KT** it is false at some world w_i of some **KT**-model (i.e., a reflexive model) $\mathcal{M}_1$. For every **KT**-model $\mathcal{M}_1 = \langle W, R, V \rangle$, we build a derived model $\mathcal{M}_2 = \langle T, \overrightarrow{R}, \overleftarrow{R}, V \rangle$ with the following properties:

(a) $W = T$.

(b) If wRw' in $\mathcal{M}_1$, then in $\mathcal{M}_2$ either $w = w'$ or $w \overrightarrow{R} w'$.

(c) $\overrightarrow{R}$ and $\overleftarrow{R}$ are converse relations.

It is obvious that $\mathcal{M}_2$ has the properties of a $\mathbf{K}_t$-model. Now we are able to prove by induction on the complexity of α that for every α, if $v(\alpha, w) = 0$ in $\mathcal{M}_1$, then $v(f(\alpha), w) = 0$ in $\mathcal{M}_2$. The induction hypothesis supports that the relation holds for any α of arbitrary complexity. The only interesting point of the proof concerns the wffs of form $\Diamond\beta$. Suppose that α of the form $\Diamond\beta$ is false at some w of $\mathcal{M}_1$, then β is false at w (since wRw) and at all other worlds R-accessible to w. By induction hypothesis, $f(\beta)$ is false at w and at all worlds w' of T in $\mathcal{M}_2$ such that either $w = w'$ or $w \overrightarrow{R} w'$. So, $\neg f(\beta)$ is true at w and, from the truth conditions of $\Box$, so is $\Box\neg f(\beta)$: therefore $f(\beta) \vee \Diamond f(\beta)$ is false at w, hence $v(f(\Diamond\beta), w) = 0$.

But if $f(\alpha)$ is refuted in some $\mathbf{K}_t$-model, it cannot be a $\mathbf{K}_t$-theorem since $\mathbf{K}_t$ is complete with respect to the the class of $\mathbf{K}_t$-models. So, if α is not a **KT**-thesis, $f(\alpha)$ cannot be a $\mathbf{K}_t$-thesis. ♦

The above result shows that the Diodorean fragment of $\mathbf{K}_t$ is precisely **KT**. By applying arguments inspired in the above treatment, we obtain several other results; the Diodorean and the Megaric fragments of the *PF*-logics mentioned in this section are:

PF–logic	Diodorean fragment	Megaric fragments
CR	S4	KTB
K_b	S4	KTB
CL	S4.3	S5
SL	S4.3	S5
PL	S4.3	S5
PCr	S5	S5

6.3. MONOMODAL FRAGMENTS OF PF-LOGICS

The system **S4.3** was mentioned at the end of Section 4.2 (page 102) as a system whose characteristic axiom is not an instance of the schema G^∞. Recall that **S4.3** is, in fact, axiomatized by adding the axiom $\Box(\Box p \supset q) \lor \Box(\Box q \supset p)$ to **S4** and that it expresses the linearity of the temporal frame. It is important to take into account that this correspondence between the additional axiom and linearity subsists only if we assume the density of temporal series (expressed by $\Diamond p \supset \Diamond\Diamond p$, see Proposition 6.2.1). If we assume the discreteness of temporal series, however, linearity will be conveyed by a different formula, called *Dummett's formula*:

(**Dum**) $\Box(\Box(p \supset \Box p) \supset p) \supset (\Diamond \Box p \supset p)$

As it can be seen from the above list, all systems expressing linearity and yielding **S4.3** as a Diodorean fragment also yield **S5** as a Megaric fragment. The presence of references to the past in the definition of such modalities makes the accessibility relation symmetric or universal.

To obtain Megaric fragments of the tense-logical systems, we will now apply a different strategy from the one used to identify Diodorean fragments. We show as an example that the Megaric fragment of K_t is **KTB** (i.e, **KT**+ $p \supset \Box\Diamond p$). The method of proof which will be used here is different from the one used for the Diodorean fragment.

Now define a translation f from **KTB** into K_t in the following way:

(i) $f(p) = p$

(ii) $f(\Box\alpha) = \boxdot f(\alpha) \land \boxminus f(\alpha) \land f(\alpha)$

(iii) $f(\alpha \# \beta) = f(\alpha) \# f(\beta)$, where # is any truth-functional binary operator

(iv) $f(\neg\alpha) = \neg f(\alpha)$

Define another translation g from K_t into **KTB** as:

(i) $g(p) = p$

(ii) $g(\boxminus\alpha) = g(\boxdot\alpha) = \Box g(\alpha)$

(iii) $g(\alpha \# \beta) = g(\alpha) \# g(\beta)$, where # is any truth-functional binary operator

(iv) $g(\neg\alpha) = \neg g(\alpha)$

Now, the following theorem can be readily proved:

Proposition 6.3.2 *(i) If α is a thesis of KTB, $f(\alpha)$ is a thesis of K_t.*
(ii) If α is a thesis of K_t, $g(\alpha)$ is a thesis of KTB.

Proof: (i) An easy induction on the length of proofs.
(ii) Also by induction on the length of proofs (Exercise 6.16). In particular, the step concerning axiom (**B**) requires us to show that $g(p \supset \boxdot \Diamond p) = p \supset \Box \Diamond p$. ♠

What remains to be proved is the following:

Proposition 6.3.3 $\alpha \equiv g(f(\alpha))$ *is a thesis of* **KTB**.

Proof: By induction on the complexity of α. It is enough to prove the following equalities and equivalences:

(i) $g(f(p)) = g(p) = p$

(ii) $g(f(\alpha \# \beta)) = g(f(\alpha) \# f(\beta)) = g(f(\alpha)) \# g(f(\beta))$

(iii) $g(f(\neg \alpha)) = g(\neg f(\alpha)) = \neg g(f(\alpha))$

(iv) $g(f(\Box \alpha)) = g(\boxdot f(\alpha)) \wedge g(\boxminus f(\alpha)) \wedge g(f(\alpha)) = \Box g(f(\alpha)) \wedge \Box g(f(\alpha)) \wedge g(f(\alpha)) \equiv \Box g(f(\alpha))$

♠

From Propositions 6.3.2 and 6.3.3, it follows that **KTB** is translatable into $\mathbf{K}_t$, and hence constitutes its modal alethic fragment.

6.4 Other temporal systems

The *PF*-logics are sometimes called tense logics, because the operators $\Diamond$ and $\Diamondminus$ and their combinations are interpreted as grammatical modifiers of the present verbal tense.

If p is "it rains", $\Diamond p$ can be read as the future "it will rain", while $\Diamond \Diamondminus p$ corresponds to the future perfect "it will have rained". But it is interesting to remark that not all temporal logics are tense logics. As a simple example, consider a primitive binary operator T such that pTq reads as "p and next q". The logic **AN** for the operator "*and next*", originally formulated by G. H. von Wright, is axiomatized as follows:

(**AN1**) $(p \vee q)T(r \vee s) \equiv (pTr) \vee (pTs) \vee (qTr) \vee (qTs)$

(**AN2**) $(pTq) \wedge (rTs) \equiv (p \wedge r)T(q \wedge s)$

(**AN3**) $p \equiv (pT(q \vee \neg q))$

(**AN4**) $\neg(pT(q \wedge \neg q))$

6.4. OTHER TEMPORAL SYSTEMS

The rules are (**US**), (**MP**) and (**Eq**).
The operators $\Box$ and $\Diamond$ can be defined in the following way:

$$\Box \alpha \stackrel{\text{Def}}{=} (p \vee \neg p) T \alpha$$

$$\Diamond \alpha \stackrel{\text{Def}}{=} \neg((p \vee \neg p) T \neg \alpha)$$

A converse operator of T, called Y, can be axiomatized by taking (**AN1**)-(**AN5**) with Y in place of T, and reading Y and T as "yesterday" and "tomorrow". On account of this extension, we can define $\boxdot$ and $\diamondsuit$ by analogy with $\Box$ and $\Diamond$. In any case, we need a *bridge axiom* (that is, an axiom joining two or more modal operators) stating that T and Y are converse operators:

(**AN5**) $\neg(pT(qY\neg p))$

T and Y can, however, be defined on the basis of an even simpler operator. Take an operator T_S (where the subscript is to remind that this logic was introduced by D. Scott) as a primitive. The intended meaning of T_S is "in the next instant of time". Consequently, the operator T can be defined as:

$$\alpha T \beta \stackrel{\text{Def}}{=} \alpha \wedge T_S \beta$$

The system which axiomatizes T_S, called **W**, results from simply adding the following axioms and rules to **PC**:

(**W1**) $T_S \alpha \equiv \neg T_S \neg \alpha$

(**W2**) $T_S(\alpha \supset \beta) \supset (T_S \alpha \supset T_S \beta)$

(**Nec**$_{T_S}$) $\vdash \alpha$ implies $\vdash T_S \alpha$

As much as T can be defined in terms of T_S, T also defines T_S, adding to **AN**:

$$T_S \alpha \stackrel{\text{Def}}{=} (\beta \vee \neg \beta) T \alpha$$

The two definitions introduced above permit us to show the equivalence between the systems **W** and **AN** (see Exercise 6.11).

The system **AN** presupposes discreteness of time as well as linearity, the latter being described by axiom (**AN2**). Suppose, in fact, by *Reductio*, that there is a branching point such that, in the next instant (in two alternative futures), p is true after a unity of time and $\neg p$ is true after another unity of time. Substitutions in the axiom (**AN2**) then lead to

$(pTp) \wedge (pT\neg p) \equiv (p \wedge p)T(p \wedge \neg p)$, which is impossible, as the second conjunct is contradictory.

Axiom (**AN2**) can be weakened in order to express a branching order of time:

(**AN2'**) $(pT(qTr)) \supset (pTr)$

A completeness result for **AN** can be given as follows. As we know, $T_S p$ can be read as "in the next unity of time p" and $\alpha T \beta$ is defined as $\alpha \wedge T_S \beta$.

Since **W** is equivalent to **AN**, any completeness result for **W** implies the completeness of **AN**. It is enough to prove what follows. Call *standard frame* the pair $\mathcal{F}^{\mathbb{N}} = \langle \mathbb{N}, \oplus \rangle$, where $\mathbb{N}$ is the set of natural numbers and $\oplus$ is the successor relation. The basic idea is that, if $T_S^m \beta$ (T_S iterated m times) is true at instant k, then β will be true in the instant $k + m$, which suggests that the validity of formulas containing T_S can be reduced to the validity of **PC**-formulas.

Proposition 6.4.1 *W is sound and complete with respect to* $\mathcal{F}^{\mathbb{N}}$.

Proof: An implicit model based on the standard frame $\mathcal{F}^{\mathbb{N}}$ is a triple $\mathcal{M}^{\mathbb{N}} = \langle \mathbb{N}, \oplus, V \rangle$, where V is a function from Var to $\wp(\mathbb{N})$ that associates a set of natural numbers $V(p_i) = P^i$ to any atomic variable p_i. Let us define truth at an arbitrary natural number n with respect to a model $\mathcal{M}^{\mathbb{N}}$ in the following way:

1. $\mathcal{M}^{\mathbb{N}}, n \vDash p_k$ iff $n \in P^k$, for p_k an atomic variable
2. $\mathcal{M}^{\mathbb{N}}, n \vDash \neg \alpha$ iff $\mathcal{M}^{\mathbb{N}}, n \nvDash \alpha$
3. $\mathcal{M}^{\mathbb{N}}, n \vDash \alpha \supset \beta$ iff, $\mathcal{M}^{\mathbb{N}}, n \nvDash \alpha$ or $\mathcal{M}^{\mathbb{N}}, n \vDash \beta$
4. $\mathcal{M}^{\mathbb{N}}, n \vDash T_S \alpha$ iff $\mathcal{M}^{\mathbb{N}}, n+1 \vDash \alpha$

The wff α will be said to be $\mathcal{F}^{\mathbb{N}}$-*valid* if α is valid in all models $\mathcal{M}^{\mathbb{N}}$ on $\mathcal{F}^{\mathbb{N}}$.

We omit the trivial proof of soundness; completeness is proved by showing that, if α is not a theorem, then α is not valid on the standard frame $\mathcal{F}^{\mathbb{N}}$. In this case, suppose α is not a theorem; in order to reduce α to a normal formula in which all T_S are placed "as most internal as possible" (see Exercise 6.10), define T_S^n recursively as:

$$T_S^0 \alpha \stackrel{\text{Def}}{=} \alpha$$

$$T_S^{i+1} \alpha \stackrel{\text{Def}}{=} T_S T_S^i \alpha$$

6.4. OTHER TEMPORAL SYSTEMS

Moreover, define a transformation that maps α (in normal form) into a formula α^* shorn of the operator T_S, where in place of each $T_S^m p_n$ (provided it is not in the scope of T_S itself) we put a new atomic variable $q^{m,n}$ univocally determined by indices m and n occurring in $T_S^m p_n$ in α.[7]

Since we are supposing that α is not a theorem of **W**, then α^* is not **PC**-valid: in fact, if it were, α would be a **PC**-theorem, and by inverse substitution in the variables $q^{m,n}$, we would have α among the theorems of **W**. We can accordingly construct a model $\mathcal{M}'^{\mathbb{N}} = \langle \mathbb{N}, \oplus, V' \rangle$ on $\mathcal{F}^{\mathbb{N}}$ which refutes α^*. Let $\langle Q^{0,1}, Q^{1,1}, Q^{0,2} \cdots \rangle$ be a sequence of values (sets of numbers) that the valuation V' in $\mathcal{M}'^{\mathbb{N}}$ assigns to atomic variables (hence, also to the new variables $q^{m,n}$ given above), and suppose $\mathcal{M}'^{\mathbb{N}}, 0 \not\models \alpha^*$. On the other hand, let $\langle P^0, P^1, \cdots P^k \cdots \rangle$ be a sequence of values that another valuation V assigns to atomic variables, such that $m \in P^n$ if and only if $0 \in Q^{m,n}$, and let $\mathcal{M}^{\mathbb{N}} = \langle \mathbb{N}, \oplus, V \rangle$.

From clause 4. of the definition of truth, we know that $m \in P^n$ iff $\mathcal{M}^{\mathbb{N}}, m \models p_n$ iff $\mathcal{M}^{\mathbb{N}}, 0 \models T_S^m p_n$. Thus, for all formulas $T_S^m p_n$ of α (provided they are not in the scope of some T_S), by the construction of $\mathcal{M}^{\mathbb{N}}$ and $\mathcal{M}'^{\mathbb{N}}$, we have that $\mathcal{M}^{\mathbb{N}}, 0 \models T_S^m p_n$ iff $\mathcal{M}'^{\mathbb{N}}, 0 \models q^{m,n}$. Hence, as it was supposed that $\mathcal{M}'^{\mathbb{N}}, 0 \not\models \alpha^*$, by an usual induction on the complexity of α we also have that $\mathcal{M}^{\mathbb{N}}, 0 \not\models \alpha$. Therefore, if α is not a theorem of **W**, α is not valid on the standard frame. ♠

The method used in the previous proposition yields the following interesting corollary:

Corollary 6.4.2 *W is decidable.*

Proof: The decision procedure is indeed implicit in the proof of Proposition 6.4.1. Since the validity of α is reducible to the validity of a certain **PC**-formula α^*, it is enough to check the validity of α^* in **PC** with usual methods.
♠

Note that, if we define a class of **W**-frames in a more general way such as to include the standard frame, the previous proposition automatically grants the completeness of **W** with respect to such class of **W**-frames.

Another more direct way to obtain a completeness result for **AN** relies on the remark that axiom (**W1**) corresponds to $\Box p \equiv \Diamond p$ (provided T_S is interpreted as $\Box$). Therefore, as (**W2**) and (**Nec**$_{T_S}$) correspond to (**K**) and (**Nec**), **AN** is actually a version of **KD** + (**F**), where (**F**) is the wff $\Diamond p \supset \Box p$ which

---

[7]For example, let α be the formula $T_S^2 p_n \supset T_S p_m$, which is equivalent to $T_S T_S p_n \supset T_S p_m$; α^* will be $q^{2,n} \supset q^{1,m}$.

expresses the functionality of accessibility relations (see Remark 3.2.15). In this way, **AN** can be straightforwardly characterized by serial and functional frames.

The so-called "temporal logic of programs" makes use of a language which extends the language of von Wright's logics. The idea is to introduce a new operator ATN with the following meaning:

- $\alpha ATN\beta \stackrel{\text{Def}}{=} \alpha$ is true in the next first instant in which β is true, if β is true in some instant of future time.

The first attempt to develop a logic of this kind requires taking ATN as a primitive connective. Let $\mathcal{M}^{\mathbb{N}} = \langle \mathbb{N}, \oplus, V \rangle$ be a model on the standard frame defined above. Then the truth-conditions for the new operator are:

- $\mathcal{M}^{\mathbb{N}}, n \models \alpha ATN\beta$ iff $\mathcal{M}^{\mathbb{N}}, m \not\models \beta$ for every $m > n$ or $\mathcal{M}^{\mathbb{N}}, l \models \alpha$ for the least $l > n$ such that $\mathcal{M}^{\mathbb{N}}, l \models \beta$.

By taking ATN as primitive, other known operators like T_S and the Diodorean box can be defined as follows:

$$T_S \alpha \stackrel{\text{Def}}{=} \alpha ATN\top$$

$$\overline{\Box}\alpha \stackrel{\text{Def}}{=} \alpha \wedge (\bot ATN \neg \alpha)$$

Yet other operators can be also defined by:

- $\beta ATN(\alpha \supset \beta)$ (α unless β)
- $\neg\beta ATN(\alpha \supset \neg\beta)$ (α while β)
- $\neg\beta ATN(\alpha \vee \beta)$ ($\neg\beta$ before α)

An axiomatization for a mixed system containing T_S, ATN and $\overline{\Box}$, which will be named **TL** here, results from subjoining the following axioms to **PC**:

(**TL1**) $\neg T_S \alpha \equiv T_S \neg \alpha$

(**TL2**) $T_S(\alpha \supset \beta) \supset (T_S \alpha \supset T_S \beta)$

(**TL3**) $\overline{\Box}\alpha \supset (\alpha \wedge T_S \overline{\Box}\alpha)$

(**TL4**) $T_S \overline{\Box} \neg \beta \supset \alpha ATN\beta$

(**TL5**) $\alpha ATN\beta \equiv T_S(\beta \supset \alpha) \wedge T_S(\neg \beta \supset \alpha ATN\beta)$

6.4. OTHER TEMPORAL SYSTEMS

The rules are:

(**MP**) $\vdash \alpha$ and $\vdash \alpha \supset \beta$ implies $\vdash \beta$

(**Nec**$_{T_S}$) $\vdash \alpha$ implies $\vdash T_S\alpha$

(**RT**$_S\overline{\square}$) $\vdash \alpha \supset \beta$ and $\vdash \alpha \supset T_S\beta$ implies $\vdash \alpha \supset \overline{\square}\beta$

The semantics for **TL**, like the one for **AN**, relies on a presupposition of discreteness, but we can disregard this presupposition and look for more general connectives.

A couple of binary operators, S and U (for "Since" and "Until"), proposed by Hans Kamp in 1968, allows us to define all the above introduced operators.

- $U(\alpha, \beta)$: to be read as "α will be true at a future time and β will be true until then " i.e., at any time between the present time and that moment (the present moment does not need to be included).

- $S(\alpha, \beta)$: to be read as "α was true at some past time and β has been true since then", i.e., at any time between that moment and the present time (again, the present moment does not need to be included).

Which is the meaning of $U(\alpha, \alpha)$? Suppose that time is discrete, hence that there exists something as "the next instant of time". If the next instant is the only instant at which α is true, then α will be true at this instant. Otherwise, α will be true from now on until some future instant at which α is true. In both cases, α will be true at the next moment of time without commitment to the discreteness of the temporal series. Therefore the following definitions may be introduced:

$$T_S\alpha \stackrel{\text{Def}}{=} U(\alpha, \alpha)$$

$$Y_S\alpha \stackrel{\text{Def}}{=} S(\alpha, \alpha)$$

The operators $\Diamond$ and $\Diamond$ are also definable in terms of U and S in the following way:

$$\Diamond\alpha \stackrel{\text{Def}}{=} U(\alpha, \top)$$

$$\Diamond\alpha \stackrel{\text{Def}}{=} S(\alpha, \top)$$

The strength of this approach can be perceived by observing that new temporal operators can now be defined besides ◻ and ◼. So, for instance:

- $\square'\alpha \stackrel{\text{Def}}{=} U(\top, \alpha)$ with the meaning of "α will be always true from now on";

- $\blacksquare'\alpha \stackrel{\text{Def}}{=} S(\top, \alpha)$ with the meaning of "α has been always true up to now".

Obviously the dual operators ◇' and ◆' can be defined in the usual way.

It is clear then that a multimodal language based on special dyadic operators permits us to define a wide variety of multimodal languages based on monadic operators.

Logics sharing the language with U and S are called *US-logics*. Frames for *US*-systems are triples $\langle T, \overrightarrow{R}, \overleftarrow{R} \rangle$ defined as the ones for *PF*-logics. The truth conditions are standard as far as truth-functional connectives are concerned, while for U and S they are as follows:

Definition 6.4.3 *Let M be a model based on a $\langle T, \overrightarrow{R}, \overleftarrow{R} \rangle$ frame. Then:*

- $M, t \models U(\alpha, \beta)$ *if and only if there exists t' such that $t \overrightarrow{R} t'$ and $M, t' \models \alpha$, and for all t'' such that $t \overrightarrow{R} t''$ and $t'' \overrightarrow{R} t'$, it holds that $M, t'' \models \beta$.*

- $M, t \models S(\alpha, \beta)$ *if and only if there exists t' such that $t \overleftarrow{R} t'$ and $M, t' \models \alpha$, and for all t'' such that $t \overleftarrow{R} t''$ and $t'' \overleftarrow{R} t'$, it holds that $M, t'' \models \beta$.*

It is worth remarking that $S(p \wedge q, \neg q)$ will mean "p was true at the last time in which q was true", while $U(p \wedge q, \neg q)$ will mean that "p will be true at the first time in which q is true". From the truth conditions for *ATN*, it turns out that:

- $M, t \models \alpha ATN \beta$ iff $M, t' \not\models \beta$ for every t' such that $t' \overrightarrow{R} t$ or else $M, t \models U(\alpha \wedge \beta, \neg \beta)$

This shows that *ATN* can be defined in terms of *U*, and it turns out that *U* is more expressive than *ATN*.

Granted that ◻ and ◼ can be defined in terms of U and S, an axiomatic basis for a standard *US*-logic (called **TLUS**) is the following extension of **PC**:

(US.1) $\square(p \supset q) \supset (((U(p, r) \supset U(q, r)) \wedge (U(r, p) \supset U(r, q)))$

(US.2) $\blacksquare(p \supset q) \supset (((S(p, r) \supset S(q, r)) \wedge (S(r, p) \supset S(r, q)))$

6.4. OTHER TEMPORAL SYSTEMS

(**US.3**) $(p \wedge U(q,r)) \supset U(q \wedge S(p,r), r)$

(**US.4**) $(p \wedge S(q,r)) \supset S(q \wedge U(p,r), r)$

The rules are (**US**), (**MP**), (**Nec**$_\square$), (**Nec**$_\boxminus$).

To appraise the gain in expressiveness reached by the new language, observe that, while intransitivity is not *PF*-definable, it is in fact *US*-definable. Indeed, a formula expressing intransitivity is simply $U(p,q) \supset U(p,r)$, as shown below:

Proposition 6.4.4 *Let* $\mathcal{F} = \langle T, \overleftarrow{R}, \overrightarrow{R} \rangle$ *be a frame for a PF-logic. Then* $\overrightarrow{R}$ *is intransitive iff* $\mathcal{F} \models U(p,q) \supset U(p,r)$.

Proof: Since intransitivity is expressed by the first-order formula:

(**Int**) $\forall t, t', t'' ((t \overrightarrow{R} t' \wedge t' \overrightarrow{R} t'') \supset \neg (t \overrightarrow{R} t''))$

then the correspondence is proved by the following argument. Let $\mathcal{F}$ be any temporal frame $\langle T, \overleftarrow{R}, \overrightarrow{R} \rangle$ and suppose that $\overrightarrow{R}$ satisfies (**Int**). For any assignment V on a model $\mathcal{M}$ over $\mathcal{F}$ and any $t \in T$, if $\mathcal{M}, t \models U(p,q)$, then there exists a $t' \in T$ such that $t \overrightarrow{R} t'$ and $V(p) = \{t'\}$. Since $\overrightarrow{R}$ satisfies (**Int**), there is no $t'' \in T$ such that $t \overrightarrow{R} t''$ and $t'' \overrightarrow{R} t'$. Then, by vacuity, we have, for this model $\mathcal{M}$ over $\mathcal{F}$ that $\mathcal{M}, t \models U(p,r)$ for each r, therefore $U(p,q) \supset U(p,r)$ is valid on $\mathcal{F}$.

For the converse direction, suppose that $\overrightarrow{R}$ does not satisfy (**Int**). Then there are $t, t', t'' \in T$ such that $t \overrightarrow{R} t'$, $t' \overrightarrow{R} t''$ and $t \overrightarrow{R} t''$. Let V be a valuation on $\mathcal{F}$ such that $V(p) = \{t''\}$, $V(q) = T$ and $V(r) = \emptyset$. It is clear that $U(p,q) \supset U(p,r)$ is false at t for this assignment, and thus it is false on $\mathcal{F}$. ♦

This gain on expressiveness, however, goes hand in hand with a proliferation of incompleteness results. We can prove for example:

Proposition 6.4.5 *Let* $\Sigma = \{U(p,q) \supset U(p,r), \Diamond\Diamond p \supset \Diamond p, \Diamond(p \vee \neg p)\}$ *be a set of formulas in the language of TLUS. Then: (i)* **TLUS**$+\Sigma \models \bot$, *but (ii)* **TLUS**$+\Sigma \nvdash \bot$.

Proof:

(i) We know that $\Diamond(p \vee \neg p)$ defines the seriality of $\overrightarrow{R}$ (it is enough to observe that it is equivalent to $\Diamond p \vee \Diamond \neg p$, thus to $\Box p \supset \Diamond p$), and, consequently, the relation $\overrightarrow{R}$ cannot be empty; we also know that $\Diamond\Diamond p \supset \Diamond p$

defines transitivity (see Section 3.3). As seen above, $U(p,q) \supset U(p,r)$ expresses intransitivity. Therefore, if there were a class of frames validating Σ, it should have both the properties of transitivity and intransitivity, which is impossible. Hence, such a class of frames coincides with the empty class, which validates the formula $\bot$.

(ii) Now, for (ii), define an "erasure" translation $\bar{e}$ from **TLUS**+Σ into **PC** in the following way:

(a) $\bar{e}(p) = p$ for p a propositional variable or $\bot$
(b) $\bar{e}(\neg \alpha) = \neg \bar{e}(\alpha)$
(c) $\bar{e}(\alpha \supset \beta) = \bar{e}(\alpha) \supset \bar{e}(\beta)$
(d) $\bar{e}(U(\alpha, \beta)) = \bar{e}(\alpha)$
(e) $\bar{e}(S(\alpha, \beta)) = \bar{e}(\alpha)$

By induction on the length of wffs we can prove that, for each α that is a theorem of **TLUS**+Σ, $\bar{e}(\alpha)$ is a classical tautology. Since $\bar{e}(\bot)$ is not a classical tautology, it follows that $\bot$ does not belong to the class of **TLUS**+Σ-theses.

♠

To give other examples, the systems obtained by adding the following sets of formulas to **TLUS** are also incomplete (see Exercise 6.15):

- $\Sigma_1 = \{U(\top, \bot), \Diamond p \supset \Diamond \Diamond p\}$
- $\Sigma_2 = \{U(\top, \bot), \Box p \supset p\}$
- $\Sigma_3 = \{U(p,q) \supset U(p,r), \Box p \supset p\}$

6.5 US-logics, metric tense logics and hybrid logics

The father of tense logic, A. N. Prior, thought that the most interesting extensions of *PF*-logics should have a metric nature, that is, they should be such as to numerically express time intervals of length n after which a certain proposition α will be (or has been) true.

By symbolizing these notions with $\Box^n \alpha$ and $\boxempty^n \alpha$ (where n is an any natural number), our language will have to be supplemented by a new formation rule, i.e.:

If $\alpha \in WFF$ and n is a natural number, then $\Box^n \alpha, \boxempty^n \alpha \in WFF$

6.5. US-LOGICS, METRIC TENSE LOGICS AND HYBRID LOGICS

$\diamondsuit^n \alpha$ and $\diamondsuit^n \alpha$ will also be suitably defined well-formed formulas.

Since we need to at least express the addition and difference of interval lengths in metric logic, we obviously need at least a fragment of Peano arithmetic. By calling this limited fragment of arithmetic **TN**, it is also necessary to admit quantification over intervals. Hence we also need the quantificational theory **QL** (see Chapter 9).

The postulates of the minimal metric tense logic, called **PF**^{n}, are the following:

(A0)　The axioms of **QL** and **TN**
(A1.1)　$\neg \diamondsuit^n \neg (p \supset q) \supset (\diamondsuit^n p \supset \diamondsuit^n q)$　(A1.2)　$\neg \diamondsuit^n \neg (p \supset q) \supset (\diamondsuit^n p \supset \diamondsuit^n q)$
(A2.1)　$\diamondsuit^n \neg \diamondsuit^n \neg p \supset p$　(A2.2)　$\diamondsuit^n \neg \diamondsuit^n \neg p \supset p$
(A3.1)　$\diamondsuit^m \exists n \diamondsuit^n p \supset \exists n \diamondsuit^m \diamondsuit^n p$　(A3.2)　$\diamondsuit^m \exists n \diamondsuit^n p \supset \exists n \diamondsuit^m \diamondsuit^n p$
(A4.1)　$\diamondsuit^n \exists n \diamondsuit^n p \supset \exists n \diamondsuit^n \diamondsuit^n p$　(A4.2)　$\diamondsuit^n \exists n \diamondsuit^n p \supset \exists n \diamondsuit^n \diamondsuit^n p$
(A5.1)　$\diamondsuit^{m+n} p \supset \diamondsuit^m \diamondsuit^n p$　(A5.2)　$\diamondsuit^{m+n} p \supset \diamondsuit^m \diamondsuit^n p$

The rules are the same as for first-order logic although extended with two new rules of temporal necessitation:

(**Nec**$_{\square^n}$)　$\vdash \alpha$ implies $\vdash \square^n \alpha$

(**Nec**$_{\boxminus^n}$)　$\vdash \alpha$ implies $\vdash \boxminus^n \alpha$

The non-metric *PF*-logic is obtained from this one via obvious definitions:

$$\diamondsuit \alpha \stackrel{\text{Def}}{=} \exists n \diamondsuit^n \alpha \qquad \diamondsuit \alpha \stackrel{\text{Def}}{=} \exists n \diamondsuit^n \alpha$$
$$\square \alpha \stackrel{\text{Def}}{=} \forall n \square^n \alpha \qquad \boxminus \alpha \stackrel{\text{Def}}{=} \forall n \boxminus^n \alpha$$

The relationship between metric logic and tense logic is interesting. Suppose for instance that we add the axiom $\square^n p \supset \diamondsuit^n p$ expressing infinity (seriality) of the future, and another axiom, $\diamondsuit^n \neg p \supset \neg \diamondsuit^n p$, expressing linearity of the future. Under such assumptions, $\square^n \alpha$ and $\diamondsuit^n \alpha$ become logically equivalent.

Then, for example, the connective T "and next" could be defined as:

$$\alpha T \beta \stackrel{\text{Def}}{=} \alpha \wedge \diamondsuit^1 \beta$$

Obviously, $\diamondsuit^1$ is identifiable with T_S.

In a parallel way, by introducing the metric operator for the past, we can identify $\diamondsuit^1$ with Y_S and $\alpha Y \beta$ with $\alpha \wedge \diamondsuit^1 \beta$.

It is obvious, vice versa, that the following definitions can also be introduced:

$$\Diamond^1 \alpha \stackrel{\text{Def}}{=} (\alpha \lor \neg \alpha) T \alpha$$

$$\Diamond^1 \alpha \stackrel{\text{Def}}{=} (\alpha \lor \neg \alpha) Y \alpha$$

But $\Diamond^1, \Diamond^2, \cdots \Diamond^n$ can also be defined in terms of T. For example, $\Diamond^2 \alpha$ can be rendered by $(\alpha \lor \neg \alpha) T((\alpha \lor \neg \alpha) T \alpha)$ etc. $\Diamond^n$ can thus be defined in terms of T or T_S if time is discrete. And considering that T_S can be defined in terms of S and U, as shown in the preceding section, these two operators turn out to be the most versatile tools in granting a basis for both metric as well as for non-metric logic.

Would it be possible to define U and S in terms of monadic operators? The answer is negative. In a model-theoretic framework, it can be proven that no modal formula can grasp their properties, i.e., they cannot be defined in the usual modal language. In fact, it can be shown[8] via bisimulations that U (and consequently also S) is not definable on arbitrary modal models. A simple proof, making use of the powerful concept of bisimulation, is given as follows.

Proposition 6.5.1 *The operators $\square$ ($\boxminus$) as axiomatized in standard normal modal logics are definable in terms of $U(p, q)$ ($S(p, q)$), but no definition of $U(p, q)$ ($S(p, q)$) is possible in terms of $\square$ ($\boxminus$).*

Proof: From Definition 6.4.3, it is easy to see that $U(\alpha, \top)$ coincides with $\Diamond \alpha$, and then $\square \alpha$ can be defined as $\neg U(\neg \alpha, \top)$. Starting from S instead, it is clear that $S(\top, \alpha)$ also defines $\boxminus \alpha$.

On the other hand, consider the two models below: From Definition 3.3.4,

it is clear that the two models are bisimilar: to see this, simply map worlds w_0 and w_1 into w. Supposing that all atomic variables are false at all worlds in both models, we see that both models assign the same values to all formulas with tense operators. Nevertheless, the models do not agree on

---

[8]See P. Blackburn and J. van Benthem [BvB07] (section 6.3).

6.5. US-LOGICS, METRIC TENSE LOGICS AND HYBRID LOGICS

the wff $U(\top, \bot)$: this wff is false in the one-world model but true on both points of the two-worlds model. Consequently, taking for granted that bisimulations preserve the satisfaction of all tensed formulas, if U were definable in terms of $\Box$ or $\Diamond$, its meaning would have to be preserved in both models. Analogous arguments hold for S. ♠

The same argument can be immediately adapted to show that U and S cannot be defined terms of the alethic operators $\Box$ or $\Diamond$.

Both metric language and US-language have been introduced in temporal logics with the aim of extending the expressive limits of the PF-language. In order to understand this requirement it is enough to remark that in ordinary language it is usual to make use of complex tenses such as the past perfect (e.g. *I had seen*), which are not expressible in PF-language: in this concern suffices it to observe that in transitive and dense temporal models an apparently obvious translation of the past perfect as $\Diamond\Diamond p$ is equivalent to the simple $\Diamond p$. As H. Reichenbach first noticed in the Forties, the problem is that in complex tenses not two but three different times are involved: the time of utterance of the statement (*point of speech*), the time of the described event (*point of event*), and a third time called *point of reference*. If i stands for such point of reference, the present tense could be represented by $i \wedge p$, the past tense by $P(i \wedge p)$ and the past perfect by $P(i \wedge Pp)$. A reasonable proposal is then to extend the PF-language with new symbols $i, k, l, \cdots$ for objects that we may call *nominals* and that we treat as propositions of a different sort. The formation rules for the language of what have been called *hybrid logics* should include rules for an operator of satisfaction @ such that $@_i \alpha$ asserts that α is satisfied at the (unique) point named by the nominal i. @ is then a two-place operator which is ruled by special axioms which include (at least) the following:

(@1) $@_i \alpha = \neg @_i \neg \alpha$

(@2) $@_i(\alpha \supset \beta) \supset (@_i \alpha \supset @_i \beta)$

The most interesting fact is that $i, k, l, \cdots$ are also symbols for propositions, so that also $@_i j$, say, should count as a well-formed formula. Asserting that the nominal j is true at i means "the point named by i is identical to the point named by j". $@_i j$ may be set to be equivalent to $i = j$ and has then the properties pertaining to the identity relation (reflexivity, symmetry, transitivity and substitution of identicals).

The enrichment gained with this extension of *PF*-language cannot be underestimate. The accessibility relations themselves may be expressed in the object language in terms of @, ◇ or ◈, by putting:

$(Def \vec{R}) : \quad t \vec{R} t' \stackrel{Def}{=} @_i \diamond j$

$(Def \overleftarrow{R}) : \quad t \overleftarrow{R} t' \stackrel{Def}{=} @_i \blacklozenge j$

Since □ is definable in terms of ◇ and ◈, also @ may be defined in terms of ◇ and ◈ in this way:

$(Def@) : \quad @_i p \stackrel{Def}{=} \Box(i \supset p)$

Many properties of the temporal series which cannot be expressed in *PF*-language are now easily rendered in the extended hybrid language. For instance:

- Irreflexivity: $@_i \neg \diamond i$
- Asymmetry: $@_i \neg \diamond \diamond i$
- Intransitivity: $\diamond \diamond i \supset \neg \diamond i$
- Trichotomy: $@_j \diamond i \vee @_j i \vee @_i j$

Quantifying over nominals yields a further enrichment and allows, for instance, translating $\Box(i \supset p)$ into $\forall i @_i(i \supset p)$. Since nominals are propositions, this kind of quantification is actually a special kind of propositional quantification (for which see Section 9.1). Hybrid logic is a way to import into modal object language important aspects of the metalanguage. From a technical viewpoint, the operator @ is a special kind of a two-place modal operator and can thus be compared with other two-place operators treated in this chapter, as the ones for "and next", "since" and "until". Modal languages whose primitives are two or more *n*-place modal operators (where $n > 1$) have been of increasing importance in the development of multimodal logics. However, a detailed analysis of them lies out of the scope of the present book.

6.6 Exercises

1. Prove the constructive completeness of the system **KT**□.

2. Show, for *s* the standard translation, that

6.6. EXERCISES

   (i) $s(p \supset \Box \Diamond p) = \forall t \forall t' (t' \overleftarrow{R} t \supset t \overrightarrow{R} t')$

   (ii) $s(p \supset \Box \Diamond p) = \forall t \forall t' (t' \overrightarrow{R} t \supset t \overleftarrow{R} t')$

3. (a) Test the formula $p \supset \Box(\Diamond p \supset \Diamond \Box \Diamond (p \lor q))$ by means of $\mathbf{K}_t$-tableaux and convert the resulting closed tableau into an equivalent tableau which is "fully corrected" in the sense defined on page 154.

   (b) Convert the tableau obtained in (a) into a proof of the formula under test.

   (c) The same as in (a) and (b), for the formula $(p \land \Diamond \Diamond q) \supset \Diamond(\Diamond p \land \Diamond p)$

4. Give a syntactic proof in $\mathbf{K}_t$ of the two formulas indicated in the previous exercise without passing through constructive completeness.

5. Prove that $\mathbf{K}_t + \Diamond \Diamond p \supset \Diamond p + \Diamond \Diamond p \supset \Diamond p$ is complete with respect of the class of transitive $\mathbf{K}_t$-frames.

6. In order to complete details of Example 6.1.14, prove that the following wffs in normalized form (i.e., after eliminating by definition every $\Diamond$ and $\Diamond$) are $\mathbf{K}^{\Box\Box}$-thesis:

   (i) $\neg \Box q \supset ((p \supset \Box \Diamond p) \supset ((p \land \Diamond q) \supset \Diamond(\Diamond p \land q)))$

   (ii) $\Box(\Diamond p \land q) \supset ((p \supset \Box \Diamond p) \supset ((p \land \Diamond q) \supset \Diamond(\Diamond p \land q)))$

   (iii) $\neg \Box \Diamond p \supset ((p \supset \Box \Diamond p) \supset ((p \land \Diamond q) \supset \Diamond(\Diamond p \land q)))$

7. Show that $\mathbf{K}_t$ is characterized by all temporal frames in which $\overleftarrow{R}$ and $\overrightarrow{R}$ are irreflexive and converse of each other and by all temporal frames in which $\overleftarrow{R}$ and $\overrightarrow{R}$ are intransitive and converse of each other.

8. Show that Dummett's formula (**Dum**), reproduced on page 165, if added to **S4** holds in all linear temporal frames provided that time order is assumed to be discrete.

9. Show that the incompleteness proof given for $\mathbf{K}_t^{\circ\circ}$ (Proposition 6.2.2) cannot be reproduced if the axioms (**McK**) and (**GL**) are subjoined to $\mathbf{K}^{\Box\Box}$.

10. Show that, in the system **W**, T_S can be distributed in such a way that sequences of T_S appear only before **PC**-formulas.

11. Show the equivalence between the systems **AN** and the fragment of **PF**^{n} in which only the operator $\lozenge^1$ occurs. Conclude from this that **AN** is equivalent to **W**.

12. After defining the non-metric future $\lozenge$ in terms of metric future as in Section 6.5, axiomatically determine the fragment of **PF**^{n} which contains all and only the theorems in which only operators $\lozenge$ occur.

13. Show that the Diodorean fragment of **CL** is exactly the system **S4.3**, while the Megaric fragment is **S5**.

14. Show that, once $\lozenge'$ and $\square'$ are defined as in Section 6.4, it is possible to express the following properties of temporal series in a synthetic way:

    (a) $\lozenge'\top$: density of $\overrightarrow{R}$.

    (b) $\square'\bot \wedge \boxempty'\bot$: discreteness of $\overleftarrow{R}$.

15. Prove the incompleteness of **TLUS**+$\Sigma_1(\Sigma_2, \Sigma_3)$.

16. Develop the inductive arguments in the proof of Proposition 6.3.2.

17. Let us call *trichotomy* the property of **K**$_t$-frames described as: for every $t, t', t = t' \vee t \overleftarrow{R} t' \vee t \overrightarrow{R} t'$. Show that no *PF*-wff can express such a property (Hint. The required *PF*-wff would take value 0 in a non-cohesive temporal frame, where $T = \{t, t'\}$, $t \overleftarrow{R} t$ and $t' \overleftarrow{R} t'$. Draw a contradiction from the supposition that the property is *PF*-definable).

6.7 Further reading

A. N. Prior is unanimously considered the founder of tense logic (*PF*-logic). His work is mainly contained in three fundamental volumes: [Pri57], [Pri67] and [Pri68]. In the Ph.D. thesis of N. B. Cocchiarella synthesized in [Coc66b], *PF*-logics receive an adequate semantical treatment using methods derived from Kripke semantics. The book by N. Rescher and A. Urquhart [RU71] goes beyond *PF*-logics, putting the tense logic within the more general class of topological logics, while the proposal of the logic of "and next" is due to G. H. von Wright (cf. [vW65]). For the completeness of **W**, see K. Segerberg [Seg68]. The introduction of the *US*-logic is due to H. Kamp in [Kam68], an analysis that gave origin to an extensive line of research (see J. P. Burgess

6.7. FURTHER READING

in [Bur84], M. Xu in [Xu88]). For an algebraic study of tense logic see R. A. Bull [Bul68].

Even when restricting attention to transitive models, to linear models and even to models isomorphic to the real numbers, U and S cannot be defined in terms of the basic modal language (cf. section 6.3 of P. Blackburn and J. van Benthem [BvB07] and chapter 7 of P. Blackburn and M. de Rijke and Y. Venema [BdRV01]). To fully understand how expressive the operators U and S are, it is advisable to resort to the so-called Kamp's theorem, one of the first purely model-theoretical results in modal logic. Kamp's theorem, originally proven in H. Kamp [Kam68] (but see D. M. Gabbay, I. Hodkinson and M. Reynolds [GHR94] for extensions and for a more clear proof), states that modal language with the operators S and U is expressively complete for the class of all flows of time isomorphic to the series of of real numbers.

The development of hybrid logic is mainly due to P. Blackburn: see for instance P. Blackburn and M. Tzakova [BT99]. However, the key idea of hybrid logic goes back to A. Prior's [Pri68]. For the analysis of complex tenses see H. Reichenbach [Rei47].

In [vB83a], van Benthem provides a synthesis of the subject of temporal logics, while for an earlier anthology in Italian, see C. Pizzi [Piz74]. For topics of temporal logic related to problems in linguistics and philosophy see D. Gabbay [Gab76]. The application of tense logic to computation theory has undergone a quick development in the last years. For the temporal logic of programs, especially see Z. Manna and A. Pnueli [MP92], and for the logic of ATN see F. Kröger [Krö87]. For a survey, see [GHR94].

Chapter 7
Epistemic logic: knowledge and belief

7.1 To know, to believe and their difficulties

In modern philosophy, the logics of belief and knowledge are both classified as *epistemic*, even though the Greek word *episteme* refers essentially just to knowledge, and the logic of belief should perhaps be more properly called *doxastic*, from the Greek *doxa*. It is not unreasonable to assert that epistemic logic is a Greek legacy, just like logic in general. In *Theaetetus*, one of his most celebrated dialogues, Plato presents a conversation in which, among other things, Socrates and the young Theaetetus strive to clarify the concept of "knowledge". Thereafter, the problem of knowledge has a central place in a large part of philosophical systems.

In the aforementioned dialogue, Socrates rejects a definition proposed by Theaetetus, according to which knowledge is a true opinion. From the discussion, a more specific hypothesis arises, according to which it is possible to define knowledge as a true opinion associated to some kind of rational explanation of the object of knowledge. It also held that we we may attain knowledge – this is a further suggestion – inasmuch as we somehow manage to grasp what unambiguously distinguishes the object of our judgment from everything else. However, Socrates maintains that to hold a true opinion means exactly to distinguish the object of our judgment from everything else. In this way, the second definition turns out not to be different from the initially rejected proposal.

The dialogue goes on up to the point at which Socrates proposes that explaining something means having a true opinion on it plus *knowing* what makes it different from everything else. Though this seems to solve the initial problem, the definition becomes circular. In fact, we resort to the notion of rational explanation to tell what knowledge is and, at the same time, resort to the notion of knowledge to tell what a rational explanation is. The dialogue thus ends in an aporia, leaving the formidable question "what is knowledge" unanswered.

Systems of epistemic logic have been developed only after the formulation of the relational (or possible-worlds) semantics in very recent times. Many important works about the foundations of epistemic logic have been produced in the second half of the 20th century, a period in which epistemic logic has become a consolidated branch of modal logic. In particular, its development led to the application of epistemic logic to Artificial Intelligence, modeling common knowledge as a phenomenon of something arising from the interaction of agents.

How can one define knowledge and how is it related to opinion or belief? This is a deep and difficult question. Here we will confine ourselves to sketch some of the main ideas with some considerations on the consequences of assuming or rejecting some basic principles.

An important feature of epistemic logic is that the reference to individuals is unavoidable. In fact, if, on the one hand, it makes sense to admit something as necessary, possible or obligatory without reference to any given individual, then, on the other hand, it is inconceivable to think of anything that is known or is believed without referring to one or more individuals who know or believe it. This point is crucial for the understanding of epistemic logic with agents as parameters, as it will be shown in Section 7.2. Although the syntactical notions will be rigorously defined in Section 7.3, it is convenient to have in mind that our language for epistemic logic is inherently multimodal, and the wffs will be closed under distinct modal operators $K_1, \ldots, K_m$ (denoting m distinct agents) such that, for $i, j \in \{1, \ldots, m\}$, $K_i \alpha$ and $K_i K_j \alpha$ will be wffs meaning, respectively, "agent i knows α" and "agent i knows that agent j knows α", etc.

Beginning from Plato's first inquiries in his *Theaetetus*, it will be herein assumed, by tacit agreement, that it is impossible to know that p unless p is in fact true. This can be seen as the first basic principle of knowledge, also called the *Knowledge Axiom*,[1] which is expressed by the formula:

(KA) $K_i \alpha \supset \alpha$

---

[1] Actually, the Knowledge Axiom is an axiom schema: it is usual in the literature of epistemic logic to use axiom schemas rather than axioms, and we conform to this practice in this chapter.

7.1. TO KNOW, TO BELIEVE AND THEIR DIFFICULTIES

where, for any wff α and any modal operator K_i, $K_i\alpha$ is a wff which means, as mentioned, that "i knows that α" (note that (**KA**) has the same logical form of axiom (**T**)). Most people accept this principle as an essential part of the concept of knowledge, but the principle is clearly insufficient to characterize knowledge. For instance, imagine that someone draws a true conclusion from a false hypothesis, or guesses the lottery numbers which will be drawn next week: would you say that such a person "knows" these facts? What is clear is that such a person believes in the facts and the facts are indeed true, but that there is something missing here. As in Socrates' argument, what is missing is the concept of rational explanation or justification. Accordingly, knowledge is defined as a *justified true belief*.

Nonetheless, beyond the difficulties found in Plato's dialogue, some objections could be proposed against this definition as well. The notion that knowledge logically implies truth is actually almost universally accepted. Obviously this does not happen with respect to belief, for someone may strongly believe that p is the case and this may be logically compatible with the actual falsity of p. In other words, nothing hinders some person from believing in false propositions. Briefly, beliefs do not imply truth, although it is acceptable that knowledge does.

An important objection to defining knowledge as justified true belief moves from the fact that we can easily identify at least two kinds of knowledge: "to know that" (as in "to know that a certain calculation is right") and "to know how" (as in "to know how to bow a tie"). Both of them result from learning; however, even if it is possible to distinguish several kinds of belief, there is not something like "to believe how". Consequently, there exists at least one kind of knowledge that is not definable as a special kind of belief.

A second possible criticism is that according to a well-known (controversial) distinction introduced by B. Russell , we should distinguish between knowledge by *acquaintance* and knowledge by *description*: the former takes place when we come to know something – for instance this table – of which we are directly aware by causal interaction, without the intermediation of any process of inference or any knowledge of truths. But knowing this table as 'the physical object which causes such-and-such sense-data' depends on the fact that we *describe* the table by means of the sense-data, so this kind of knowledge belongs to the latter category.

Other objections are directed against identification of knowledge with justified true belief (as in the so called "Gettier's problem",) but we shall skip the details here. Russell's distinction also has no counterpart in some parallel distinction of beliefs.

As belief is concerned, most treatments of epistemic logic only assume that one concept of belief should be analyzed. Some authors, however, claim that there are at least two distinct notions of belief: a "strong" one and a "weak" one, as we will stress in Section 7.6.

Besides the knowledge axiom (**KA**), and in correspondence with the previously mentioned problem of knowledge, there are many other issues which have been discussed at length in the literature. Two of them are the problem of positive introspection and of negative introspection, which will be examined in next section.

Beyond (**KA**), there are two other controversial principles usually assumed in epistemic logics. The first is the generalization of axiom (**K**) and the second is the generalization of (**Nec**), both introduced in Chapter 2. The generalization of (**K**), in few words, amounts to the following condition which must hold for each K_i:

(**KK$_i$**) $(K_i(\alpha \supset \beta) \wedge K_i\alpha) \supset K_i\beta$,

whose meaning is that the collection of all facts known by some agent i is deductively closed. Although epistemic closure in this way has some *prima facie* plausibility, some puzzling consequences arise especially in conjunction with the epistemic version of the Necessitation Rule. The generalized Necessitation Rule (**Nec**) says that, for each K_i:

(**Nec$_{K_i}$**) $\vdash \alpha$ implies $\vdash K_i\alpha$

which means that, if α is a thesis, then every individual must know α.

For example, if a logician knows the axioms of Peano arithmetic (s)he should know all their consequences, including the answers to all the open questions of number theory in Peano arithmetic, such as the famous conjecture formulated by C. Goldbach in 1742 (which states that every even number greater than 2 is the sum of two prime numbers). In this way, although Goldbach's conjecture has been an open question for more than two centuries and nobody knows whether or not it is provable in Peano arithmetic, if Goldbach's conjecture is a theorem, then all logicians must already know it, and if they do not know it, this would mean that it is not provable.

Difficulties such as the one outlined above are aspects of the problem which has been called the *problem of logical omniscience*, and has led some authors to propose that knowledge should be treated as a primitive and undefined concept, which would only be partially characterized by axioms.

No doubt, omniscience is a source of predicaments if epistemic logic is conceived as a class of systems whose aim is to describe the actual behavior of rational subjects. It is not such, however, if it is conceived as a class of normative systems of rational thought whose aim is to guide the behavior of (human and non-human) agents in the matter of knowledge and belief.

In 1963, F. Fitch proved a puzzling theorem about the combination of alethic and epistemic operators, which became known as "the paradox of knowability". Fitch's theorem proves that, by accepting some minimal presuppositions about knowledge as expressed in the system $\mathbf{KT}^m$ (see Section 7.4) and alethic necessity as expressed in the system $\mathbf{K}$, there must be an unknowable assertion (Exercise 7.4). This seems to be paradoxical in view of the apparently reasonable "Principle of Knowability" (**KP**) (also known as the "Verificationist Principle") which says that any truth is (in principle) knowable, or possible to be known:

(**KP**) $\alpha \supset \Diamond K\alpha$, where K here stands for any operator K_i.

A deeper negative result stems from the joint acceptance, under the same assumptions, of (**KP**) and what is called the *Anti-Omniscience Principle* i.e. the idea that there is some unknown truth:

(**AO**) There exists α_0 such that $\alpha_0 \wedge \neg K\alpha_0$

A contradiction in fact follows from the conjunction of (**KP**) and (**AO**) (see Exercise 7.5). Thus, philosophers supporting (**KP**) are forced to assert the negation of (**AO**) and so that all truths are actually known: $\alpha \supset K\alpha$. Given that $K\alpha$ implies α, this amounts to the collapse of $K\alpha$ over α. This puzzling result has been the source of a wide literature offering a varied range of answers to the problem (from proposing the revision of the underlying logic, to investigating the background of communication and learning). To confine attention to a simple remark on the strategy of revising the underlying logic, let us observe that the intuitionistic antirealist may unproblematically rely on (**KP**) for the reason that (s)he does not accept the principle that $\neg\neg\alpha$ implies α so (s)he cannot pass from $\neg(\alpha \wedge \neg K\alpha)$ to $\alpha \supset^* K\alpha$ (where $\supset^*$ is the intuitionist implication) but simply to $\alpha \supset^* \neg\neg K\alpha$, which intuitionistically asserts that there is no way to find truths that will never be known.

7.2 Knowledge, belief and agents

The issue of characterizing knowledge, besides its intrinsic philosophical interest, appears to be crucial in understanding how certain systems work; this is the case of multi-agent and distributed systems where the computers or programs themselves exchange messages that modify the state of the knowledge thereof. A distributed system is basically composed of a set of processors or agents (either machines or programs) interconnected via a communication web. Distributed systems not only have a great impact on applications that are structurally cooperative, such as for instance message exchanging, but they also allow the implementation of error tolerance due to the duplication of processes by distinct agents (computation unities). They are of great interest in Artificial Intelligence (henceforth AI) and introduce theoretical issues of higher complexity. No matter how developed hardware is, the major difficulty rests on the software inasmuch as it depends on theoretical presuppositions: every processor is always in one state that varies in dependence of the initial state, received messages, etc. In what follows, we shall see how relational frames can be associated to these systems in a quite natural manner. Hence a large part of the issues concerning distributed systems may be seen as issues treatable in multimodal logic endowed with intrinsic philosophical importance.

We have to remark that many languages are able to express subtle differences between the meaning of the verbs "to know" (French *connaître*, Italian *conoscere*, Portuguese *conhecer*, German *zu kennen*) and "to wit" (French *savoir*, Italian *sapere*, Portuguese *saber*, German *zu wissen*). Such differences are perceived when, for instance, someone says 'I know the Amazon River, but I do not wit its exact size'. It seems that this distinction adds to Russell's distinction between knowledge by acquaintance and knowledge by description. Nevertheless, the logics of knowledge, which will be henceforth discussed, are not able to handle such distinction.

As already said, among philosophers there is no agreement about what should be the fundamental properties of knowledge: questions like those below may be answered either affirmatively or negatively, and various principles may be ruled out or included among the axioms of various logical systems.

Q_1: Can any knower know something that is not true?

Q_2: Must the knower necessarily know what (s)he knows?

Q_3: Must the knower necessarily know what (s)he does not know?

The negative answer to the first question is given by the previously mentioned knowledge axiom (**KA**) which, according to several authors, clearly distinguishes "knowledge" from "opinion", and so knowledge from belief.

The concept of an agent is a central notion for the logic of knowledge and belief, whereby an agent may be regarded as a particularization or individuation of the knowledge state. The key idea of epistemic logics rests on introducing an arbitrary number of modal operators into the language, one for each agent, so that the statement "agent i knows α" is interpreted as "α is true in all worlds that agent i holds to be plausible". (Here the term "plausible" is preferred over "possible" in order to avoid confusion with the notion of possibility studied in the previous chapters).

Given the above mentioned relation of plausibility P, we may bring in different axioms that meet the foregoing questions Q_1–Q_3:

- If P is a reflexive relation, then what an agent knows is true (answer to question Q_1).

- If P is transitive, then an agent knows what (s)he knows (answer to question Q_2).

- If P is transitive and symmetric, then an agent knows what (s)he does not know (answer to question Q_3).

Of course, where only one agent exists, the epistemic logic, whose semantics is endowed with a reflexive plausibility relation, corresponds to the system **KT**. Alternatively, a system of epistemic logic corresponds to **S4** if its underlying plausibility relation is both reflexive and transitive, and to **S5**, if the mentioned relation is reflexive, symmetric and transitive. But which system would be preferable in the case of many agents or in the specific case of one single agent? If, for instance, **S4** proves to be the most adequate system for the notion of knowledge, would it really be adequate if bounded to a single human agent?

We are now ready to show that, given the approach assumed herein, even the system of minimal logic for knowledge contains some quite controversial assumptions (cf. Section 7.1).

7.3 The minimal logic of knowledge

Our goal here is to provide the basis for every logical system suitable to formalizing the various notions of knowledge, which will be herein called *minimal logic of knowledge*. We shall employ the same propositional language used in the previous chapters, expanding it with a finite number of modal operators $K_1, \ldots, K_m$ related to a group G of m agents[2] (usually referred to as "knowers"). Of course, all the propositional definitions given in Section 1.2 still apply.

Let $A = \{p, q, r, \ldots\}$ be a set of atomic statements and $O = \{K_1, \ldots K_m\}$ be a set of modal operators indexed by agents (intuitively associating their knowledge to each of them): the formula $K_i\alpha$ will mean "agent i knows α", and the knowledge of the agents represented in O based on A, $\mathcal{K}_O(A)$, is defined as the smallest set of formulas containing A which is closed under $\bot$ and $\supset$ (a fortiori, under $\neg$, $\wedge$ and $\vee$) and under the modal operators in O. Accordingly, if α, β belong to $\mathcal{K}_O(A)$, then $\neg\alpha$, $\alpha \wedge \beta$, $\alpha \vee \beta$, $\alpha \supset \beta$, $K_i\alpha$ belong to $\mathcal{K}_O(A)$ (for every $K_i \in A$).

The resulting language is remarkably expressive; for instance, $K_iK_j\alpha \in \mathcal{K}_O(A)$, that which also permits us to express complex statements like $K_i\neg K_j K_i p$, meaning "i knows that j does not know that i knows p".

Intuitively, the knowledge of an agent corresponds to her/his capacity to answer questions about some specific world. At a given world, we associate to each agent the set of worlds that this agent holds plausible. Consequently, an agent knows a fact α iff α is true at all worlds of that set or, in other words, (s)he fails to know a fact α iff there exists a world (s)he deems plausible, where α does not hold.

We define a *(relational) multi-agent frame* as an $(m + 1)$-tuple $\mathcal{F} = \langle W, P_1, \ldots, P_m \rangle$, where:

1. W is a non-empty set of worlds.

2. $P_1, \ldots, P_m$ are binary plausibility relations between worlds (as usual, we often write wP_iw' instead of $\langle w, w' \rangle \in P_i$).

A *multi-agent model* based on $\mathcal{F}$ is an $(m+2)$-tuple $\mathcal{M} = \langle W, P_1, \cdots P_m, V \rangle$, where $\mathcal{F} = \langle W, P_1, \ldots, P_m \rangle$ is a multi-agent frame and V is a function from the propositional variables of A to the subsets of W. It is convenient here to recall the definition of implicit models introduced in Definition 3.2.1.

An *epistemic state* is a pair $\langle \mathcal{M}, w \rangle$, $\mathcal{M}$ being a multi-agent model and w a world in $\mathcal{M}$. The relation P_i captures the notion of "plausibility with respect to an agent i" to the effect that, if wP_iw', then for the agent i the state $\langle \mathcal{M}, w' \rangle$ is plausible from $\langle \mathcal{M}, w \rangle$. When there is no risk of misunderstanding, we simply write $\mathcal{M}, w$ instead of $\langle \mathcal{M}, w \rangle$, and $w \vDash \alpha$ instead of $\mathcal{M}, w \vDash \alpha$ when $\mathcal{M}$ is contextually presupposed; we also sometimes refer to the world w as a *state*.

---
[2] We realistically suppose here that the cardinality of the set of agents is finite, but there is no substantial difficulties in working out a theory with infinitely many agents.

7.3. THE MINIMAL LOGIC OF KNOWLEDGE

The notion of *satisfiability* in an epistemic state is formally defined as follows:

1. $\mathcal{M}, w \not\models \bot$.

2. $\mathcal{M}, w \models p$ iff $w \in V(p)$ for $p \in A$.

3. $\mathcal{M}, w \models \alpha \supset \beta$ iff $\mathcal{M}, w \not\models \alpha$ or $\mathcal{M}, w \models \beta$.

4. $\mathcal{M}, w \models K_i \alpha$ iff $\mathcal{M}, w' \models \alpha$, for every w' such that $wP_i w'$ for each $K_i \in A$.

From such conditions, it obviously follows that:

5. $\mathcal{M}, w \models \neg \alpha$ iff $\mathcal{M}, w \not\models \alpha$.

6. $\mathcal{M}, w \models \alpha \wedge \beta$ iff $\mathcal{M}, w \models \alpha$ and $\mathcal{M}, w \models \beta$.

Notice that clause 4 from the above definition expresses the intuition that an agent i in an epistemic state $\langle \mathcal{M}, w \rangle$ knows a fact α iff α is true in all epistemic states that the agent i holds to be plausible with respect to $\langle \mathcal{M}, w \rangle$.

The notions of validity on a frame and of validity in the reference logic are defined similarly as in Section 3.3.

Given a distributed system, according to what we have discussed in the previous section, we may associate a frame $\mathcal{F}$ to it as follows: let $A = \{K_1, K_2, \cdots K_m\}$ be a set of indexed knowledge operators and W be an appropriate set of worlds which describes all possible configurations of the system; we define a relation P_j between worlds as $wP_j w'$ iff the "mental state" of an agent or processor j does not change in the configurations w and w'.

In order to better clarify these points and to illustrate the expressive capacity of this language, even before assuming any axioms, let us consider the following example: let $\mathcal{F}$ be a frame containing three agents a, b and c and three worlds w_1, w_2, w_3, and let A consist of one single atomic formula p, meaning "it is raining in São Paulo". Assume that the following are the plausibility relations: $P_a = \{\langle w_1, w_2 \rangle, \langle w_2, w_2 \rangle\}$, $P_b = \{\langle w_1, w_3 \rangle, \langle w_3, w_3 \rangle\}$ and $P_c = \{\langle w_2, w_3 \rangle, \langle w_1, w_1 \rangle\}$, and assume that p is true in w_1 and w_3, but not in w_2.

We can always represent frames by means of directional multi-labeled graphs, which proves to be especially useful in the case of finite frames. The states of the frame are represented as nodes on the graph, and two nodes w and w' are interconnected by an arrow with a label P_j iff $wP_j w'$. Accordingly, the preceding example is represented by Figure 7.1 below.

CHAPTER 7. EPISTEMIC LOGIC: KNOWLEDGE AND BELIEF

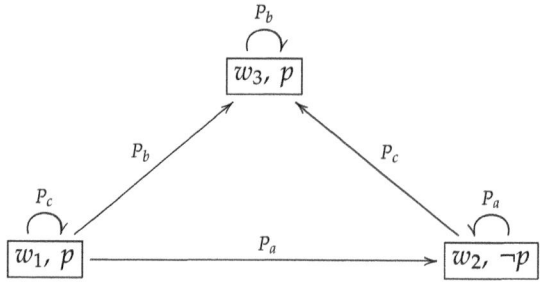

Figure 7.1: A toy example

In order to make the example clearer, let us explore some valid fundamental formulas:

1. $\mathcal{M}, w_1 \vDash p$ [see Figure 7.1]
2. $\mathcal{M}, w_2 \vDash \neg p$ [see Figure 7.1]
3. $\mathcal{M}, w_3 \vDash p$ [see Figure 7.1]
4. $\mathcal{M}, w_1 \vDash K_a \neg p \wedge \neg K_a p$ [from 1 and 2, since $w_1 P_a w_2$ and $\mathcal{M}, w_2 \nvDash p$]
5. $\mathcal{M}, w_1 \vDash K_b p$ [from 3, since $P_b = \{\langle w_1, w_3 \rangle, \langle w_3, w_3 \rangle\}$]
6. $\mathcal{M}, w_1 \vDash K_c p$ [from 1, since $P_c = \{\langle w_2, w_3 \rangle, \langle w_1, w_1 \rangle\}$]
7. $\mathcal{M}, w_2 \vDash K_b \neg p$ [by vacuity]
8. $\mathcal{M}, w_3 \vDash K_a p$ [by vacuity]

Therefore, it becomes clear that at the state w_1:

- According to 1, it is raining in São Paulo but, by 4, the agent a does not know it: $\mathcal{M}, w_1 \vDash p$ and $\mathcal{M}, w_1 \vDash \neg K_a p$.

- On the other hand, by 5, b knows whether it is raining or not, and it is easy to see, by 5 and 7, that a knows that b knows it: $\mathcal{M}, w_1 \vDash (K_b p \vee K_b \neg p)$ and $\mathcal{M}, w_1 \vDash K_a(K_b p \vee K_b \neg p)$.

- So b knows that it is raining in São Paulo, but b does not know that a does not know it: $\mathcal{M}, w_1 \vDash K_b p$ and $\mathcal{M}, w_1 \vDash \neg K_b \neg K_a p$.

- c knows all the following information: $\mathcal{M}, w_1 \vDash K_c K_b p$ and $\mathcal{M}, w_1 \vDash K_c \neg K_b \neg K_a p$ etc.

Through similar analysis, we shall obtain formulas depicting what the whole system knows. Even in toy models like this, we appreciate the fact that, by starting from a small number of sentences (in this case, just one) we may express sophisticated information about the state of knowledge of the agents.

7.3. THE MINIMAL LOGIC OF KNOWLEDGE

It is easy to remark that moving from some basic knowledge (i.e., the knowledge of one single proposition) any agent may know an infinite amount of statements, though this may seem to contradict the fact that, initially, each agent knew *only* one proposition.

It is also quite interesting to observe that, as we have seen, though p is true in w_1, a does not know it:

$$\mathcal{M}, w_1 \vDash p \land \neg K_a p$$

And even if we wished to inform agent a of this fact (s)he could not know it, which is to say:

$$\mathcal{M}, w_1 \nvDash K_a(p \land \neg K_a p)$$

In other words, even if the fact $p \land \neg K_a p$ were consistent with the epistemic world of agent a, it would cease to be so after a had been informed!

This sort of "blocked learning" situation legitimates the introduction of temporal operators together with other modal operators if we want to improve our description of knowledge. Though temporal logic has already been introduced in Chapter 6, this multimodal aspect will be treated in the next chapter.

Obviously, the following properties will hold for any logic which relies on the above semantic definitions, and they will therefore be maintained in any logic which tries to give a formal treatment of knowledge:

Proposition 7.3.1 *jhfgkjfhjk*

(a) *All instances of propositional tautologies are valid in the logic of knowledge, and Modus Ponens preserves this property.*

(b) *For every formula α, β in $\mathcal{K}_O(A)$ and every K_i, the axiom* (KK_i) *is valid.*

(c) *For every formula α in $\mathcal{K}_O(A)$ and every K_i, the rule* (Nec_{K_i}) *preserves validity.*

Proof: Exercise 7.1. ♦

The preceding result means that some basic properties, i.e. (KK_i) and (Nec)$_{K_i}$, must always hold semantically, regardless of the plausibility relations that are involved. Nevertheless, properties (a) and (b) of Proposition 7.3.1 are questionable for they characterize logical omniscience, which is to say, as seen in Section 7.1, that every agent must know all propositional theorems and all logical consequences of what (s)he knows. This is of course too much even for artificial agents.

In the next section, we shall examine the minimal systems of the logic of knowledge.

7.4 The systems K^m, KT^m, $S4^m$ and $S5^m$

The system K^m of minimal logic of knowledge (where m is the number of the knowing agents) consists in adding some principles and rules (as in all systems defined in this book) to **PC**, which we describe in detail for clarity:

(KK$_i$) $(K_i(p \supset q) \wedge K_i p) \supset K_i q$ is a K^m-thesis for each $1 \leq i \leq m$

(MP) If $\vdash_{K^m} \alpha$ and $\vdash_{K^m} \alpha \supset \beta$, then $\vdash_{K^m} \beta$

(Nec$_{Ki}$) If $\vdash_{K^m} \alpha$, then $\vdash_{K^m} K_i \alpha$, for every $1 \leq i \leq m$

The family of axioms (KK$_i$) is a direct generalization of the axiom (K) introduced in Chapter 3. In fact, if $m = 1$ (i.e., if there exists only one agent), K_1 may be identified with the notion of necessity of the system **K**. The notion of derivation is the familiar one and we could, alternatively, as usual, introduce the uniform substitution (US) rule and consider axioms instead of axiom schemas.

The way to prove the completeness of K^m with respect of the class of multi-agent frames is parallel to the completeness proof for $K+G^\infty$ in Chapter 4. Also, completeness for K^m, in turn, is a particular case of the general completeness theorem for standard multi-modal systems which will be proved in next chapter. The notion of a canonical multi-agent frame is defined along the lines already given in Chapter 4, replacing clause (2) for canonical models (see Definition 4.2.6) by the following (2') for each modal operator K_i:

(2') $wP_i w'$ iff $Den_{K_i}(w) \subseteq w'$, where w and w' are maximal consistent extensions of K^m and $Den_{K_i}(w) = \{\alpha : K_i \alpha \in w\}$.

Under the definition already given in Section 1.2 (page 11), we recall the notion of a formula α being consistent with respect to a system **S**.

Proposition 7.4.1 *Let $\mathcal{F}$ be a canonical multi-agent frame: then, for every state w in a canonical model $\mathcal{M}$ based on $\mathcal{F}$, we have that $\mathcal{M}, w \models \alpha$ iff $\alpha \in w$.*

Proof: ($\Leftarrow$) By induction on the complexity of formulas.

(i) If α is atomic, the result follows from the definition of canonical model.

(ii) If α is of the form $\bot$ or $\beta \supset \gamma$, the proof is simple and is left to the reader.

(iii) If α is of the form $K_i\beta$ and $\alpha \in w$, then $\beta \in Den_{K_i}(w)$ and the definition of P_i implies that, if wP_iw', then $\beta \in w'$. By inductive hypothesis, $M, w' \models \beta$ holds for every w' such that wP_iw'. Consequently, $M, w \models K_i\beta$. This completes the first part of the proof.

($\Rightarrow$) Assume that $M, w \models K_i\beta$ (let us consider just the modal case). It follows that $Den_{K_i}(w) \cup \{\neg\beta\}$ is necessarily inconsistent. Suppose, in fact, that it is not inconsistent. In such a case, by reasoning as in Lemma 4.2.11, this set has a maximal consistent extension w', and, consequently, wP_iw' holds. Since $\neg\beta \in w'$, by the first part of the proof $M, w' \not\models \beta$, so $M, w \not\models K_i\beta$, which contradicts the initial hypothesis.

Thus $Den_i(w) \cup \{\neg\beta\}$ is inconsistent, which implies that at least some of its subsets is inconsistent, say: $\{\alpha_1, \alpha_2, \cdots \alpha_r, \neg\beta\}$. Notice that $\neg\beta$ may be included in the aforementioned set without loss of generality. So, the formula $\neg(\alpha_1 \wedge \alpha_2 \wedge \cdots \wedge \alpha_r \wedge \neg\beta)$ in particular is provable, or equivalently, $\vdash \alpha_1 \supset (\alpha_2 \supset (\cdots \supset (\alpha_r \supset \beta) \cdots))$, and, by (**Nec**$_{K_i}$), it follows that $\vdash K_i(\alpha_1 \supset (\alpha_2 \supset (\cdots \supset (\alpha_r \supset \beta) \cdots)))$.

Given that a maximal consistent set is deductively closed, $K_i(\alpha_1 \supset (\alpha_2 \supset (\cdots \supset (\alpha_r \supset \beta) \cdots))) \in w$. As $\alpha_1, \alpha_2, \cdots \alpha_r \in Den_{K_i}(w)$, then $K_i\alpha_1, K_i\alpha_2, \cdots K_i\alpha_r \in w$. By iterating applications of (**MP**), (**KK**$_i$) and finally (**Nec**$_{K_i}$), we obtain $K_i\beta \in w$. ♠

From the above result, one obtains:

Proposition 7.4.2 K^m *is sound and complete with respect to the class of relational multi-agent frames.*

Proof: On the one hand, Proposition 7.3.1 grants the soundness of K^m. On the other hand, if $\not\vdash_{K^m} \alpha$, then $\neg\alpha$ can be included in a maximal consistent set w within the canonical multi-agent frame for K^m; therefore by Proposition 7.4.1, $M, w \not\models \alpha$, where M is the canonical model, that which by contraposition proves completeness. ♠

We may now consider some axiom schemas concerning questions Q_1, Q_2 and Q_3 already mentioned in Section 7.2, more specifically:

(**TK**$_i$) $K_i\alpha \supset \alpha$, for each $1 \leq i \leq m$ Knowledge axiom

(**4K**$_i$) $K_i\alpha \supset K_iK_i\alpha$, for each $1 \leq i \leq m$ Positive introspection axiom

(**5K**$_i$) $\neg K_i\alpha \supset K_i\neg K_i\alpha$, for each $1 \leq i \leq m$ Negative introspection axiom

We may then define the systems:

- **KT**^{m}=**K**^{m}+(**T**K_i) for $1 \leq i \leq m$
- **S4**^{m}=**KT**^{m}+(**4**K_i) for $1 \leq i \leq m$
- **S5**^{m}=**S4**^{m}+(**5**K_i) for $1 \leq i \leq m$

It is clear that (**T**K_i), i.e., $K_i\alpha \supset \alpha$, implies the epistemic counterpart of the deontic axiom (**D**) $K_i\alpha \supset \neg K_i \neg \alpha$, for each $1 \leq i \leq m$ (Exercise 7.2).

Considering adequate plausibility relations, analogue to the accessibility relations of **KT**, **S4** and **S5** studied in previous chapters, we can now show the following:

Proposition 7.4.3 *(i)* **KT**^{m} *is sound and complete with respect to the class of multi-agent frames wherein all plausibility relations are reflexive.*

(ii) **S4**^{m} *is sound and complete with respect to the class of multi-agent frames wherein all plausibility relations are reflexive and transitive.*

(iii) **S5**^{m} *is sound and complete with respect to the class of multi-agent frames wherein all plausibility relations are reflexive, symmetric and transitive.*

Proof: The proofs are analogous to the one given in Chapter 4, with suitable modification as in Propositions 7.3.1 and 7.4.1 (see Exercise 7.3). ♠

Bearing in mind the classical conception of knowledge as justified true belief, axiom (**5**K_i) does not seem totally coherent with our usual view about knowledge (even if it has sometimes been defended by certain philosophers) for it requires that the agents be aware of their own ignorance. There are many other possible ways to tackle this question, but we shall not examine them here.

7.5 Common knowledge and implicit knowledge

The notions of *common knowledge*, *common belief* and *implicit knowledge* are of great interest for the study of the internal logic of a group (of agents or processors, for instance) and suggest an interesting philosophical problem which may explain why the notion of common knowledge was first proposed in philosophical terms (involving the analysis of social conventions).

7.5. COMMON KNOWLEDGE AND IMPLICIT KNOWLEDGE

Given a fact α and a group G, α is *common knowledge* among the agents in the group G iff all members of G know α and all of them know that they all know α, and so forth. In other words, common knowledge turns out to be "what everyone knows" and will be formally defined as follows.

First, let us introduce an operator E, meaning "all the m agents know that", defining it as:

$$E\alpha \stackrel{\text{Def}}{=} K_1\alpha \wedge K_2\alpha \wedge \cdots \wedge K_m\alpha$$

Common knowledge may, thus, be inductively defined in terms of the above operator as the conjunction of all the infinite formulas defined as follows:

$$E^1\alpha \stackrel{\text{Def}}{=} E\alpha$$
$$E^{i+1}\alpha \stackrel{\text{Def}}{=} E(E^i\alpha)$$

In other words, common knowledge is represented by an infinitary formula C:

$$C\alpha \stackrel{\text{Def}}{=} E\alpha \wedge E^2\alpha \wedge \cdots \wedge E^n\alpha \wedge \cdots$$

Notice that the above definitions are relevant only if at least two agents are involved, for should there be one single agent, then $E\alpha \equiv K_1\alpha$ and $E^n\alpha \equiv K_1^n\alpha$.

But just two agents may provide interesting cases of common reasoning. When, for instance, a couple (man and wife) lose each other in a department store without any prior agreement on where to meet, they will reason with coinciding interests, since one wants to find the other. The best strategy for the man, for example, is not to think "What would I do if I were she?" but "What would I do if I were she wondering what she would do if she were wondering what I would do if I were she $\cdots$?"

It is not difficult to see that, for the operator E, the following holds (see Exercise 7.6):

$\mathcal{M}, w \models E\alpha$ iff $\mathcal{M}, w' \models \alpha$ for every w' such that $\langle w, w'\rangle \in P_1 \cup P_2 \cup \cdots \cup P_m$

The definition of a multi-agent frame may, under $C\alpha$, be interpreted as follows: let us say that a state w' is *reachable* from w if there is a sequence $w = w_1, w_2, \ldots, w_n = w'$ such that $w_1 P_{j_1} w_2, w_2 P_{j_2} w_3, \cdots w_{n-1} P_{j_{n-1}} w_n$. Consequently, the following holds:

$\mathcal{M}, w \models C\alpha$ iff $\mathcal{M}, w' \models \alpha$ for every w' reachable from w

Interestingly enough, as shown in the above definition, even if, on the one hand, common knowledge is a syntactically infinitary notion, on the other hand it turns out to be simple from the semantical perspective. Actually, we do not use the infinitary formula $C\alpha$, but rather its finitary approximations $C^k\alpha = E\alpha \wedge E^2\alpha \wedge \cdots \wedge E^r\alpha \wedge \cdots \wedge E^k\alpha$, of which $C\alpha$ is the limiting case.

In spite of its infinitary character, $C\alpha$ may be finitely axiomatized. Let us define the system $\mathbf{CK}^m$ by adding to $\mathbf{K}^m$ just one axiom and the Common Knowledge Rule (**CKr**):

(**C**) $C\alpha \supset E(\alpha \wedge C\alpha)$

(**CKr**) $\vdash \alpha \supset E(\alpha \wedge \beta)$ implies $\vdash \alpha \supset C\beta$

Recalling the definition of E, the notion of C treated axiomatically grants a certain "fixed point solution" approach to common knowledge, in the sense that C is a solution of the equation $X = E(\alpha \wedge X)$. This helps to explain that E and C are two "virtual" modal operators which do not interfere with the conditions on the plausibility relations.

From $\mathbf{CK}^m$, one obtains $\mathbf{CKT}^m$, $\mathbf{CS4}^m$ and $\mathbf{CS5}^m$ by adding the axiom (**C**) and the rule (**CKr**), respectively, to the systems **T**, **S4** and **S5**. The following characterization can be proved (we do not give a proof here, since axiom (**C**) falls outside the main axiom schemas we treat in Chapter 8, but the interested reader can find references in the "Further reading" section of this chapter):

Proposition 7.5.1 *(i)* $\mathbf{CK}^m$ *is sound and complete with respect to the class of multi-agent frames.*

(ii) $\mathbf{CKT}^m$ *is sound and complete with respect to the class of multi-agent frames wherein all plausibility relations are reflexive.*

(iii) $\mathbf{CS4}^m$ *is sound and complete with respect to the class of multi-agent frames wherein all plausibility relations are reflexive and transitive.*

(iv) $\mathbf{CS5}^m$ *is sound and complete with respect to the class of multi-agent frames wherein all plausibility relations are reflexive, symmetric and transitive.*

The "King's puzzle" is an example that clearly illustrates the potentialities of applying the notion of common knowledge. The story is as follows.

Once upon a time in a kingdom, there lived equally intelligent subjects who could not see the colors of their own eyes due to the lack of mirrors in the kingdom. Of course, as they saw each other, each of them knew the colors of the other subjects' eyes. One day, the king began to suspect that all blue-eyed subjects were planning an uprising, and he decided to call his court and order all the blue-eyed citizens to leave his kingdom.

"Some people in this room have blue eyes and must go immediately abroad!" says the King. He keeps repeating his command various times till

7.5. COMMON KNOWLEDGE AND IMPLICIT KNOWLEDGE

the moment in which all blue-eyed people simultaneously stand up and go away. Considering that each of them had previously ignored the colour of his own eyes and that they did not have any mutual interchange of information, how did they reason? (See Exercises 7.7 and 7.8).

In fact, it can be proven that where there are exactly k blue-eyed persons, the common knowledge of the group of k agents, expressed by $C^k\alpha$ (α meaning "some people in this room have blue eyes") suffices to lead people to the conclusion that they also have blue eyes, while $C^{k-1}\alpha$ does not.

If we think common knowledge as corresponding to what "any fool" knows, the notion of *distributed knowledge*, in comparison, would correspond to what a "wise man" knows: distributed knowledge is a kind of meta-knowledge which does not correspond to the knowledge of any individual member, nor to the common knowledge of a group, but to the cooperative knowledge of all members together. As an example, let a and b be two agents. Suppose that a knows α and b knows $\alpha \supset \beta$. Together they may deduce β, while singularly taken, they could not do it; so the knowledge of β is distributed among them.

The extension of the language with a new operator D becomes appropriate to define distributed knowledge. This new operator can be semantically characterized as follows:

$$\mathcal{M}, w \vDash D\alpha \text{ iff } \mathcal{M}, w' \vDash \alpha \text{ for every } w' \text{ such that } \langle w, w' \rangle \in P_1 \cap P_2 \cap \cdots \cap P_m$$

Let us emphasize that distributed knowledge does not correspond to the conjunction of states of knowledge (as in the definition of $E\alpha$), but rather to the articulated knowledge of the agents. Again, where only one agent gets involved, then $D\alpha \equiv K_1\alpha$.

We define the system $\mathbf{DK}^m$ by adding the following axiom schemas to the system $\mathbf{K}^m$:

(**DK1**) $K_i\alpha \supset D\alpha$ for each $1 \leq i \leq m$

(**DK2**) $(D(\alpha \supset \beta) \land D\alpha) \supset D\beta$

Alternatively (**DK2**) can be replaced by the following rule of *distributed knowledge*

(**DKm**) $\vdash_{DKm} (K_1\alpha_1 \land \cdots \land K_m\alpha_m) \supset \beta$ implies $\vdash_{DKm} (K_1\alpha_1 \land \cdots \land K_m\alpha_m) \supset D\beta$

It is possible to prove by propositional arguments that (**DKr**) is inferred from both (**DK2**) and ($\mathbf{K}_m$), and vice versa (See Exercise 7.9).

We also define the systems $\mathbf{DT}^m$, $\mathbf{DS4}^m$ and $\mathbf{DS5}^m$ by adding axioms (**DK1**) and (**DK2**), respectively, to the systems $\mathbf{KT}^m$, $\mathbf{S4}^m$ and $\mathbf{S5}^m$. In a way similar to the case of common knowledge, it is possible to prove what follows (but the proof is not provided here, again because axiom (**DK2**) falls outside the main axiom schemas treated in Chapter 8; references in the "Further reading" of this chapter):

Proposition 7.5.2 *(i) $\mathbf{DK}^m$ is sound and complete with respect to the class of multi-agent frames.*

(ii) $\mathbf{DKT}^m$ is sound and complete with respect to the class of multi-agent frames wherein all plausibility relations are reflexive.

(iii) $\mathbf{DS4}^m$ is sound and complete with respect to the class of multi-agent frames wherein all plausibility relations are reflexive and transitive.

(iv) $\mathbf{DS5}^m$ is sound and complete with respect to to the class of multi-agent frames wherein all plausibility relations are reflexive, symmetric and transitive.

We may also introduce the operator S, meaning "somebody knows"

$$S\alpha \stackrel{\text{Def}}{=} K_1\alpha \vee K_2\alpha \vee \cdots \vee K_m\alpha$$

Thus, all the defined operators are part of a hierarchy with the following property:

$$C\alpha \supset \cdots \supset (C^k\alpha \cdots \supset (E\alpha \supset (S\alpha \supset (D\alpha \supset \alpha)\cdots)))$$

We have already argued that there are cases where $C^k\alpha$, unlike $C^{k+1}\alpha$, is insufficient to lead to any action, as in the case of the King's puzzle. Now, there are also cases in which no $C^k\alpha$ is sufficient and it would be necessary to use $C\alpha$.

7.6 The logic of belief

It important to remark that $K_i\alpha \supset \alpha$ implies the epistemic counterpart of the deontic (**D**): $K_i\alpha \supset \neg K_i\neg\alpha$ (Exercise 7.2). We may read $\neg K_i\neg\alpha$ as "i's knowledge does not exclude α" and introduce a new symbol B_i for "Belief" such that $B_i\alpha \stackrel{\text{Def}}{=} \neg K_i\neg\alpha$.

It cannot be denied that $K_i\alpha \supset \neg K_i\neg\alpha$ is an intuitive epistemic principle, which might accepted in place of the stronger and more controversial (**TK$_i$**). Since we know that $B_i(\alpha \vee \neg\alpha)$ is equivalent to $K_i\alpha \supset B_i\alpha$ (see Exercise 7.10), it is natural to rule out the axiom (**T$_m$**) and to endorse a new axiom schema:

7.6. THE LOGIC OF BELIEF

($\mathbf{B}_i$) $\neg B_i \bot$ for each $1 \le i \le m$

The intuitive meaning of belief axiomatized here is that i believes α iff i has *some* degree of belief in α. However, treating belief as some kind of "not-knowledge" entails, for example, that, if you do not know that Bishkek is *not* the capital of Kyrgyzstan, you must believe it is, even if you never heard about Bishkek (actually, it is the capital, so you are not believing in any contradiction). But this has the unacceptable effect of making you believing compulsorily. So we may take a new postulate for the effect that, for a given agent i, 'the act of believing must be known": $B_i(p) \supset K_i(B_i(p))$. However, this derives $B_i(K_i(p)) \supset K_i(p)$, which seems plainly improper.

As a matter of fact, as anticipated in Section 7.1, it is possible to argue that there are several kinds of belief. For instance, one can distinguish *strong* beliefs from *weak* ones: the former include ideas that cannot be dropped without traumatic consequences, as in "*a* thinks that paradise and hell are the same" or "*a* believes in the existence of God", while the latter are viewpoints that admit possible refutations which the believer may be prepared to accept, as in "*a* thinks that it will rain tomorrow" or "*a* believes in a certain hypothesis". It is also possible to treat hypotheses as special cases of beliefs.

Also, by treating belief as a dual of knowledge based on certain notions of belief may yield counterintuitive consequences. For instance, among the intuitive postulates for the notion of belief in the strong sense, we should admit the notion of a *conjunctive belief*:

($\mathbf{CB}_i$) $(B_i\alpha \wedge B_i\beta) \supset B_i(\alpha \wedge \beta)$

But as a special instance of ($\mathbf{CB}_i$), we would have $(B_i\alpha \wedge B_i\neg\alpha) \supset B_i(\alpha \wedge \neg\alpha)$, so on the ground of the axiom ($\mathbf{B}_i$) above, by contraposition we derive $\neg(B_i\alpha \wedge B_i\neg\alpha)$, i.e., $B_i\alpha \supset \neg B_i\neg\alpha$. But $\neg B_i\neg\alpha$ equals $K_i\alpha$, which means that belief collapses on knowledge. The distributivity of B_i on conjunction suggests that B_i has some properties in common not with the $\Diamond$-operators but with the $\Box$-operators, even if, of course, $B_i\alpha \supset \alpha$ is unacceptable.

By changing the operator K_i to B_i, dropping ($\mathbf{TK}_i$) and adding ($\mathbf{B}_i$) in the system $\mathbf{S5}^m$, we obtain a system which is called $\mathbf{KD45}^m$.

As far as accessibility relations are concerned, the following proposition may be proven (either directly or as special cases of the multimodal systems to be analyzed in the next chapter, or by appeal to the properties of transitivity, seriality and euclideanity):

Proposition 7.6.1 $\mathbf{KD45}^m$ *is sound and complete with respect to the class of multi-agent frames which are euclidean, transitive and serial.*

Proof: The proof is analogous to the ones in Chapter 4 (Exercise 7.11). ♦

In the light of such considerations, there is a wide agreement on the view that doxastic logic has different properties from that of standard epistemic logic, or that we also need a plurality of doxastic logics to mirror the plurality of currently used notions of belief.

Likewise, it has already been stressed that we should take into account the existence of several kinds of knowledge, which leads to the possibility of an essentially multimodal theory of epistemic and doxastic logic, which so far as until now, does not seem to have been worked out. Doxastic logic can also be combined with other systems, yielding numberless multimodal systems which, given certain conditions, can be proven to be complete with respect to classes of multimodal frames. As a matter of fact, we can treat several of these systems and their combinations homogeneously, as we shall see in next chapter.

7.7 Exercises

1. Prove Proposition 7.3.1.

2. (i) It may be expected that for belief we have: $B_i(B_i p \supset p)$. Prove that this is indeed a **KT**^{m}-thesis if B_i is taken as the dual of K_i.
   (ii) Prove that $K_i \alpha \supset \alpha$ implies $K_i \alpha \supset \neg K_i \neg \alpha$ (Hint: recall that (**T**) implies (**D**); see page 30).

3. Prove Proposition 7.4.3.

4. Show that, by accepting the epistemic system **KT**^{m} together with the alethic system **K** for $\Box$, we cannot know whether or not there is some unknown fact, thus going against the Principle of Knowability (**KP**). (Hint: suppose that $K(\alpha \wedge \neg K\alpha)$. Using (**KK**$_i$) and (**TK**$_i$), obtain $K\alpha \wedge \neg K\alpha$, a contradiction, thus $\neg K(\alpha \wedge \neg K\alpha)$ is a **KT**^{m}-thesis. Now apply rule Nec for $\Box$ in **K**).

5. Show, under the same assumptions of the Exercise 7.4, that (**KP**) and (**AO**) cannot be accepted simultaneously. (Hint: consider the following instance of (**KP**): $(\alpha \wedge \neg K\alpha) \supset \Diamond K(\alpha \wedge \neg K\alpha)$. From Exercise 7.4, $\neg K(\alpha \wedge \neg K\alpha)$ is a **KT**^{m}-thesis. Obtain $\alpha \supset K\alpha$, which together with (**TK**$_i$), entails the collapse of the knowledge operator.)

6. For $E = K_1 \alpha \wedge \cdots \wedge K_m \alpha$, show that $\mathcal{M}, w \models E\alpha$ iff $\mathcal{M}, w' \models \alpha$ for every s' such that $\langle s, s' \rangle \in P_1 \cup \cdots \cup P_m$.

7. Show by arithmetical induction that, in the King's puzzle, if there exist k blue-eyed persons, only at the kth time at which the King orders them to leave, can they *prove* that they have blue eyes.

8. Consider the King's puzzle once more: should there be just one blue-eyed person in a room, then the king, by telling this person that someone in the room has blue eyes, would be telling something genuinely new to this person. On the other hand, if the number of blue-eyed individuals is $k > 1$, the King would apparently be telling the group something they already know. Show that the subjects would not be able to reach the desired conclusion if the King had not imparted the mentioned information.

9. Prove that the Distributed Knowledge Rule (**DKr**) is deducible from (**DK2**) and that, the other way around (**DK2**) may also be deduced from the former and other axioms of $\mathbf{K}^m$.

10. Prove that $B_i(\alpha \vee \neg\alpha)$ is equivalent to $K_i\alpha \supset B_i\alpha$ (Hint: recall the distributivity of $\square$ over conjunction; see Proposition 2.3.11).

11. Sketch the main steps of the proof of Proposition 7.6.1.[3]

7.8 Further reading

In Plato's *Theaetetus*, one finds Socrates' thought-provoking challenge together with the basics of epistemic logic derived from the Greek heritage. From the standpoint of modern logic, it was J. Hintikka in [Hin62] who first tackled the issues mentioned in Section 7.2. J. McCarthy and P. Hayes in [MH69] and Hintikka in [Hin63] have found some connections with Artificial Intelligence. Another important reading is W. Lenzen [Len78], which sums up all the literature on epistemic logic after Hintikka in [Hin62] and includes more than four hundred bibliographical references. See also [Len79] and [Len80]. Hintikka in [Hin63] defends the idea that **S5** is the most adequate epistemic system.

F. Fitch proved the theorem stated at page 187 in [Fit63]. More information and references on recent developments can be found in the encyclopedia entry under B. Brogaard and J. Salerno [BS04]. For the intuitionistic attitude towards the problem see especially T. Williamson [Wil92] and chapter 12 of [Wil00].

---
[3]A complete proof can be found in R. Fagin, J. Y. Halpern and M. Y. Vardi [FHV92], Theorem 2.5.

The example of the department store appears in T. Schelling [Sch60]. The concept of common knowledge was first introduced in the philosophical literature by D. Lewis in [Lew69]; Lewis defended the idea that common knowledge is essential for defining the the notion of convention among the members of a group. Common knowledge was also studied in the context of probability in the seminal paper of R. J. Aumann [Aum76], where he proved that agents having the same prior probability distribution over the states of the world cannot agree to disagree.

Epistemic logic has been extensively studied in Computer Science; good references are J. Y. Halpern and Y. Moses [HM85] and [HM92] (where a proof of Proposition 7.5.1 can be found at Section 4, Theorem 4.3) and R. Fagin, J. Y. Halpern and M. Y. Vardi in [FHV92] (where a proof of Proposition 7.5.2 can be found) and [FHMV95]. The last text contains introductory material and sophisticated applications to computation, especially concerning distributed systems. Formal analysis of issues concerning knowledge about knowledge are proposed there in terms of possible-worlds semantics, as in the case of the puzzle of kids with muddy foreheads, which is structurally analogous to the King's puzzle. [FHMV95] also introduces combinations of epistemic and temporal operators, which allows treating time combined with common knowledge in synchronic systems, with important applications to protocols for message exchanging such as the so-called "simultaneous attack puzzle".

For a recent survey about doxastic logic, see J. J. Meyer [Mey03]. The theory of belief revision, which studies the ways in which a rational subject dynamically modifies her/his beliefs, has been formulated by PC. Alchourrón, P. Gärdenfors and D. Makinson (the so-called AGM theory: see [Gär88]). Though, strictly speaking, it does not belong to modal logic, K. Segerberg showed in [Seg95] that it is possible to reconstruct the theory of belief revision within the object language of doxastic logic.

Chapter 8

Multimodal logics

8.1 What are multimodalities?

It is known that, from a philosophical point of view, it is not easy to give a satisfactory definition of modality. Moreover, we do not know whether for every modality which is expressed in natural language there is a modal operator that can represent it: for instance, in Chapter 7 we have treated the modality "to know that" with a certain degree of success from a logical point of view. But can we treat the modality "to know how" from a logical point of view with the same success?

Though the questions above may remain unanswered, we can consider multimodal systems as extensions of propositional logic whose language is equipped with one or more non-truth-functional operators that are interpretable via some sort of semantics, such as relational semantics. Some simple multimodal systems have been examined in Chapters 6 and 7. Modal systems as such are largely employed in the representation of knowledge, belief and other propositional attitudes (intentions, desires and obligations), especially in computer science and also in other fields such as the philosophy of language. One of the main reasons for the interest in multimodal systems rests on the possibility of modeling several scenarios by which an agent may reason (i.e. operate deductions), through which (s)he interacts with other agents, which may produce changes in the scenarios themselves. In specific cases, this involves either the representation of the dynamic aspects of the agents or the reasoning about their actions in time.

In monomodal systems, the operators allow us to treat modal concepts one by one as notions of possibility and necessity in **K**, **KT**, **S4**, **S5**, etc. More complex systems, as the temporal and epistemic logics introduced in Chapters 6 and 7, can treat several subsystems simultaneously and allow them to interact. In the cases where accessibility relations share similar properties, the systems are called *homogeneous*. In comparison, we call *heterogeneous* those systems where each operator has an independent behaviour, which may also be related to the behaviour of other operators (see Section 8.4). The epistemic logic systems examined in Chapter 7, for instance, are cases of homogeneous systems. By contrast, a system which deals jointly with temporal operators and other modalities would be an instance of a heterogeneous system. Of course, it is easy to see that there are infinitely many heterogeneous systems with different levels of complexity. This chapter aims to introduce a general theory of multimodality to cover a wide class of the aforementioned systems (albeit, certainly, not all possible systems, as it will be seen). Therefore, it is necessary to introduce a formal language that allows the use of arbitrary modalities on the one hand and, on the other hand, permits to express general axiom schemas that govern both the modalities and the relations between them.

The interest in this kind of assemblage of modalities has led to the development of proof methods that allow for a comprehensive treatment of a wide class of multimodal systems which is also available for applications. We shall, nevertheless, be herein concerned only with the logical and definitional aspects of multimodal systems, rather than their algorithmic side.

Given the exorbitant varieties of multimodal logics, one of the first difficulties faced is to obtain adequate formulations leading to a systematic treatment of multimodal systems paralleling the case of the systems $G^{k,l,m,n}$ for monomodal logics seen in Chapter 4.

8.2 Multimodal languages

Before we face the task of defining multimodal languages formally, it is useful to emphasize the flexible nature of modal operators, which confers a special philosophical interest to them. Even a single modal operator like $\Box \alpha$ can get, as we have already discussed, different interpretations: we may interpret it alethically, i.e., "α is necessarily true", temporally, i.e., "α will always be true", epistemically (with respect to an agent i), i.e., "i knows that α", or dynamically, i.e., "after the execution of a certain process α will be true". Nothing prevents that such distinct interpretations might be represented within the same system. For example, the epistemic systems treated in Chapter 7 axiomatize an arbitrary number of epistemic notions, with respect to different agents.

8.2. MULTIMODAL LANGUAGES

A *multimodal propositional language* is defined as the standard modal propositional language extended with a (fixed) set Φ_0 of *atomic modal parameters*, generalizing what was done in Section 2.2. There is a major difference with respect to the extant modal systems however: the multimodal systems we treat here deeply generalize all the previously treated ones since they allow a calculus of the modal operators themselves, whereby complex modal operators are defined in terms of atomic parameters. This treatment will be granted by a parametric representation defined below. As we shall see, this representation allows us to simultaneously define and treat a wide class of modal operators starting from atomic operators.

Accordingly, besides the propositional variables contained in *Var* and the symbols $\bot$ and $\supset$, our basic alphabet includes a class Φ_0 of atomic modal parameters $a, b, c \cdots$ intended to behave as indices for modal operators. Two special parameters will be the *null parameter*, denoted by 0, and the *identity parameter* denoted by 1. As expected, the propositional connectives are defined in terms of $\bot$ and $\supset$ as in Section 1.2.

The class Φ of *modal parameters over* Φ_0 is defined from Φ_0 by closure under two *formation operators*, viz., $\cup$ and $\odot$:

1. If $a \in \Phi_0$, then $a \in \Phi$.

2. If $a, b \in \Phi$, then $a \cup b \in \Phi$ and $a \odot b \in \Phi$.

Finally, Θ, the class of *modal operators* indexed by modal parameters, is defined as follows:

- If $a \in \Phi$, then $[a] \in \Theta$

We recall the definition of modal language **ML** in Section 2.3; that definition is generalized here considering a multimodal language as formally defined as a quadruple **MML** $= \langle Var, \bot, \supset, \langle \Phi_0, \odot, \cup \rangle \rangle$, where $\Phi_0, \odot$ and $\cup$ are as above. **ML** is recovered from **MML** by taking Φ_0 as $\{a, 0, 1\}$, where a is the only non-null and non-identity atomic modal parameter. Details are given on page 214.

An intuitive interpretation of this apparatus is provided by regarding the elements of Θ as modal operators labeled by combined agents whose mutual attitudes will be governed by such multimodal operators (determined by appropriate axioms). For instance, we may interpret the multimodal operator $[a \cup b]$ as the "knowledge", the "necessity" or the "obligations" shared by agents a and b. Similarly, $[a \odot b]$ may represent the "knowledge", the "necessity" or the "obligations" of an agent a relative to (or depending upon) b.

The interpretations above, however, are not the only ones. Multimodal operators may, according to an alternative approach, be regarded as labels for processes. The process interpretation provides an elegant formalization of multimodal logics as logics of *serial processes* (in the case of $[a \odot b]$) and of *parallel processes* (in the case of $[a \cup b]$); under this interpretation [1] can be seen as the *identity process*, and [0] as the *halting clause* in a process. In this way, it is possible to ascribe a procedural meaning to multimodal systems, as it appears in the so-called dynamic logic.

The multimodal well formed formulas are defined in the usual manner (cf. Section 2.2), as far as truth-functional operators are concerned, by adjoining the following clause:

- If $\alpha \in WFF$ and $[a] \in \Theta$, then $[a]\alpha \in WFF$

For any modal operator $[a]$, we define $\langle a \rangle$ as follows:

$$\langle a \rangle \alpha \stackrel{\text{Def}}{=} \neg [a] \neg \alpha.$$

The key idea that will be used to define the semantics of multimodal logics is that a relation R_a is associated to every modal parameter a, in such a way that the satisfiability of the modal formulas $[a]\alpha$ and $\langle a \rangle \alpha$ at a world $w \in W$ in a certain model $\mathcal{M}$ is determined by two conditions:

- $\mathcal{M}, w \vDash [a]\alpha$ iff for all $w' \in W$, wR_aw' implies $\mathcal{M}, w' \vDash \alpha$

- $\mathcal{M}, w \vDash \langle a \rangle \alpha$ iff there exists $w' \in W$ such that wR_aw' and $\mathcal{M}, w' \vDash \alpha$

This is a direct generalization of the conditions we introduced for modal formulas in the previous chapters. But, in this way, it is more important that we can associate new binary relations (derived from the initial ones) with complex modal parameters (i.e., those obtained via the formation operations $\cup$ and $\odot$). Consequently, it becomes clear that multimodal logics, as we treat them here, lead to an algebra or calculus of accessibility relations.

In the following, we shall show how the better known systems can be dealt with as special instances of multimodal systems and how the relational calculus helps to represent them.

8.3 The elementary multimodal systems

As it will be shown a little later, the most simple instances of multimodal systems involve the combination of modalities, as in the case of epistemic logics and of deontic logics. More complex systems involve combinations of modal operators of a distinct kind and require bridge axioms governing the interaction between these operators, as illustrated in the cases below.

(a) **Epistemic-doxastic logics**. As seen in Chapter 7, from the perspective of multimodal logic the usual systems of epistemic logic are obtained simply by gluing n "copies" of the usual modal systems.

The usual epistemic and doxastic systems comprise simple multimodal systems requiring intuitively clear bridge principles connecting the respective modal operators. There exist several epistemic-doxastic systems in the literature that combine knowledge (K_i) with belief (B_i), such as the systems characterized by the axioms below:

- $K_i p \supset B_i p$
- $B_i p \supset K_i B_i p$

(b) **Deontic logics**. Monomodal deontic logics have already been mentioned at the end of Section 2.1. In order to treat multimodal deontic logics, it is better to use the symbols O (obligation) and P (permission). The operator F (forbiddance) can be defined as $F\alpha \stackrel{\text{Def}}{=} O\neg\alpha$. One may obtain deontic multimodal systems by assuming a family of operators $O_1, \ldots, O_n$, which may correspond to obligation under different kinds of norms, namely juridical, religious, civil, or other. While it makes sense to require, from a philosophical viewpoint, that $O_i p \supset P_i p$ and $O_i(O_i p \supset p)$ hold for a given i, it is conceivable that an act may be allowed under a certain normative code, but forbidden under another. Thus we may, for instance, formalize the interaction between codes via the following bridge axioms:

1. $\neg(O_i p \land O_j \neg p)$
2. $O_i p \supset O_j p$
3. $O_i O_j p \supset O_i p$

The requirement for compatibility of normative codes is captured by the principle in (1), which means that an act cannot be simultaneously obligatory under one code and forbidden under another. The principle in (2) ranks two normative codes, meaning that, if an act is obligatory under code i, it will continue to be obligatory under j, while the principle in (3) expresses the transitivity of obligations, in the sense that, if code i imposes the obligatoriness of p upon code j, then p is obligatory under i.

(c) **Dynamic logics**. Dynamic logics, which have been introduced to account for the logic of computer programs, may be regarded as privileged systems in the study of multimodality due to the uniform way its operators can be represented and to the application of operations defined upon them.

A typical example of an axiom in dynamic logic is the following, intended to codify the behavior of abstract processes (in computer programs, engineering, linguistics, etc.) where $[a]$ and $[b]$ denote the actions of processes a and b and $[a \odot b]$ denotes the *sequential* action of processes a and b:

$$[a \odot b]p \equiv [a][b]p$$

This axiom says that, if performing a and then b must produce p, then a must result in a situation in which b must result in p. So, for instance, if I cook and eat my dinner, then cooking must bring about a situation in which my dinner can be eaten. This principle will be taken as our "Axiom of serial processes" to be discussed below.

(d) **Temporal logics**. Temporal logics (already seen in Chapter 6) may be looked upon as the most typical and oldest multimodal systems. In a sense, they are essentially multimodal, because it is convenient to reason about time by employing various distinct modal operators.

Some bridge axioms have already been studied in the logical tradition, such as (for the meaning of ◈, ▣, ◇ and □, see Chapter 6):

- ◈$p \supset$ ▣◈p: *Descending linearity* (if p has occurred, then p has always been possible in the past).

- ◇$p \supset$ □◇p: *Ascending linearity* (if p will occur, then p will always be possible in the future).

- □(▣$p \supset$ ◇▣p) $\supset$ (▣$p \supset$ □p): *Continuity* (if the fact that p has always been the case implies that it will occur that p has been the case is a necessary fact, then the fact that p has always been the case implies that p will always be the case).

8.3. THE ELEMENTARY MULTIMODAL SYSTEMS

As completeness and incompleteness of several temporal systems have been studied in Chapter 6, we shall herein present two examples of interaction between temporal, deontic and epistemic logics.

(e) **Deontic temporal logics.** It is interesting to combine the temporal modalities ◇, □, ⧫ and ⊡ with the deontic modalities O and P. An instance is given by a logic that contains the following principles:

- $O\boxdot(\diamondsuit p \supset \neg O\boxdot\neg p)$
- $O\boxdot p \supset O\boxdot Op$

Throughout axioms that allow the interaction between obligation and historical necessity (□), modal axioms that combine O and P with □ and ◇ can be proposed in other systems:

(a) $\quad \Box p \supset Op$
(b) $\quad \Box p \equiv O\Box p$
(c) $\quad \diamondsuit p \equiv O\diamondsuit p$
(d) $\quad (\Box p \vee \Box\neg p) \supset (p \equiv Op)$

Axiom (1) is a principle meaning that whatever is historically necessary is obligatory as well; (2) and (3) mean that obligation is irrelevant or useless if applied to what is historically necessary or possible; and (4) states that obligation has no effect over (does not apply to) what is historically not contingent.

(f) **Epistemic temporal logics.** It is also interesting to combine epistemic operators K_i, belief operators B_i and the temporal operator T_s (see D. Scott's T_s in Section 6.4). Such combinations suggest the following axioms:

1. $B_i T_s p \supset K_i T_s \neg K_i \neg p$
2. $B_i T_s p \supset T_s B_i p$
3. $B_i T_s p \supset B_i T_s B_i p$

As one may notice, the above axioms reflect some kind of strong belief of the kind mentioned in Chapter 7, namely a kind of belief the believer is unwilling to give up. Here are other weaker versions of belief:

4. $B_i p \supset (T_s B_i p \vee T_s K_i \neg p)$
5. $B_i p \supset (T_s B_i p \vee T_s B_i \neg p)$

which may be respectively interpreted by the following (4') and (5'):

4'. If today i believes p, then (s)he will believe so tomorrow unless (s)he will know tomorrow that p is false.

5'. If today i believes p, then (s)he will believe so tomorrow unless (s)he is going to change her/his mind.

Thus, we account for the fact that within a comprehensive outlook of multimodalities it is possible to formulate principles with a specific philosophical interest.

(g) **Other multimodal systems.** Several authors have proposed more or less direct generalizations of the systems **KD**, **KT**, **KB**, **S4** and **S5**. A very plausible proposal consists in taking into account several modal operators $\Box_1, \Box_2, \cdots \Box_n$ and $\Diamond_1, \Diamond_2 \cdots \Diamond_m$ and introducing axiom schemas that generalize those of **KD**, **KT**, **KB**, **S4** and **S5**, using strings of the above mentioned operators. For instance, the following would be generalizations of axiom (**B**):

$$p \supset \Box_1 \Box_3 \cdots \Box_{2r+1} \Diamond_2 \Diamond_4 \cdots \Diamond_{2s} p$$
$$\Diamond_1 \Diamond_3 \cdots \Diamond_{2r+1} \Box_2 \Box_4 \cdots \Box_{2s} p \supset p$$

These principles can be variously instantiated in distinct contexts. For instance, historical necessity could be expressed by means of the first one by $p \supset \boxed{P}\Box\Diamond p$, and $\Diamond K_i p \supset p$ would, from the second one, express a sort of strengthening of the "Knowledge Axiom" (see Chapter 7): "if it is possible that i knows p, then p is true".

The completeness proof for such systems results from an adequate generalization of canonical models, which we shall talk about later on. Another interesting application of multimodalities consists in representing provability properties via modal notions. As already remarked, if the notorious axiom (**GL**) ($\Box(\Box p \supset p) \supset \Box p$) is added to **S4**, a logic is obtained which formalizes provability in Peano arithmetic (see Section 4.4), whereby it is also possible to express the modal version of Gödel's Second Incompleteness Theorem. On the other hand, following C. Smorynski [Smo85] (see Section 8.8), it is possible to introduce a modal operator, say $\Box^d$, which represents provability in relation to some different theory, and so to obtain a multi modal system, the completeness of which is proven via a multi-relational possible worlds-semantics. Such a system may also be axiomatized along the lines we are going to present in the next sections.

8.4 Axioms for multimodal logics

The general approach to multimodal systems which we shall adopt here is a generalization of the axioms belonging to the schema $G^{k,l,m,n}$ (i.e., $\Diamond^k \Box^l p \supset \Box^m \Diamond^n p$) studied in Section 4.2. Our approach includes three classes of multimodal systems of growing complexity:

1. The *basilar systems* $\mathbf{G}^{\langle a,b,c,d \rangle}$, characterized by the bridge axioms $G(a,b,c,d)$

2. The *affirmative systems* $\mathbf{G}^{\langle a,b,\varphi \rangle}$, containing the so-called $G(a,b,\varphi)$ axioms as bridge axioms, which are generalizations of the preceding ones

3. The *Catach-Sahlqvist systems* $\mathbf{G}^{\langle \varphi,\psi \rangle}$, which contain an even wider class of bridge axioms, namely the *Catach-Sahlqvist's axioms* $G(\varphi, \psi)$

Definition 8.4.1 *A multimodal system S^Φ based upon a collection Φ of modal parameters is a collection of (multi)modal wffs containing all the tautologies of **PC** and closed under the classical propositional rules (**MP**) (Modus Ponens) and (**US**) (Uniform Substitution).*

When there is no risk of misunderstanding, we may drop the reference to Φ and simply write **S** instead of $\mathbf{S}^\Phi$. We suppose that all multimodal systems are governed by the following basic axioms:

Definition 8.4.2 *Let S^Φ be a multimodal system; then S^Φ is said to be:*

(a) *Normal if it satisfies, for each atomic modal parameter $a \in \Phi_0$, variables p and q and wff α, the **Normality Axiom** ($K_{[a]}$):*

- $[a](p \supset q) \supset ([a]p \supset [a]q)$

*and the **Necessitation Rule** ($Nec_{[a]}$):*

- *If $\vdash \alpha$, then $\vdash [a]\alpha$*

(b) *Standard if it satisfies, for every $a, b \in \Phi$ and variable p, the **Axioms for Multimodal Operators**:*

(**MM1**)	$[a \cup b]p \equiv [a]p \wedge [b]p$	(Axiom of parallel processes)
(**MM2**)	$[0]p \equiv \top$	(Halting axiom)
(**MM3**)	$[a \odot b]p \equiv [a][b]p$	(Axiom of serial processes)
(**MM4**)	$[1]p \equiv p$	(Axiom of neutral process)

Here we deal with standard normal multimodal logics only. The particular case of dynamic logic provides an intuitively clear interpretation for the standard multimodal operators:

1. The parallel processes axiom $[a \cup b]p \equiv [a]p \wedge [b]p$ may be understood as "p will be true after $a \cup b$ has been carried out iff p will be true after a has been carried out and p will be true after b has been carried out".

2. The serial processes axiom $[a \odot b]p \equiv [a][b]p$ may be understood as "p will be true after a has been carried out followed by the execution of b iff after a has been carried out, it will be true that, after b is carried out, p will be true".

The great advantage of this parametric notation with regard to standard multimodal systems is that complex combinations of multimodal operators can be compressed into a single operator by means of the algebraic operations $\cup$ and $\odot$[1]: for instance, $[a]([b]p \wedge p)$ can be represented as $[a \odot (b \cup 1)]p$.

Basilar systems $G^{\langle a,b,c,d \rangle}$

A normal and standard multimodal system $\mathbf{S}^\Phi$ is *basilar* if all its multimodal axioms are *bridge axioms* of the form $G(a,b,c,d) \stackrel{\text{Def}}{=} \langle a \rangle [b] p \supset [c] \langle d \rangle p$ for a, b, c, d arbitrary parameters in Φ.

It is interesting to notice that the bridge axioms satisfy the equivalence below:

$$G(a, b, c, d) \equiv G(c, d, a, b)$$

which is easily obtained by contraposition from the definition of $\langle a \rangle$.

In order to make ideas clearer, we show that the schema $G^{k,l,m,n}$ studied in Section 4.2 is a special instance of $G(a, b, c, d)$, remarking that in the former formula the parameters are natural numbers, while in the latter they are formal modal parameters. Indeed, it is sufficient to take an atomic parameter a, in addition to 0 and 1, and to interpret $[a]$ as the usual modal operator $\Box$; then $[a \odot \cdots \odot a]p = \Box^n p$ just by iterating the operator $\odot$ n times.[2] In this way, $G^{k,l,m,n}$ is obtained in terms of a and $\odot$. Thus all systems axiomatized by instances of $G^{k,l,m,n}$ are really particular cases of the basilar systems.

Now, it may happen that for two distinct modal parameters a and b one may prove, in a given system $\mathbf{S}$, the equivalence $\vdash [a]p \equiv [b]p$, in which

---

[1] The multimodal operators closed under $\cup$ and $\odot$ form a Kleene algebra, a fundamental structure in computer science which generalizes Boolean algebras (cf. J. H. Conway [Con71]).

[2] Actually, this amounts to writing the natural numbers in unary notation.

8.4. AXIOMS FOR MULTIMODAL LOGICS

case one of them would be redundant. To simplify matters and get rid of redundancy, we may define an equivalence relation $\simeq$ on the collection Φ by stipulating that $a \simeq b$ iff $\vdash [a]p \equiv [b]p$, and replacing Φ with its quotient $\Phi/\simeq$. We suppose this reduction has been performed when referring to Φ.

Even if the Normality Axiom and the Necessitation Rule are introduced for atomic modal parameters, it is easy to prove that such properties are extensible to all modal parameters (hence to all multimodal operators). It is, firstly, convenient to observe that the notion of deducibility for multimodal logics will be analogous to the one given for monomodal logic and, in particular, that the Deduction Theorem (recall Definition 2.3.6) will also hold in our multimodal setting: for the atomic modal parameters in Φ_0 the proof is basically the same as for monomodal logics, and for general modal operators in Φ we just use the axioms for multimodal operators.

Proposition 8.4.3 *For every multimodal system S and for every c in Φ the following holds:*

(i) $[c](p \supset q) \supset ([c]p \supset [c]q)$
(ii) If $\vdash \alpha$, then $\vdash_S [c]\alpha$

Proof:

(i) By induction on the complexity of the multimodal operators:

- If c is atomic, the result holds by definition.
- If c is $a \cup b$:
    1. $[a \cup b](p \supset q)$ [Hyp.]
    2. $[a \cup b](p \supset q) \equiv [a](p \supset q) \wedge [b](p \supset q)$ [(MM1)]
    3. $[a](p \supset q) \wedge [b](p \supset q)$ [(MP) in 1, 2]
    4. $[a](p \supset q)$ [PC in 3]
    5. $[b](p \supset q)$ [PC in 3]
    6. $[a]p \supset [a]q$ [Ind. Hyp in 4]
    7. $[b]p \supset [b]q$ [Ind. Hyp in 5]
    8. $([a]p \wedge [b]p) \supset [a]p$ [PC]
    9. $([a]p \wedge [b]p) \supset [a]q$ [PC 8 and 6]
    10. $([a]p \wedge [b])p \supset [b]p$ [PC]
    11. $([a]p \wedge [b]p) \supset [b]q$ [PC 10 and 7]
    12. $([a]p \wedge [b]p) \supset ([a]q \wedge [b]q)$ [PC 9 and 11]
    13. $[a \cup b]p \supset [a \cup b]q$ [(MM1) and PC in 12]

- If c is $a \odot b$, the argument is analogous.

(ii) The case of the Necessitation Rule, i.e., $\vdash_S \alpha$ implies $\vdash_S [c]\alpha$, is similar.

♠

Proposition 8.4.4 *The axioms for multimodal operators (in Definition 8.4.2) are equivalent to the following:*

(i) $\langle a \cup b \rangle p \equiv \langle a \rangle p \vee \langle b \rangle p$

(ii) $\neg \langle 0 \rangle p \equiv \top$

(iii) $\langle a \odot b \rangle p \equiv \langle a \rangle \langle b \rangle p$

(iv) $\langle 1 \rangle p \equiv p$

Proof: Directly from the axioms for multimodal operators and the definition of $\langle a \rangle p$ as $\neg [a] \neg p$. ♠

The following are some examples of particular basilar systems in terms of the axioms $G(a, b, c, d)$ which cover several interaction axioms. If we assume two distinct modal operators $\Box_1 = [a]$ and $\Box_2 = [b]$, then we get:

$(\mathbf{K}_{1,2})$:	$\Box_2 p \supset \Box_1 p$	i.e. $G(1, b, a, 1)$
$(\mathbf{D}_{1,2})$:	$\Box_2 p \supset \Diamond_1 p$	i.e. $G(1, b, 1, a)$
$(\mathbf{B}_{1,2})$:	$p \supset \Box_1 \Diamond_2 p$	i.e. $G(1, 1, a, b)$
$(4_{1,2})$:	$\Box_1 p \supset \Box_2 \Box_1 p$	i.e. $G(1, a, (b \odot a), 1)$
$(5_{1,2})$:	$\Diamond_1 p \supset \Box_2 \Diamond_1 p$	i.e., $G(a, 1, b, a)$
$(\mathbf{SC}_{1,2})$:	$\Box_2 \Box_1 p \supset \Box_1 \Box_2 p$	i.e. $G(1, (b \odot a), (a \odot b), 1)$
$(\mathbf{I}_{1,2})$:	$\Box_2 p \supset (\Box_1 p \supset \Box_1 \Box_2 p)$	i.e. $G(1, (b \cup a), (a \odot b), 1)$

Despite the fact that the $G(a, b, c, d)$ systems are reasonably general, they are unable to account for axiom schemas like the following, which is a generalization of the known McKinsey axiom (see page 67):

1. $[a]\langle b \rangle p \supset \langle c \rangle [d] p$

Several extensions of the $G(a, b, c, d)$ axioms are possible. In the following section we shall approach one of the most useful of these generalizations.

8.4. AXIOMS FOR MULTIMODAL LOGICS

Affirmative systems $G^{\langle a,b,\varphi \rangle}$

The notion of affirmative and negative occurrences of a variable p in a formula is inductively defined as follows:

1. p has no occurrence neither in $\bot$ nor in any other variables distinct from p.

2. The occurrence of p in p is affirmative.

3. If the occurrence of p is affirmative in α, then it is affirmative in $\beta \supset \alpha$, negative in $\neg \alpha$[3] and negative in $\alpha \supset \beta$ (on the other hand, if the occurrence of p is negative in α, then it is negative in $\beta \supset \alpha$, affirmative in $\neg \alpha$ and affirmative in $\alpha \supset \beta$).

4. If the occurrence of p is affirmative in α, then it is affirmative in $Q\alpha$, where Q is any modal operator (on the other hand, if the occurrence of p is negative in α, then it is negative in $Q\alpha$, where Q is any modal operator).

For example, if p and q are atomic, the occurrence of q is affirmative and of p is negative in $p \supset q$. Similarly, the first occurrence of p is affirmative and the second one is negative in $p \supset p$. In contrast, both occurrences of p and q are affirmative in $p \wedge q$ and in $p \vee q$ (just recall that $\alpha \wedge \beta \stackrel{\text{Def}}{=} \neg(\alpha \supset \neg \beta)$ and $\alpha \vee \beta \stackrel{\text{Def}}{=} \neg \alpha \supset \beta$).

A formula α is *affirmative (negative) in the variable* p if all the occurrences of p in α are affirmative (negative). If all the occurrences of its propositional variables in a formula are affirmative (negative), the formula as a whole is said to be *affirmative (negative)*. For instance, $p \wedge q$, $p \vee q$ and $\langle a \rangle p$ are affirmative formulas, just like the formulas that begin with sequences of modal operators such as $Q_1, Q_2 \cdots Q_n$, every Q_i being either $[a_i]$ or $\langle a_i \rangle$.

It is easy to see that, if α is affirmative (negative), then $\neg \alpha$ is negative (affirmative) and that, if α is negative (affirmative) and β is affirmative, then $\alpha \supset \beta$ is affirmative (negative). Consequently, if α and β are both affirmative (both negative), then $\alpha \wedge \beta$ and $\alpha \vee \beta$ are affirmative (negative). Also, if α is affirmative (negative) then so is $[a]\alpha$ or $\langle a \rangle \alpha$ for any atomic modal parameter a.

It can be easily seen (cf. Exercise 8.7) that a wff α is affirmative iff α is obtained from propositional variables, $\bot$ and $\top$ under $\wedge$, $\vee$, $[a]$ and $\langle a \rangle$ for any atomic modal parameter a.[4]

---
[3] The clause for negation is, strictly speaking, unnecessary since $\neg \alpha$ is by definition $\alpha \supset \bot$.
[4] Note that, by definition, $\bot$, and hence $\top$, are not negative; it is technically convenient to consider them as affirmative.

A much wider extension of the foregoing class may be axiomatized by means of axioms of the following form:

$G(a, b, \varphi): \quad \langle a \rangle [b] p \supset \varphi$

where a, b are arbitrary modal parameters and φ is an affirmative formula which has p as its only propositional variable.

Of course, $G(a, b, \varphi)$ generalizes $G(a, b, c, d)$ just by taking $\varphi = \langle c \rangle [d] p$ (which is an affirmative formula, as seen before).

Some properties of certain notable cases of the schema $G(a, b, \varphi)$ are the following:

- $G(0, b, \varphi): \langle 0 \rangle [b] p \supset \varphi$ is valid, as $\langle 0 \rangle p$ is a contradiction (cf. Proposition 8.4.4).

- $G(a, b, \top): \langle a \rangle [b] p \supset \top$ is a tautology.

- $G(1, 0, \varphi)$ is equivalent to φ and, thus, every multimodal affirmative formula is itself an axiom of the form $G(a, b, \varphi)$, as for instance $\langle a \rangle [b] p \wedge q$ and $[b]\langle a \rangle p \vee q$.

The wff $\langle a \rangle [b] \neg p \supset \varphi(\neg p)$ is an instance of $G(a, b, \varphi)$ (recall that φ is an affirmative formula which has p as its only propositional variable). By contraposition, this wff is equivalent to $\neg \varphi(\neg p) \supset [a]\langle b \rangle p$. Because φ is affirmative, $\neg \varphi(\neg p)$ (called the *dual* of φ) is also affirmative. Hence $G(a, b, \varphi)$ also covers wffs of form $\varphi \supset [a]\langle b \rangle p$ (which we call *dual axiom*), where φ is affirmative.

Some interesting classes of axioms covered by $G(a, b, \varphi)$ are the following:

(i) $\langle a \rangle [b] p \supset \langle c \rangle [d] p$ (since $\langle c \rangle [d] p$ is affirmative).

(ii) $[a]\langle b \rangle p \supset [c]\langle d \rangle p$ (dual of (i)).

(iii) $\langle a \rangle [b] p \supset Q_1 \cdots Q_n p$, where each Q_i is either $[a_i]$ or $\langle a_i \rangle$ (since $Q_1 \cdots Q_n p$ is affirmative).

(iv) $Q_1 \cdots Q_n p \supset [a]\langle b \rangle p$, where each Q_i is either $[a_i]$ or $\langle a_i \rangle$ (dual of (iii)).

(v) $\langle a \rangle [b] p \supset (\langle c_1 \rangle [d_1] p \vee \cdots \vee [d_n]\langle c_n \rangle)$ (since $\langle c_1 \rangle [d_1] p \vee \cdots [d_n]\langle c_n \rangle$ is affirmative).

Item (ii) above generalizes axioms of the form $\Box^k \Diamond^l p \supset \Box^m \Diamond^n p$ studied in the literature but not expressible in terms of $G(a, b, c, d)$. Interesting instances of $G(a, b, \varphi)$ of this kind are some epistemic axioms involving mixed knowledge of agents, as in the case of the so-called "mutual negative introspection axiom": $K_i \neg K_j p \supset K_i p$, meaning "if a_i knows that a_j does not know p, then a_i knows p". Indeed, this axiom is equivalent to $\neg K_i \neg K_j p \vee K_i p$, which is an affirmative formula of the form $\langle i \rangle [j] p \vee [i] p$.

Item (v) covers, for instance, the axiom $\Diamond p \supset (p \vee \Box \Diamond p)$ and its equivalents $p \vee (\Diamond p \supset \Box \Diamond p)$ and $p \supset (\Diamond \Box p \supset \Box p)$. Since the system **S4.4** is known to be characterized[5] by adding the latter axiom to **S4**, **S4.4** is then an affirmative system.

The stated properties show that the $G(a, b, \varphi)$ axioms are much more comprehensive than the $G(a, b, c, d)$ axioms. Their generality notwithstanding, they do not cover some cases such as:

(vi) $[a]\langle b \rangle \alpha \supset \langle c \rangle [d] \alpha$, for arbitrary α (non necessarily affirmative), which generalizes McKinsey axiom (see Section 5.1) and does not correspond to any first-order property.

(vii) $[a]([a]p \vee p) \supset \langle a \rangle ([a]p \wedge p)$.

(viii) The induction axiom $[b](p \supset [a]p) \supset ([a]p \supset [b]p)$, which equally do not correspond to any first-order property.

Catach-Sahlqvist systems $G^{\langle \varphi, \psi \rangle}$

Now the $G(a, b, \varphi)$ axioms may be generalized via an even broader class of the so-called *(multimodal) Catach-Sahlqvist axioms* $G(\varphi, \psi)$, which provide a generalization of the known Sahlqvist's monomodal axioms mentioned at the end of Section 4.2, defined as:

$$G(\varphi, \psi) : [a]^n (\varphi \supset \psi),$$

where a is a non-null modal parameter whatsoever, φ and ψ are affirmative and φ is such that it satisfies certain conditions (particularly satisfied by $\langle a \rangle [b] p$). Thus, $[c](\langle a \rangle [b] p \supset \psi)$ is a specific instance of the Catach-Sahlqvist axioms, where c is a non-null arbitrary parameter. In this way, we see that $G(\varphi, \psi)$ generalizes $G(a, b, \psi)$, for $[1](\langle a \rangle [b] p \supset \psi)$ is equivalent to $\langle a \rangle [b] p \supset \psi$.

Although the axioms above are pretty general, yet they do not allow us to express the cases (vi) and (vii) mentioned above, and we shall not study them here. The reasons for focusing on $G(a, b, \varphi)$ instead are the following: firstly $G(a, b, \varphi)$ axioms suffice to cover a wide portion of the multimodal systems of philosophical interest and of relevance in computer science and

---
[5]See [Che80], page 146.

information theory. In fact, until the present time there is not a systematic survey of the philosophical or linguistic issues that may be approached via such axioms. Secondly, the completeness of the affirmative systems $G(a, b, \varphi)$ is a very interesting and instructive issue, inasmuch as the relational equation method is employed. We shall show in detail in Section 8.6 that, if a $G(a, b, \varphi)$ formula of the form $\langle a \rangle [b] p \supset \varphi$ is an axiom of a multimodal system **S**, then its canonical multi-model satisfies the equation $R_a \subseteq F^\varphi R_b^{-1}$, where F is an operation that depends on φ defined over the set of binary relations.

The completeness of the Catach-Sahlqvist's multimodal systems can be proved without great difficulty by extending the completeness proof of the Sahlqvist's monomodal systems, but we shall not deal with this problem.

8.5 Multimodal systems and strict implication

An interesting question concerning multimodalities is to what extent we can overlap several multimodal systems without loosing their specificity. This issue matters because, if multimodal combinations do not collapse, we can, for instance, recover singular modalities that occur in the multimodal systems. Such is the case of the multimodal systems whose fundamental modalities are *separable*, which is to say that the modalities are mutually independent. In this way, it is possible to control their interaction by adequately combining the axioms by means of adequate bridge principles.

Let **S** be a multimodal system and Ax its collection of axioms and rules, and suppose that $\Box \in \Theta$ and $L(\Box)$ is a sublanguage of the language of **S**, whose modal symbols are limited to the simple $\Box$. Thus, we define:

- $Th(\Box) \stackrel{\text{Def}}{=} \{\alpha \in L(\Box) : \alpha \text{ is a theorem of } \mathbf{S}\}$

- $Ax(\Box) \stackrel{\text{Def}}{=} \{\alpha \in L(\Box) : \alpha \in Ax\}$

An axiomatization (in $\Box \in \Theta$) for **S** is *separable* if $Th(\Box) = \{\alpha : Ax(\Box) \vdash \alpha\}$.

In other words, the class of theorems containing $\Box$ does not change if the axioms governing $\Box$ are extended with other axioms. This property can be of course extended to the case of several modal operators.

Example 8.5.1 *The multimodal system below is a case of non-separable axiomatization (where a and b are distinct modal parameter):*

1. $[a](p \supset q) \supset ([a]p \supset [a]q)$

2. $[a]p \supset p$

8.5. MULTIMODAL SYSTEMS AND STRICT IMPLICATION

3. $[b](p \supset q) \supset ([b]p \supset [b]q)$

4. $[b]p \supset [a]p$

Notice that $[b]\alpha \supset \alpha$ is a theorem of the system above which is in $L([b])$, but it is not derivable by $Ax([b])$, since $Ax([b])$ contains only axiom 3.

A separable multimodal system **S** is *simple* if its axiomatization consists of the juxtaposition of the axiomatization of $Ax(\square)$ for any subsystem $L(\square)$ and *homogeneous* if, additionally, the modalities are of a similar kind.

Example 8.5.2 *The epistemic (knowledge and belief) logic systems approached in Chapter 7 are all homogeneous. Of course, elementary monomodal systems are homogeneous too, including the 15 normal modal systems studied in Chapter 3 (**KD**, **KT**, **KT4**, **KTB4=S5**, **KD45**, etc.).*

When the axiomatization of a separable multimodal system **S** includes axioms that mix different multimodal operators, such a system is said to be *heterogeneous*. The following are some typical cases of bridge axioms of type $G(a, b, c, d)$:

Example 8.5.3 *Some examples of axioms (particular cases of $G(a, b, c, d)$) which characterize heterogeneous systems and receive suggestive names in the literature are the following:*

1. $[a]p \supset [b]p$ *Inclusion*

2. $p \supset [a]\langle b\rangle p$ *Semi-adjunction*

3. $[a][b]p \supset [b][a]p$ *Semi-commutativity*

4. $[a]p \equiv [b][c]p$ *Composition*

Separable systems are interesting because they make it possible to control the role of the different modalities that make up a multimodal system. Yet, on many occasions, a non-separable axiomatization turns out to be more economic, for it allows a more sinthetic representation of the interactions between the modalities. Though it is possible to prove that the schemas $G(a, b, c, d)$ and $G(a, b, \varphi)$ grant separable systems, this does not preclude the possibility of there being other equivalent axiomatizations available, which may be more succinct for those systems.

Another interesting aspect of multimodal logics is that they allow for generalization of strict implication. In fact, it is well-known that Lewis has introduced modal logic with the aim of studying the aforesaid strict implication, which has already been defined as $\alpha \prec \beta \stackrel{\text{Def}}{=} \Box(\alpha \supset \beta)$ (see Chapter 2).

By generalizing this idea within a multimodal approach, strict implication operators $\prec_a$ may be defined for any modal operator of the type $[a]$ as:

$$\alpha \prec_a \beta \stackrel{\text{Def}}{=} [a](\alpha \supset \beta)$$

On the other hand, the modal operator $[a]$ is also definable from $\prec_a$, as in the definition below:

$$[a]\alpha \stackrel{\text{Def}}{=} \top \prec_a \alpha$$

Some equivalences can be easily proven in any multimodal system (see Exercise 8.1), viz.:

1. $\alpha \prec_a \beta \equiv \neg\langle a\rangle(\alpha \wedge \neg\beta)$
2. $\alpha \prec_a (\beta \wedge \gamma) \equiv (\alpha \prec_a \beta) \wedge (\alpha \prec_a \gamma)$
3. $\alpha \prec_1 \beta \equiv [1]\alpha \supset \beta \equiv \alpha \supset \beta$
4. $\alpha \prec_{a \cup b} \beta \equiv (\alpha \prec_a \beta) \wedge (\alpha \prec_b \beta)$
5. $\alpha \prec_{a \odot b} \beta \equiv \top \prec_a (\alpha \prec_b \beta)$

We may, thus, alternatively consider the multimodal logics as systems that axiomatize families of strict implication operators, going back to the original motivation of the founding fathers of modal logic.

8.6 Multimodal models and completeness

At this point, two facts should be intuitively clear: firstly, that multimodal systems should contain several accessibility relations and, secondly, that the accessibility relations should be combined to make up new relations. We now turn to the ideas outlined at the end of Section 8.2. Consequently, operations over accessibility relations are naturally taken into account as well: it suffices to define certain fundamental operations and some special relations, as we shall show.

8.6. MULTIMODAL MODELS AND COMPLETENESS

Considering W a set of worlds (sometimes called *indices* in the multimodal context), some distinguished relations in the cartesian product $W \times W$ are the following:

(i) $0 = \emptyset$ \hfill The empty relation

(ii) $1 = W \times W$ \hfill The universal relation

(iii) $Id = \{\langle w, w \rangle : w \in W\}$ \hfill The identity or diagonal relation

Additionally, the following are important *operations over relations*:

1. $R \cup S = \{\langle w, w' \rangle : wRw' \lor wSw'\}$ \hfill Union

2. $R \cap S = \{\langle w, w' \rangle : wRw' \land wSw'\}$ \hfill Intersection

3. $R \odot S = \{\langle w, w' \rangle : \exists w''(wRw'' \land w''Sw')\}$ \hfill Relative product[6]

4. $R \oplus S = \{\langle w, w' \rangle : \forall w''(wRw'' \lor w''Sw')\}$ \hfill Relative addition

5. $\overline{R} = \{\langle w, w' \rangle : \langle w, w' \rangle \notin R\}$ \hfill Complementation

6. $R^{-1} = \{\langle w, w' \rangle : \langle w', w \rangle \in R\}$ \hfill Inverse

7. $R \Rightarrow S = \{\langle w, w' \rangle : \forall w''(wRw'' \supset w''Sw')\}$ \hfill Relative implication

8. $R \triangleright S = \{\langle w, w' \rangle : wRw' \land \exists w''(w''Sw)\}$ \hfill Restriction

Two additional operations are the *right residuation* S/R and *left residuation* $R \backslash S$ of two given relations R and S, introduced as solutions to the relational equations $X \odot R = S$ and $R \odot X = S$. Since such solutions do not necessarily exist in all cases, the best approximation is to consider the greatest element of each set $\{X : X \odot R \subseteq S\}$ and $\{X : R \odot X \subseteq S\}$. As such elements always exist and are well-defined, we may form the new operations S/R (read as S *over* R) and $R \backslash S$ (read as R *within* S) and characterized by the following equivalences, for any relations R, S and T:

$T \odot R \subseteq S$ iff $T \subseteq S/R$
$R \odot T \subseteq S$ iff $T \subseteq R \backslash S$

The above operations enable us to operate on accessibility relations and will play a crucial role in the completeness proof for the systems. To be more specific, relative implication and restriction are essential in the

---
[6] Also called "composition" and sometimes denoted by $R | S$ in the literature.

proof of completeness of the affirmative systems, which are axiomatized by instances of the schema $G(a, b, \varphi)$.

All the operations are definable from $\odot$ plus the usual set-theoretical operations $\cap$, $\cup$ and $^-$ (Exercise 8.2), and several operations are interdefinable, as proposed in the Exercises 8.2–8.6.

Although a single proof of completeness encompassing both the basilar systems and the affirmative systems can be given at once (since the completeness of the basilar systems turns out to be a particular case of the one for the affirmative systems), we deem it convenient to sketch a separate proof of the former systems as a preparatory step for the latter.

Such a proof for the basilar systems is an almost straight generalization of the proof for the systems $G^{k,l,m,n}$ treated in Section 4.2, but it makes use of an operation ρ that associates modal parameters with operations over relations. A proof for the schema $G(a, b, \varphi)$ is more sophisticated and uses an operation F^φ of a similar kind, but it now associates affirmative formulas with operations over relations.

The operation ρ may be derived from the operation F^φ, as shown below, and this fact helps to make clear how the completeness proof of the basilar systems can be seen as a particular case of the completeness proof for the affirmative systems.

A *multi-relational frame* (or *multi-frame*) is a pair $\mathcal{F} = \langle W, \Omega \rangle$, where W is a set of worlds and Ω is a set of binary relations over W.

Definition 8.6.1 *Let S^Φ be a basilar system; $\mathcal{F} = \langle W, \Omega \rangle$ is a multi-frame for S^Φ if there exists a function $\rho : \Phi \longrightarrow \Omega$ associating a relation R_a to every modal parameter a in Φ (where Φ are the modal parameters over Φ_0) satisfying the following conditions:*

1. $\rho(1) = Id$ (i.e. $R_1 = Id$)   Identity relation

2. $\rho(a \cup b) = \rho(a) \cup \rho(b)$ (i.e. $R_{a \cup b} = R_a \cup R_b$)   Union

3. $\rho(a \odot b) = \rho(a) \odot \rho(b)$ (i.e. $R_{a \odot b} = R_a \odot R_b$)   Composition

4. $\rho(0) = 0$ (i.e. $R_0 = \emptyset$)   Empty relation

To better clarify the issue, notice that the symbols $\cup$ and $\odot$ are being used with two distinct meanings: while $a \cup b$ and $a \odot b$ denote operations over the modal parameters, $\rho(a) \cup \rho(b)$ and $\rho(a) \odot \rho(b)$ denote operations over the relations (union and composition of relations, respectively), as defined above. The algebraically inclined reader will notice that the function $\rho : \Phi \longrightarrow \Omega$

8.6. MULTIMODAL MODELS AND COMPLETENESS

is just a homomorphism between the algebraic structures defined by the modal parameters and the relations, respectively.

Remark 8.6.2 *To be more precise, ρ is a homomorphism between the Kleene algebras (see footnote on page 214) defined by the modal parameters Φ with 0 and 1 closed under $\cup$ and $\odot$, and the relations Ω with $\varnothing$ and Id closed under union $\cup$ and composition $\odot$. The role of Kleene algebras is crucial to understand the issue of algebraic semantics for multimodal logics, as done in Section 5.2 for monomodal logics (in fact, an analogous of Proposition 5.2.4 can be proved for multimodal logics, but we do deal in detail with this topic here).*

The maximal consistent sets of formulas are defined just as in the monomodal case, for obviously the notion of maximal consistency is independent from the logical system to which it applies. Likewise, their properties are analogous to the properties provable in the monomodal case (cf. Lemma 4.2.10):

Proposition 8.6.3 *Every consistent set of multimodal formulas w can be extended to a maximal consistent set w'.*

Proof: Analogous to the monomodal case (see Exercise 8.8). ♠

From a multi-frame $\mathcal{F} = \langle W, \Omega \rangle$, one obtains a *multi-model* $\mathcal{M} = \langle W, \Omega, v \rangle$ by adding a valuation assignment v to the multi-frame $\mathcal{F}$ in the same way as in the cases studied in previous chapters.

Definition 8.6.4 *The canonical multi-model for a basilar system S^Φ is a triple $\mathcal{M} = \langle W_S, \Omega_S, v_S \rangle$, where:*

1. *W_S is the class of maximal consistent extensions of S*

2. *The relations $R_{S_a} \in \Omega_S$ are defined, for every $a \in \Phi$ as:*

   $$wR_{S_a}w' \text{ iff } Den_a(w) \subseteq w', \text{ where } Den_a(w) = \{\alpha : [a]\alpha \in w\}$$

3. *For every atomic variable p,*

   $$v_S(p, w) = \begin{cases} 1 & \text{if } p \in w \\ 0 & \text{if } p \notin w \end{cases}$$

Notice that the second condition amounts to a generalization of Definition 4.2.9. Also, $Den_a(w)$ can be seen as a notion of multi-modal "denecessitation", to which corresponds an equivalent notion of "possibilitation" as in Corollary 4.2.14 (see Exercise 8.9).

The notion of *truth of a formula at a world of a multi-frame* $\mathcal{M}, w \vDash \alpha$ is defined in the usual fashion by adding the clause:

$$\mathcal{M}, w \vDash [a]\alpha \text{ iff } \mathcal{M}, w' \vDash \alpha \text{ for every } w' \text{ such that } \langle w, w' \rangle \in \rho(a).$$

The fundamental property of the canonical models as enunciated above has already been used in the previous completeness proofs and is the basis of the one developed here. We first have to prove a result analogous to that of Lemma 4.2.12:

Lemma 8.6.5 *If w is a maximal consistent extension of S containing $\neg[a]\alpha$, then $Den_a(w) \cup \{\neg\alpha\}$ is a consistent set.*

Proof: The proof follows by contraposition, much the same as in Lemma 4.2.12. ♠

It still remains to prove that the canonical multi-frame of the canonical multi-model for a basilar system $\mathbf{S}^\Phi$ satisfies the condition required in Definition 8.6.1; this is a consequence of the following lemma:

Lemma 8.6.6 *Let $\mathcal{F} = \langle W_S, \Omega_S \rangle$ be a canonical frame and $\rho_S : \Phi \longrightarrow \Omega_S$ be defined as $\rho_S(a) = R_{S_a}$. Then ρ_S satisfies the following conditions:*

1. $\rho_S(1) = Id$
2. $\rho_S(a \cup b) = \rho_S(a) \cup \rho_S(b)$
3. $\rho_S(a \odot b) = \rho_S(a) \odot \rho_S(b)$
4. $\rho_S(0) = 0$

Proof: Exercise 8.10. ♠

This lemma shows that the canonical multi-frame satisfies the requirement of Definition 8.6.1. Then we have:

Proposition 8.6.7 (*Fundamental Theorem for Basilar Systems*). *If $\mathcal{M} = \langle W_S, \Omega_S, v_S \rangle$ is the canonical model for a basilar system S, $w \in W_S$ and α is a multimodal formula, then $\mathcal{M}, w \vDash \alpha$ iff $\alpha \in w$.*

8.6. MULTIMODAL MODELS AND COMPLETENESS

Proof: See Exercise 8.11. ♠

Departing from such property, the general method to prove completeness of a particular system **S** (as seen in Chapter 3) via canonical models consists in determining a certain class of models, using properties of classes of relations (which is to say, through a certain condition **C** to be defined), such that two properties of **S** can be proven:

I. Its soundness: every theorem α of **S** is valid in all models that satisfy condition **C**.

II. Its completeness: if α is valid in all models that satisfy condition **C**, then α is a theorem of **S**.

The condition **C** for multimodal basilar systems $\mathbf{G}^{\langle a,b,c,d \rangle}$ is a direct generalization of the corresponding condition for $\mathbf{G}^{k,l,m,n}$. Given a bridge axiom of type $G(a,b,c,d)$, we say that the *condition of* (a,b,c,d)-*interaction* holds iff:

$$\rho(a)^{-1} \odot \rho(c) \subseteq \rho(b) \odot \rho(d)^{-1}$$

That is, for every w, w' and w'', if $\langle w, w' \rangle \in \rho(a)$ and $\langle w, w'' \rangle \in \rho(c)$, there exists w''' such that $\langle w', w''' \rangle \in \rho(b)$ and $\langle w'', w''' \rangle \in \rho(d)$.

It is perfectly perceivable that the condition of (a,b,c,d)-interaction is a direct generalization of the "diamond property" $\mathbf{C}^{k,l,m,n}$ introduced in Section 4.2: indeed, it is easy to see that $\mathbf{C}^{k,l,m,n}$ holds iff $R^{-m} \odot R^k \subseteq R^n \odot R^{-l}$ (or equivalently $R^{-k} \odot R^m \subseteq R^l \odot R^{-n}$, which by its turn is equivalent to $R^k \subseteq R^m \Rightarrow (R^n \odot R^{-l})$).

Proposition 8.6.8 (*Completeness of the basilar multimodal systems* $\mathbf{G}^{\langle a,b,c,d \rangle}$). *If* **S** *is a normal multimodal system with axioms* $G(a,b,c,d) = \langle a \rangle [b] p \supset [c] \langle d \rangle p$, *then* **S** *is sound and complete with respect to the multi-frames that satisfy the corresponding condition of* (a,b,c,d)-*interaction.*

Proof: The proof consists of an almost immediate generalization of the method of canonical models: let $\mathcal{F}^* = \langle W^*, \Omega^* \rangle$ be the canonical multi-frame defined as follows:

1. W^* is the class of maximal consistent extensions of **L**.

2. For every $a \in \Phi$, w, w' in W^*, $\langle w, w' \rangle \in \rho(a)$ iff $\{\alpha : [a]\alpha \in w\} \subseteq w'$.

As soundness is concerned, it is not difficult to show that, if $G(a,b,c,d) = \langle a \rangle [b]p \supset [c]\langle d \rangle p$ is an axiom (or theorem) of $\mathbf{S}^\Phi$, then $\mathcal{F}^*$ satisfies the condition of (a,b,c,d)-interaction. For the converse, if α is valid on all frames for $\mathbf{S}^\Phi$, then α is a theorem of $\mathbf{S}^\Phi$ extended with $G(a,b,c,d)$; the proof is analogous to the argument of Section 4.2, using Proposition 8.6.7. ♦

A similar proof of additive completeness as obtained in Proposition 4.2.19 holds for basilar systems, but we skip the details here.

The completeness proof for affirmative systems $G(a,b,\varphi)$ is structurally similar to the one for the basilar systems $G(a,b,c,d)$, though more elaborate. We shall in the following examine the details of this proof and explain how completeness for $G(a,b,c,d)$ systems is obtained as a particular case from completeness of $G(a,b,\varphi)$ systems.

For each affirmative formula φ, we inductively define an operation $F^\varphi : \Omega \longrightarrow \Omega$ based on φ over the class Ω of binary relations as follows:

Definition 8.6.9 1. If φ is p, then $F^p(R) = R$, where p is a propositional variable.

2. If φ is $\bot$, then $F^\bot(R) = 0$.

3. If φ is $\top$, then $F^\top(R) = Id$.

4. If φ is $\alpha \wedge \beta$, then $F^{\alpha \wedge \beta}(R) = F^\alpha(R) \cap F^\beta(R)$.

5 If φ is $\alpha \vee \beta$, then $F^{\alpha \vee \beta}(R) = F^\alpha(R) \cup F^\beta(R)$.

6. If φ is $\langle a \rangle \alpha$, then $F^{\langle a \rangle \alpha}(R) = R_a \odot F^\alpha(R)$.

7. If φ is $[a]\alpha$, then $F^{[a]\alpha}(R) = R_a \Rightarrow F^\alpha(R)$.

As an illustration of the definition above, let φ be the formula $\langle a \rangle \top$. Hence, by Definition 8.6.9(6), $F^\varphi(R) = F^{\langle a \rangle \top}(R) = R_a \odot F^\top(R) = R_a \odot Id = R_a$.

The operation ρ may be seen, as already mentioned, as a particular instance of the operation F^φ, and this fact results from the properties described in the Proposition 8.6.10, proved below.

Given an affirmative axiom $G(a,b,\varphi)$, we say that the *condition of* (a,b,φ)-*interaction* holds if $R_a \subseteq F^\varphi(R_b^{-1})$.

We now see that, as a consequence of Proposition 8.6.13, the condition of (a,b,φ)-interaction generalizes the condition of (a,b,c,d)-interaction.

8.6. MULTIMODAL MODELS AND COMPLETENESS

Proposition 8.6.10 Let φ be an affirmative formula. Then the following equalities hold for the operation F^φ:

(i) $F^{\langle 1 \rangle \varphi}(R) = F^\varphi(R)$

(ii) $F^{[1]\varphi}(R) = F^\varphi(R)$

(iii) $F^{\langle a \cup b \rangle \varphi}(R) = (R_a \cup R_b) \odot F^\varphi(R)$

(iv) $F^{[a \cup b]\varphi}(R) = (R_a \cup R_b) \Rightarrow F^\varphi(R)$

(v) $F^{\langle a \odot b \rangle \varphi}(R) = (R_a \odot R_b) \odot F^\varphi(R)$

(vi) $F^{[a \odot b]\varphi}(R) = (R_a \cup R_b) \Rightarrow F^\varphi(R)$

(vii) $F^{[0]\varphi}(R) = 0 \Rightarrow F^\varphi(R) = 1$

(viii) $F^{\langle 0 \rangle \varphi}(R) = 0 \odot F^\varphi(R) = 0$

Proof: The equalities above follow from the properties of composition (or relative product) $\odot$ and from the relative implication $\Rightarrow$ (see Exercise 8.12). ♦

Recall that $F^\top(R) = Id$ and notice that $R \Rightarrow Id = R$ and $R \odot Id = 1$, from which other equalities are easily obtained:

(i) $F^{\langle 1 \rangle \top}(R) = Id$

(ii) $F^{[1]\top}(R) = Id$

(iii) $F^{\langle a \cup b \rangle \top}(R) = (R_a \cup R_b)$

(iv) $F^{[a \cup b]\top}(R) = (R_a \cup R_b)$

(v) $F^{\langle a \odot b \rangle \top}(R) = (R_a \odot R_b)$

(vi) $F^{[a \odot b]\top}(R) = (R_a \odot R_b)$

(vii) $F^{[0]\top}(R) = 0$

(viii) $F^{\langle 0 \rangle \top}(R) = 0$

It follows from Proposition 8.6.10 that the operation ρ (see Definition 8.6.1) can be defined from F as $\rho(a) = F^{[a]\top}(R)$ (or equivalently, as $\rho(a) = F^{\langle a \rangle \top}(R)$). Indeed, it is clear that (i) and (ii) define $\rho(1) = Id$ (identity), (iii) and (iv) define $\rho(a \cup b) = \rho(a) \cup \rho(b)$ (union), (v) and (vi) define $\rho(a \odot b) = \rho(a) \odot \rho(b)$

(composition), and (vii) and (viii) define $\rho(0) = 0$. Consequently, ρ can be obtained as a special case of the definition of F.

The notion of a canonical model and canonical frame for affirmative systems is the same as in Definition 8.6.4.

Proposition 8.6.11 (*Fundamental Theorem for Affirmative Systems*). *Let φ be an affirmative formula depending on p, and let $\varphi(\alpha)$ denote the substitution $\varphi[p/\alpha]$, w and w' be maximal consistent extensions and $a \in \Phi$. Then the following equivalence holds:*

$$\langle w, w' \rangle \in F^\varphi(R_a^{-1}) \text{ iff } \{\varphi(\alpha) : [a]\alpha \in w'\} \subseteq w,$$

where F^φ is the above defined operation.

Proof: By induction over the complexity of φ. As we have already observed, φ may be re-written in an affirmative form, i.e., without $\neg$ or $\supset$.

Case 1. $\varphi = \top$. On the one hand, $\langle w, w' \rangle \in F^\varphi(R_a^{-1})$ iff $\langle w, w' \rangle \in 1$. On the other hand, given that $\top \in 1$, $\{\varphi(\alpha) : [a]\alpha \in w'\} = \{\top\} \subseteq w$. Therefore, both sides are true and the equivalence holds.

Case 2. $\varphi = \bot$. The argument is analogous to the previous one, with the difference that the equivalence holds because both sides are false.

Case 3. $\varphi = \varphi_1 \wedge \varphi_2$. We have to show that:

$$\langle w, w' \rangle \in F^{\varphi_1 \wedge \varphi_2}(R_a^{-1}) \text{ iff } \{\varphi_1(\alpha) \wedge \varphi_2(\alpha) : [a]\alpha \in w'\} \subseteq w.$$

From Definition 8.6.9, $F^{\varphi_1 \wedge \varphi_2}(R_a^{-1}) = F^{\varphi_1}(R_a^{-1}) \cap F^{\varphi_2}(R_a^{-1})$. By induction assumption, we suppose the equivalence holds for φ_1 and φ_2; then:

$$F^{\varphi_1}(R_a^{-1}) \text{ iff } \{\varphi_1(\alpha) : [a]\alpha \in w'\} \subseteq w \text{ and}$$

$$F^{\varphi_2}(R_a^{-1}) \text{ iff } \{\varphi_2(\alpha) : [a]\alpha \in w'\} \subseteq w.$$

It suffices to show, thus, that:

$$\{\varphi_1(\alpha) : [a]\alpha \in w'\} \subseteq w \text{ and } \{\varphi_2(\alpha) : [a]\alpha \in w'\} \subseteq w$$

iff

$$\{\varphi_1(\alpha) \wedge \varphi_2(\alpha) : [a]\alpha \in w'\} \subseteq w$$

But this follows immediately from the fact that, if $[a]\alpha \in w'$, $\varphi_1(\alpha) \in w$ and $\varphi_2(\alpha) \in w$ iff $\varphi_1(\alpha) \wedge \varphi_1(\alpha) \in w$, since w is maximal consistent.

8.6. MULTIMODAL MODELS AND COMPLETENESS

Case 4. $\varphi = \varphi_1 \vee \varphi_2$. Again, by induction hypothesis, let us assume that the above mentioned equivalence holds for φ_1 and φ_2. As, by Definition 8.6.9, it suffices show that:

$$\{\varphi_1(\alpha) : [a]\alpha \in w'\} \subseteq w \quad \text{or} \quad \{\varphi_2(\alpha) : [a]\alpha \in w'\} \subseteq w$$

iff

$$\{\varphi_1(\alpha) \vee \varphi_2(\alpha) : [a]\alpha \in w'\} \subseteq w.$$

($\Rightarrow$)
From left to right, it is clear that, if $[a]\alpha \in w'$, then $\varphi_1(\alpha) \in w$ or $\varphi_2(\alpha) \in w$ implies $\varphi_1(\alpha) \vee \varphi_2(\alpha) \in w$, since w is maximal consistent.
($\Leftarrow$)
Conversely, assume by *Reductio ad Absurdum* that

$$\{\varphi_1(\alpha) \vee \varphi_2(\alpha) : [a]\alpha \in w'\} \subseteq w$$

but that there exist certain α_1 and α_2 (they do need to be equal) such that $[a]\alpha_1 \in w'$, $[a]\alpha_2 \in w'$ but $\varphi_1(\alpha_1) \notin w$ and $\varphi_2(\alpha_2) \notin w$. Considering $\beta = \alpha_1 \wedge \alpha_2$, it can be shown (see Exercise 8.14) that $[a]\beta \in w'$, but $\varphi_1(\beta) \vee \varphi_2(\beta) \notin w$, contradicting the hypothesis.

Case 5. $\varphi = \langle b \rangle \psi$. In such case, by Definition 8.6.9 $\langle w, w' \rangle \in F^{\langle b \rangle \psi}(R_a^{-1})$ iff $\langle w, w' \rangle \in R_b \odot F^{\psi}(R_a^{-1})$ iff $\exists w''(wR_bw'' \wedge \langle w'', w' \rangle \in F^{\psi}(R_a^{-1}))$. Therefore, from this equivalence and by inductive hypothesis, we must show that: $\exists w''(wR_bw'' \wedge \langle w'', w' \rangle \in F^{\psi}(R_a^{-1}))$ iff $\{\langle b \rangle \psi(\alpha) : [a]\alpha \in w'\} \subseteq w$
($\Rightarrow$)
From left to right, suppose $\exists w''(wR_bw'' \wedge \langle w'', w' \rangle \in F^{\psi}(R_a^{-1}))$ and consider an element $\langle b \rangle \psi(\alpha)$ in the set $A = \{\langle b \rangle \psi(\alpha) : [a]\alpha \in w'\}$; we have to show that $\langle b \rangle \psi(\alpha) \in w$.

If $\langle w'', w' \rangle \in F^{\psi}(R_a^{-1})$, then by induction hypothesis we have $\{\psi(\alpha) : [a]\alpha \in w'\} \subseteq w''$. Since $\langle b \rangle \psi(\alpha) \in A$, then it holds $[a]\alpha \in w'$, hence $\psi(\alpha) \in w''$ given that $\{\psi(\alpha) : [a]\alpha \in w'\} \subseteq w''$. As by hypothesis wR_bw'', then from $\psi(\alpha) \in w''$ it follows $\langle b \rangle \psi(\alpha) \in w$. Therefore $A \subseteq w$.
($\Leftarrow$)
For the converse, consider the following set:

$$\Gamma = \{\alpha : [b]\alpha \in w\} \cup \{\psi(\alpha) : [a]\alpha \in w'\}$$

We will show that Γ is included in a maximal consistent extension w''.

(a) Γ is consistent. Indeed, if not, there exist $\alpha_1, \ldots, \alpha_m$ and $\beta_1, \ldots, \beta_n$ such that $[b]\alpha_i \in w$ for $1 \leq i \leq m$, $[a]\beta_j \in w'$ for $1 \leq j \leq n$ and $\alpha_1 \wedge \cdots \wedge \alpha_n \wedge \psi(\beta_1) \wedge \cdots \wedge \psi(\beta_m) \equiv \bot$.
Let $\alpha = \alpha_1 \wedge \cdots \wedge \alpha_n$ and $\beta = \beta_1 \wedge \cdots \wedge \beta_m$. Because $\beta \supset \beta_j$ is a theorem and ψ is affirmative, it can be proven by an easy induction that $\psi(\beta) \supset \psi(\beta_j)$ is also a theorem. Consequently, $\psi(\beta) \supset (\psi(\beta_1) \wedge \cdots \wedge \psi(\beta_m))$ is a theorem as well.
It is easy to check that $([b]p \wedge \langle b \rangle q) \supset \langle b \rangle (p \wedge q)$ is a theorem of any multimodal logic (Exercise 8.15), and obviously $\Diamond \bot \supset \bot$ is also a theorem of any multimodal logic. Consequently, $\alpha_1 \wedge \cdots \wedge \alpha_n \wedge \psi(\beta_1) \wedge \cdots \wedge \psi(\beta_m) \equiv \bot$ entails $\langle b \rangle (\alpha_1 \wedge \cdots \wedge \alpha_n \wedge \psi(\beta_1) \wedge \cdots \wedge \psi(\beta_m)) \equiv \bot$, which by its turn entails $\langle b \rangle (\alpha \wedge \varphi_1(\beta)) \equiv \bot$.
From the observation about $([b]p \wedge \langle b \rangle q) \supset \langle b \rangle (p \wedge q)$, it follows that $[b]\alpha \wedge \langle b \rangle \psi(\beta) \equiv \bot$, a contradiction since by hypothesis $[b]\alpha \in w$ and $\langle b \rangle \psi(\beta) \in w$, and w is consistent.

(b) Now, as Γ is consistent, by Proposition 8.6.3 Γ can be extended to a maximal consistent set w''.

It remains to be shown that this w'' is appropriate. In fact, we have $wR_b w''$ since $\{\alpha : [b]\alpha \in w\} \subseteq \Gamma \subseteq w''$ and $\langle w'', w' \rangle \in F^\psi(R_a^{-1})$ since $\{\psi(\alpha) : [a]\alpha \in w''\} \subseteq \Gamma \subseteq w''$.

Case 6. $\varphi = [b]\psi$. In this case, from Definition 8.6.9, $\langle w, w' \rangle \in F^{[b]\psi}(R_a^{-1})$ iff $\langle w, w' \rangle \in (R_b \Rightarrow F^\psi(R))$ iff $\forall w''(wR_b w'' \supset \langle w'', w' \rangle \in F^\psi(R_a^{-1}))$.

Since by induction hypothesis

$\langle w'', w' \rangle \in F^\psi(R_a^{-1})$ iff $\{\psi(\alpha) : [a]\alpha \in w'\} \subseteq w''$

we have to show that:

$\forall w''(wR_b w'' \supset \{\psi(\alpha) : [a]\alpha \in w'\} \subseteq w'')$ iff $(\{[b]\psi(\alpha) : [a]\alpha \in w'\} \subseteq w)$.

($\Rightarrow$)
Supposing the left sentence, we must show that $[b]\psi(\alpha) \in w$ under the condition that $[a]\alpha \in w'$. Clearly, $[b]\psi(\alpha) \in w$ iff $\forall w''(wR_b w'' \supset \psi(\alpha) \in w'')$. But from the left sentence, under the condition that $[a]\alpha \in w'$, $wR_b w''$ implies $\psi(\alpha) \in w''$.

($\Leftarrow$)
Under the condition that $[a]\alpha \in w'$, $[b]\psi(\alpha) \in w$. Supposing $wR_b w''$, $[b]\psi(\alpha) \in w$ implies $\psi(\alpha) \in w''$. This completes the proof. ♠

8.6. MULTIMODAL MODELS AND COMPLETENESS

Corollary 8.6.12 *Let $G(a,b,\varphi) = \langle a \rangle [b] p \supset \varphi$ (where a and b are arbitrary parameters and φ is an affirmative formula depending on p) be an axiom (or a theorem) of S. Then $R_a \subseteq F^\varphi(R_b^{-1})$ holds in the canonical multi-frame for S.*

Proof: Consider $\langle w, w' \rangle \in R_a$; we have to show that $\langle w, w' \rangle \in F^\varphi(R_b^{-1})$. From Proposition 8.6.11, this is equivalent to showing that $\{\varphi(\alpha) : [b]\alpha \in w'\} \subseteq w$. But if $[b]\alpha \in w'$, then $\langle a \rangle [b]\alpha \in w$, since $wR_a w'$ in the canonical multi-frame. Therefore $\varphi(\alpha) \in w$, since w is maximal consistent and $G(a,b,\varphi)$ holds in S. ♠

We are now ready to establish the most important result of this chapter, namely, that affirmative systems axiomatized by instances of the schema $G(a,b,\varphi) = \langle a \rangle [b] p \supset \varphi$ are correct and complete with respect to multimodal frames satisfying the relational condition $R_a \subseteq F^\varphi(R_b^{-1})$. It is advisable to recall the discussion in Section 4.1.

Proposition 8.6.13 *(Completeness for affirmative systems $G^{\langle a,b,\varphi \rangle}$). Let S be a multimodal system axiomatized by instances of the schema $G(a,b,\varphi) = \langle a \rangle [b] p \supset \varphi$, where a, b are arbitrary parameters, and let φ be an affirmative formula. Then S is sound and complete with respect to the class of multi-frames that verify the condition $R_a \subseteq F^\varphi(R_b^{-1})$.*

Proof: The soundness direction is provable by *Reductio*, arguing as in Proposition 4.2.1 (see Exercise 8.15). For completeness, we must prove that, if α is valid on all multi-frames satisfying the condition $R_a \subseteq F^\varphi(R_b^{-1})$, then α must be a theorem of S (axiomatized by instances of the schema $G(a,b,\varphi) = \langle a \rangle [b] p \supset \varphi$). Suppose that α is not a theorem of S; then there exists a maximal consistent extension w of S containing $\neg \alpha$, and, consequently, the canonical multi-frame for S validates $\neg \alpha$. But by Corollary 8.6.12, $R_a \subseteq F^\varphi(R_b^{-1})$ holds in the canonical multi-frame for S, but by hypothesis α is also valid on the canonical multi-frame, a contradiction. This concludes the proof. ♠

By applying the result of Proposition 8.6.13 to formulas of form $G(a,b,c,d) = \langle a \rangle [b] p \cdots [c] \langle d \rangle p$, we obtain $R_a \subseteq F^{[c]\langle d \rangle p}(R_b^{-1})$ which, by the definitions of $F^{[c]\langle d \rangle p}$, amounts to

$$R_a \subseteq (R_c \Rightarrow (R_d \odot R_b^{-1})).$$

This is readily seen to be equivalent to the condition of (a,b,c,d)-interaction $\rho(a)^{-1} \odot \rho(c) \subseteq \rho(b) \odot \rho(d)^{-1}$ since the above inclusion is equivalent both to $R_c^{-1} \odot R_a \subseteq R_d \odot R_b^{-1}$ (see Exercise 8.13 (a)) and to $R_a^{-1} \odot R_c \subseteq R_b \odot R_d^{-1}$

(see Exercise 8.13 (d)). It becomes thereby clear that Proposition 8.6.8 is derivable from Proposition 8.6.13.

Likewise, Proposition 8.6.13 shows that the models of the affirmative systems are first-order definable, a fact that emphasizes the important parallelism between the logic of multimodal operators and the theory of their relations.

In fact, Proposition 8.6.13 can be seen as a generalization of the procedures of Correspondence Theory seen in Section 3.3, to the extent that it shows that the formula $G(a, b, \varphi) = \langle a \rangle [b] p \supset \varphi$, where a and b are arbitrary parameters and φ is an affirmative formula, corresponds to the first-order formula

$$\forall x \forall y (x R_a y \supset x F^\varphi (R_b^{-1}) y)$$

given that the preceding formula corresponds to $R_a \subseteq F^\varphi(R_b^{-1})$.

Taking φ as $[c]\langle d \rangle p$, we directly find that the formula $G(a, b, c, d) = \langle a \rangle [b] p \supset [c] \langle d \rangle p$ corresponds to the first-order formula

$$\forall x \forall y (x R_a y \supset \exists z (x R_c z \wedge \forall t z R_d t \supset y(R_b t)))$$

This is schematically depicted as:

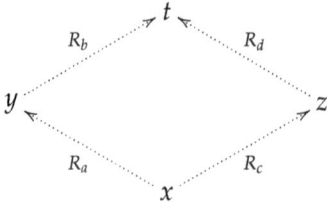

As an illustration, we re-examine the systems with interaction axioms of the kind $G(a, b, c, d)$ introduced in Section 8.3 and show how to obtain the corresponding relational equations using $R_a \subseteq (R_c \Rightarrow (R_d \odot R_b^{-1}))$; recalling that in such case $\Box_1 = [a]$ and $\Box_2 = [b]$ and that $R_1 = 1$, we have the following correspondences:

($K_{1,2}$):	$\Box_2 p \supset \Box_1 p$	$G(1, b, a, 1)$	$R_a \subseteq R_b$
($D_{1,2}$):	$\Box_2 p \supset \Diamond_1 p$	$G(1, b, 1, a)$	$1 \subseteq R_a \odot R_b^{-1}$, i.e., $R_b \subseteq R_a$
($B_{1,2}$):	$p \supset \Box_1 \Diamond_2 p$	$G(1, 1, a, b)$	$R_a \subseteq R_b^{-1}$
($4_{1,2}$):	$\Box_1 p \supset \Box_2 \Box_1 p$	$G(1, a, b \odot a, 1)$	$R_b \odot R_a \subseteq R_a$
($5_{1,2}$):	$\Diamond_1 p \supset \Box_2 \Diamond_1 p$	$G(a, 1, b, a)$	$R_a \subseteq R_b \Rightarrow R_a$
($SC_{1,2}$):	$\Box_2 \Box_1 p \supset \Box_1 \Box_2 p$	$G(1, b \odot a, a \odot b, 1)$	$R_a \odot R_b \subseteq R_b \odot R_a$

To sum up, the affirmative systems may be described in first-order language and are legitimate propositional modal systems. This is also the case of the Catach-Sahlqvist axioms $G(\varphi, \psi)$, but we refer to L. Catach [Cat89] (which generalizes results of H. Sahlqvist [Sah75]) for details.

Several authors have studied these issues and have additionally investigated other kinds of models for multimodalities, especially those of algebraic nature. It is also possible to study first-order systems by applying the present method. Nevertheless, this method is not applicable when certain axioms, such as the induction axioms on page 219, are considered. In that case, it is compelling to use more general methods, such as the method of filtrations.

8.7 Exercises

1. Show that the following equivalences are provable in any basilar system:

   A1. $\alpha \dashv_a \beta \equiv \neg\langle a \rangle (\alpha \wedge \neg \beta)$
   A2. $\alpha \dashv_a (\beta \wedge \gamma) \equiv (\alpha \dashv_a \beta) \wedge (\alpha \dashv_a \gamma)$
   A3. $\alpha \dashv_1 \beta \equiv \alpha \supset \beta$
   A4. $\alpha \dashv_{a \cup b} \beta \equiv (\alpha \dashv_a \beta) \wedge (\alpha \dashv_b \beta)$
   A5. $\alpha \dashv_{a \odot b} \beta \equiv \top \dashv_a (\alpha \dashv_b \beta)$

2. Show that the operations $\oplus, \Rightarrow, \triangleright, /$ and $\backslash$ can be defined in terms of $\odot, \cap, \cup$ and $^-$, i.e.:

   (a) $R \oplus S = \overline{\overline{R} \odot \overline{S}}$
   (b) $R \Rightarrow S = \overline{R \odot \overline{S}}$
   (c) $R \triangleright S = R \cap \overline{S}$
   (d) $R/S = \overline{R \odot S^{-1}}$
   (e) $R \backslash S = \overline{R^{-1} \odot \overline{S}}$

3. Show that relative product and relative implication are interdefinable, that is: $R \odot S = \overline{(R \Rightarrow \overline{S})}$.

4. Show that the two operations of residuation are interdefinable:

   (a) $R/S = \overline{R^{-1} \backslash S^{-1}}$
   (b) $R \backslash S = \overline{R^{-1} / S^{-1}}$

5. Show that the relative product is interdefinable with respect to each operation of residuation, that is:

   (a) $R \odot S = \overline{\overline{R}/S^{-1}}$
   (b) $R \odot S = \overline{R^{-1}\backslash \overline{S}}$

6. Show that the relative implication is interdefinable with respect to each operation of residuation, that is:

   (a) $R \Rightarrow S = \overline{\overline{R}/S^{-1}}$
   (b) $R \Rightarrow S = R^{-1}\backslash S$

7. Prove that α is affirmative iff α is obtained from propositional variables, $\bot$ and $\top$ under $\wedge$, $\vee$, $[a]$ and $\langle a \rangle$ for any atomic modal parameter a. (Hint: To the one direction, show that these are all affirmative. To the other direction use induction).

8. Prove Proposition 8.6.3.

9. Show that $Den_a(w) \subseteq w'$ if and only if $Poss_a(w') \subseteq w$, where $Poss_a(w') = \{\langle a \rangle \alpha : \alpha \in w'\}$, and thus in the canonical models $wR_a w'$ iff $Poss_a(w') \subseteq w$.

10. Prove Lemma 8.6.6. (Hint: for $\rho_S(a \cup b)$, show that $\langle w, w' \rangle \in R_{Sa \cup b}$ iff $\langle w, w' \rangle \in R_{Sa} \cup R_{Sa}$. Use Exercise 8.9, Proposition 8.4.4 and the properties of disjunction in maximal consistent sets. The other cases are similar).

11. In order to complete the proof of Proposition 8.6.7, show that the accessibility relations for the canonical models of systems $G^{\langle a,b,c,d \rangle}$ satisfy the condition of (a,b,c,d)-strict interaction; in other words, if $G(a,b,c,d) = \langle a \rangle[b]p \supset [c]\langle d \rangle p$ is an axiom (or a theorem) of a system L, then $\mathcal{F}^*$ satisfies the condition $\rho(a)^{-1} \odot \rho(c) \subseteq \rho(b) \odot \rho(d)^{-1}$.

12. Complete the proof of Proposition 8.6.10. Hints:

    (a) Prove that $Id \odot R = R$ and, from this result, that items 1 and 2 are derivable.
    (b) Prove that $(R \cup S) \odot T = (R \odot T) \cup (S \odot T)$ and $(R \cup S) \Rightarrow T = (R \Rightarrow T) \cap (S \Rightarrow T)$ and, from this result, that items 3 and 4 are derivable.
    (c) Prove that $(R \odot S) \odot T = R \odot (S \odot T)$ and $(R \odot S) \Rightarrow T = R \Rightarrow (S \Rightarrow T)$ and, from this result, that items 5 and 6 are derivable.
    (d) Prove that $0 \Rightarrow R = 1$ and $0 \odot R = 0$ for items 7 and 8.

13. Show that:

    (a) $S^{-1} \odot R \subseteq T$ iff $R \subseteq (S \Rightarrow T)$. Use this equivalence to show that $R_a \subseteq (R_c \Rightarrow (R_d \odot R_b^{-1}))$ is equivalent to $R_c^{-1} \odot R_a \subseteq R_d \odot R_b^{-1}$.

    (b) $(S \odot R)^{-1} = R^{-1} \odot S^{-1}$.

    (c) $S \subseteq R$ iff $S^{-1} \subseteq R^{-1}$.

    (d) Use the equivalences above to show that $R_c^{-1} \odot R_a \subseteq R_d \odot R_b^{-1}$ is equivalent to $R_a^{-1} \odot R_c \subseteq R_b \odot R_d^{-1}$.

14. Complete the proof of the converse of case 4 in Proposition 8.6.11. (Hint: Considering $\beta = \alpha_1 \wedge \alpha_2$, from the fact that $\beta \supset \alpha_1$ and $\beta \supset \alpha_2$ are S-theorems in any multi-modal system, show that so are $\varphi_1(\beta) \supset \varphi(\alpha_1)$ and $\varphi_2(\beta) \supset \varphi(\alpha_2)$ since φ_1 and φ_2 are affirmative.) As w is maximal consistent, taking the hypothesis into account conclude $\varphi_1(\beta) \notin w$ and $\varphi_2(\beta) \notin w$, thus $\varphi_1(\beta) \vee \varphi_2(\beta) \notin w$, a contradiction.

15. Show that, for any multimodal system S, $\vdash_S ([b]p \wedge \langle b \rangle q) \supset \langle b \rangle (p \wedge q)$. (Hint: generalize Exercise 2.2.(ii)).

16. Define, for R any binary relation, the *cut* of R with respect to X by $\langle R \rangle X = \{x : (\exists y) x R y \wedge y \in X\}$. Prove that:

    (a) $\langle R \odot S \rangle X = \langle R \rangle X \odot \langle S \rangle X$.

    (b) $\langle R \cup S \rangle X = \langle R \rangle X \cup \langle S \rangle X$.

    (c) Use the previous items to give an alternative proof of the properties of ρ_S in Lemma 8.6.6, defining $\langle a \rangle \alpha \in w$ iff $\alpha \in \langle \rho(a) \rangle w$.

8.8 Further reading

In a sense, investigation on multimodality can be seen as a reply to the critiques against monomodalism put forth more than 30 years ago. In fact, D. Scott in [Sco70] wrote: "Here is what I consider one of the biggest mistakes of all in modal logic: concentration on a system with just one modal operator".

Whatever the reason may be, several authors devoted their studies to the combination of modal systems, aiming to formalize discursive aspects of natural language (e.g. J. van Benthem in [vB83b] and [vB83a]) or to examine problems in the field of Artificial Intelligence, such as in the mentioned dynamic logics, which are concerned with the formalization of the properties of computational programs (e.g. K. Segerberg in [Seg77]).

The abstract calculus of binary relations, which plays an important role in combining modalities, was introduced in the middle of the 19th century in a paper by Augustus De Morgan; later on Charles Peirce and Ernst Schröder turned to it, but the subject has been neglected till 1940 when it was revived by Alfred Tarski. Today the general calculus of relations is a formidable source of research and a sharp tool for expressing a number of mathematical concepts (see V. R. Pratt [Pra92]).

Dynamic algebra and dynamic logic are kindred, and dynamic algebra can be used to give an algebraic interpretation to propositional dynamic logic (PDL) as much as Boolean algebra is connected to classical propositional logic. For instance, a duality between some topological Kripke models and dynamic algebras (analogous to the well-known duality between Boolean algebras and their Stone spaces) can be proved (see [Koz80]).

Even if they are so intimately connected, the concepts and the origins of dynamic logic and dynamic algebra can be separated. The notion of dynamic logic is due to V. R. Pratt [Pra76] where the idea of multimodal logic, as presented here in general terms, was developed. The notion of dynamic algebra was introduced by D. Kozen in [Koz79]. An historical account about dynamic logic and dynamic algebra can be found in D. Harel, D. Kozen and J. Tiuryn's [HKT00]. For dynamic logic as logic of programs see D. Harel [Har79].

From the proof-theoretical viewpoint, deviating from the Hilbertian approach we are taking here, some multimodal systems may be usefully axiomatized via the analytic tableau methods. This approach, when possible, brings advantages for the applications and provides the possibility of uniformly treating certain classes of systems. For analytic tableau systems for multimodal systems, see L. Catach in [Cat91] and M. Baldoni, L. Giordano and A. Martelli in [BGM98].

Multimodalities in logic programming have been investigated by P. Enjalbert and L. Fariñas del Cerro in [EdC79] with emphasis on modal resolution.

M. K. Rennie in [Ren70] and H. Sahlqvist in [Sah75] are important works on the systematic formalization of multimodal system, whose findings were later generalized in L. Catach [Cat88] and [Cat89] and M. Kracht [Kra93]. Here we have followed the general lines of these three approaches, proposing both an abstract definition of the modal operators and a calculus on the accessibility relations, the foundations of which are due to A. Tarski in [Tar41], which enables us to prove a uniform theorem of completeness for multimodalities. For applications of multimodal systems to the logic of provability, we recommend the pioneering work by G. Boolos in [Boo79]

8.8. FURTHER READING

and especially its successor [Boo93], as well as R. Solovay in [Sol76] and C. Smorynski in [Smo85].

Modalities, negation and non-classical logics intermingle in various aspects. Negations can be seen as modalities, as for example in C. W. Harvey and J. Hintikka [HH91], but also negated modalities may behave as non-classical negations, as shown in J.-Y. Béziau [B́05], where it is argued that $\neg\Box\alpha$, for $\Box$ in **S5**, has some characteristic properties of paraconsistent negation. Many-valued modal logics have been considered e.g. by K. Segerberg [Seg67] and M. C. Fitting [Fit91] and [Fit92]; some paraconsistent modal logics are treated in C. McGinnis [McG06], and intuitionistic modal logics are treated, for instance in D. Gabbay, A. Kurucz, F. Wolter and M. Zakharyaschev [GKWZ03].

The notion of bridge principle lies in the realm of the combination of languages, and many bridge principles can be defined and combined within modal logics in general. Several of the constructions in this book can be readily recast as *fusions* and *products* of normal modal logics. There are, however, several other methods not only to combine but to also decompose modal logics: for a comprehensive treatment of the subject, see W. A. Carnielli, M. E. Coniglio, D. Gabbay, P. Gouveia, and C. Sernadas [CCG$^+$07]. The area of combination of logics has gained a strong impetus due to its range of applications, which go from problems in computer science to purely philosophical questions. For an online reference, see W. A. Carnielli and M. E. Coniglio [CC07].

Chapter 9

Towards quantified modal logic

9.1 Propositional quantifiers

As is well known, the logics of quantification may be built on the ground of languages of different expressive power. While first-order predicate logic – which in this book we conventionally call **QL** – is based on a language which admits quantification over individual variables, second-order predicate logic is based on a language which admits quantification over predicate variables. The traditional distinction between first and second-order logic, however, may lead to overlook that also propositional variables are full-blooded variables, so they may also in principle be submitted to quantification. In this section, we try to outline the way in which, moving from **QL**, it is possible to build a formal theory of propositional quantification.

The language of **QL** consists of the propositional connectives $\bot$ and $\supset$ introduced in Chapter 1, plus a new symbol ∀ for the *universal quantifier* and the following sets of primitive symbols:

Ind: $\{x, y, z, \cdots, x_1, y_1, z_1, \cdots, x_2, y_2, z_2, \cdots\}$ (denumerable set of individual variables)

Const: $\{a, b, c, \cdots\}$ (denumerable set of names or individual constants)

Pred: $\{P_1^0, P_2^0, \cdots, P_1^1, P_2^1, \cdots P_1^k, P_2^k \cdots\}$ (denumerable set of predicate letters)

Fun: $\{f_1^0, f_2^0, \cdots, f_1^1, f_2^1, \cdots f_1^k, f_2^k \cdots\}$ (denumerable set of function letters)

Parentheses will also be employed as auxiliary symbols in the usual way. In each of the letters contained in **Pred** and **Fun**, the superscript indicates the number of places of the predicate or function (or *arity*), while the subscript indicates the order place.[1]

The *terms* of our language are so defined:

(i) Every member of **Ind** or **Const** is a term.

(ii) If $t_1 \cdots t_n$ are terms and $f^n \in$ **Fun**, $f^n(t_1 \cdots t_n)$ is a term.

(iii) Nothing else is a term.

The set *WFF* of well-formed formulas (wffs) is defined as follows:

(a) $\bot \in$ *WFF*.

(b) If P^n is an n-place predicate letter and $t_1, \cdots, t_n$ are terms, $P^n(t_1, \cdots, tn) \in$ *WFF*.

(c) If $\alpha \in$ *WFF*, $\beta \in$ *WFF* and $x \in$ **Ind**, then $\alpha \supset \beta \in$ *WFF* and $\forall x \alpha \in$ *WFF*.

(d) Nothing else belongs to *WFF*.

We remark that the superscript n in any predicate letter P^n may be omitted without ambiguity in any wff, as n is also the number of the terms which follow P^n. The elimination of parentheses will be made according to standard rules.

The definitions of auxiliary symbols are the same as in Section 1.2, with the addition of the following symbol for existential quantifier:

$$\exists x \alpha \stackrel{\text{Def}}{=} \neg \forall x \neg \alpha$$

If $\forall x \alpha$ or $\exists x \alpha$ is a wff, α is said to be the *scope* of $\forall$ or $\exists$. A variable x is said to be *bound* when it is in the scope of a quantifier $\forall x$ or $\exists x$, otherwise it is said to be *free*.

We will call an *interpretation* of **QL**[2] a couple $\mathcal{M} = \langle D, V \rangle$ where $D \neq \emptyset$ and V is a function defined as follows:

(i) If t is a term, $V(t) \in D$.

(ii) $V(f^n)$ is a function from D^n to D, i.e., is a set of $n+1$-tuples of elements of D.

---

[1]Predicate and function letters represent variables, not constants. For sake of simplicity, the distinction between variables and constants is here applied only to symbols representing individuals.

[2]An interpretation of **QL** may also be called a **QL**-*model*; the term "interpretation" is here reserved to non-modal models.

9.1. PROPOSITIONAL QUANTIFIERS

(iii) $V(P^n)$ is a relation in D^n, i.e. a subset of D^n, where D^n is a set of ordered n-tuples of elements of D.

The notion of truth with respect to a given interpretation $\mathcal{M} = \langle D, V \rangle$ is defined as follows, where $\mathcal{M}^{x/d} = \langle D, V^{x/d} \rangle$ denotes the *alternative interpretation* such that $V^{x/d}$ differs from V only for having $V(x) = d$:

(iv) $\mathcal{M} \vDash P(t_1, \cdots, t_n)$ if and only if $\langle V(t_1), \cdots, V(t_n) \rangle \in V(P^n)$.

(v) $\mathcal{M} \vDash \neg \alpha$, $\mathcal{M} \vDash \alpha \wedge \beta$, $\mathcal{M} \vDash \alpha \vee \beta$, $\mathcal{M} \vDash \alpha \supset \beta$ are defined as in standard propositional logic.

(vi) $\mathcal{M} \vDash \forall x \alpha$ iff $\mathcal{M}^{x/d} \vDash \alpha$ for every interpretation $\mathcal{M}^{x/d} = \langle D, V^{x/d} \rangle$.

A wff is **QL**-*valid* (in symbols: $\vDash \alpha$) if and only if, for every $\mathcal{M}$, $\mathcal{M} \vDash \alpha$.
We write $[x/t]$ to indicate that the term t takes the place of x, and we say that t is *free* for x when it does not contain variables which turn out to be bound after the substitution of x by t.

The axiom schemas of **QL** are the following:

(QL1) All the theorem schemas and rules of the propositional calculus **PC**;

(QL2) $\forall x \alpha \supset \alpha[x/t]$, provided t is free for x in α.

The only additional rule is:

(R1) $\vdash \alpha \supset \beta$ implies $\vdash \alpha \supset \forall x \beta$ (if x is not free in α)

The following derived rules are usefully employed in many proofs of **QL**-theorems:

(E∀) $\dfrac{\forall x \alpha}{\alpha[x/t]}$, x free for t in α (∀-Elimination)

(I∀) $\dfrac{\alpha[x]}{\forall x \alpha}$, x not free in α (∀-Introduction)

(E∃) $\dfrac{\exists x \alpha}{\alpha[x/t]}$, x not free in α (∃-Elimination)

(I∃) $\dfrac{\alpha[x/t]}{\exists x \alpha}$, x free for t in α (∃-Introduction)

In second-order predicate logic, quantification is allowed not only on individual variables, but on predicate variables as well, so the set *WFF* should be extended to admit wffs whose form is $\forall P_i^m \alpha$ and $\exists P_i^m \alpha$ for arbitrary m and i, preserving, *mutatis mutandis*, the distinction between free and bound variables. The semantics for the second-order quantified wffs parallels the one for first-order ones with an obvious modification in the notion of alternative interpretation, which is now of form $\mathcal{M}^{p^n/-}$.

The expressive enrichment obtained by quantifying over predicates cannot be undervalued. As we saw in Chapter 3, for instance, the correspondence between modal formulas and formulas expressing the relational properties of the frames is essentially a correspondence between modal formulas and second-order formulas, some of which are not reducible to equivalent first-order formulas.

Second-order logic has a wide number of interesting subsystems, one of which may be interpreted as a system for propositional quantification. Let us call **Pred**^{0} the subset of **Pred** which contains only 0-place predicate variables: $\{P^0_1, P^0_2, \cdots\}$. We may then identify a language, much like the language of **QL**, with the following supplementary properties:

(i) **Pred** − **Pred**^{0} = ∅

(ii) **Const** = **Ind** = ∅

(iii) **Fun** = ∅

It is now natural to identify **Pred**^{0} with the set $Var = \{p_1, p_2, \cdots\}$ of propositional variables of Section 2.3. In other words, the propositional variables previously introduced may be intended as notational variants for 0-place predicate variables by identifying any symbol P^0_i with the related propositional symbol p_i.

Quantifiers binding propositional variables will be called *propositional quantifiers*. The class of the wffs, *mutatis mutandis*, is then **QL** with the following additional clause:

(c) If $\alpha \in WFF$ and $p \in Var$, $\forall p\alpha \in WFF$.[3]

Extending the set of logical symbols with the necessity operator □, a further clause is needed (below) for the language of a modal logic with propositional quantifiers:

(d) if $\alpha \in WFF$, then $\Box\alpha \in WFF$

A useful remark concerns condition (ii) above. In place of putting **Const** = ∅, in fact, a less drastic alternative is available. Just as we distinguish among

---

[3]When convenient we shall write $\alpha(p)$ to indicate that p is part of α.

9.1. PROPOSITIONAL QUANTIFIERS

individual variables and individual constants in **QL**, we may draw a distinction between propositional variables and propositional constants – a distinction which is encouraged by the fact that $\bot$ may be properly classified as a propositional constant.[4]

A proposal which has been advanced in the field of deontic logic, for instance, is to introduce, besides the propositional variables, a propositional constant Q_a (to be read as "the moral/legal code is applied") or, alternatively, a constant Q_b (to be read as "a sanction is applied"). Q_a may be governed by the axiom $\Diamond Q_a$, and Q_b by $\neg \Box Q_b$. With the aid of Q_a, the deontic operator O could be so introduced in the following way:

$$O\alpha \stackrel{\text{Def}}{=} \Box(Q_a \supset \alpha)$$

while, with the aid of Q_b, a definition could be:

$$O\alpha \stackrel{\text{Def}}{=} \Box(\neg \alpha \supset Q_b)$$

By generalizing this strategy, one could propose the introduction of an unlimited number of symbols for propositional constants $Q_a, Q_b, Q_c \cdots$, to be added to the set of propositional variables. A possible drawback of this extension is that the use of propositional constants may yield self-referential situations which may be conceptually puzzling. Suppose that in ordinary language each proposition is symbolized by some constant, and let, for instance, Q_z be the symbol for the proposition "all the propositions read today by me are false", i.e. $\forall p((p \text{ is read today by me}) \supset \neg p)$. Then we can produce the following semi-formal argument:

1. Q_z [Hyp.]
2. $\forall p((p \text{ is read today by me}) \supset \neg p)$ [1]
3. $Q_z \text{ is read today by me} \supset \neg Q_z$ [($E_\forall$) in 2]
4. $Q_z \supset \neg(Q_z \text{ is read today by me})$ [(PC) in 3]
5. $\neg(Q_z \text{ is read today by me})$ [(MP) in 1 and 4]

But (5) is cogently false,[5] so (1) is false and Q_z is true. This argument then gives a logical proof of the fact that some proposition read today by

---
[4]It should be remarked that having at our disposal propositional quantifiers makes the basic set of connectives introduced here redundant. In fact, $\bot$ might be introduced by definition as $\forall p p$.

[5]This blatant falsity is of the kind which is sometimes called a *pragmatic contradiction*. It is to be noted, however, that suitable postulates for Q_z (for instance, "Q_z belongs to a class of propositions read today by me") may convert it into a logical contradiction.

me is true. But the conclusion is questionable, as it is logically possible that all propositions read today by me are false.

In what follows, we will ignore propositional constants but call attention to the suggestion that they could be interestingly used to improve the resources of modal language. Even without such tools, however, it is clear that modal language extended with propositional quantifiers has a relevantly high expressive power. To give an example, the two principles (**KP**) and (**AO**) involved in Fitch's Theorem (see page 187) may be fully expressed in object language, respectively, as $\forall p(p \supset \Diamond Kp)$ and $\exists p(p \land \neg Kp)$.

This enriched language is also able to express properties of the accessibility relation which are not modally definable. Let us recall that the truth of α on a frame $\mathcal{F}$ means that every substitution instance of α is true on $\mathcal{F}$: so, if p, q, r, say, are the atomic variables of α, this amounts to saying that $\forall p \forall q \forall r \alpha$ is true. One can prove, for instance, that irreflexivity of frames, which is not modally definable, may be expressed by the wff $\exists p(\Box p \land \neg p)$, while the existence of more than one accessible world is expressed by $\exists p(\neg \Box p \land p)$ (see Exercise 9.8). The application of propositional quantification to hybrid languages (see Section 6.5, especially page 177) also yields a progressive enrichment, which however will not be treated here.

The first step in axiomatizing a modal system of propositional quantification **QP** is to build a propositional quantificational basis in the following way:

(**QP0**): Axioms and rules for **PC** as on page 6;

(**QP1**): $\forall p\alpha \supset \alpha[p/\beta]$, where β is a **PC**-wff free for p in α;

(**QP2**): $\forall p(\alpha \supset \beta) \supset (\forall p\alpha \supset \forall p\beta)$;

(**UQ**): $\vdash \alpha$ implies $\vdash \forall p\alpha$.

Note that the propositional rule of Uniform Substitution (**US**) which is a primitive rule in **PC** can be proved here in the form: $\vdash \alpha(p)$ implies $\vdash \alpha[p/\beta]$, under the proviso that β is a **PC**-wff free for p in α:

1. $\vdash \alpha(p)$ [hypothesis]
2. $\vdash \forall p\alpha(p)$ [(**UQ**) in 1]
3. $\forall p\alpha(p) \supset \alpha[p/\beta]$ [(**QP1**), under proviso]
4. $\vdash \alpha[p/\beta]$ [(**MP**) in 2 and 3, under proviso]

The difference with standard substitution is that now its application may involve modifications inside universal and existential quantifiers. For

9.1. PROPOSITIONAL QUANTIFIERS

instance, it will be possible to perform derivations as in the following example[6]:

1. $q \supset q$ [PC]
2. $\forall q(q \supset q)$ [(UQ) in 1]
3. $\exists q(q \supset q)$ [⊢ $\forall p\alpha \supset \exists p\alpha$ in 2]
4. $\exists r \exists p((r \vee p) \supset (r \vee p))$ [(US), $[q/(r \vee p)]$ in 3]

At step (4) the wff $r \vee p$ has been substituted to q by (US), but this involves a substitution of the quantifiers which after substitution will have to bind not one but two variables.[7]

Let us now have a look at the axiom schemas for the modal extensions of **QP**. We will stipulate that, if **S** is an arbitrary normal modal system, **S**π^* will be the name of **S** extended with **QP** and with the following wff (often named "Barcan Formula"):

(**BF**) $\forall p \Box \alpha \supset \Box \forall p \alpha$

The name "Barcan Formula" is sometimes applied to the equivalent wff (expressed in terms of $\exists$ and $\Diamond$):

(**BF'**) $\Diamond \exists p \alpha \supset \exists p \Diamond \alpha$

It is to be remarked that a stronger version of axiom (**QP1**) is the axiom (**QP1**$^+$) $\forall p \alpha(p) \supset \alpha[p/\beta]$, where β is an arbitrary **QP**-formula free for p.

It may be then useful to admit, for every modal system **S**, a variant of **S**π^* – let us call it **S**π – which is like **S**π^* except for the fact that (**QP1**$^+$) takes the place of (**QP1**).

As a matter of fact, the addition of (**BF**) to every extension of **KB**π (where, as we recall, (**B**) is the axiom $p \supset \Box \Diamond p$) is redundant, as both (**BF**) and (**BF'**) can be proved in **KB**π in the following way:

1. $\forall p \Box \alpha \supset \Box \alpha$ [(**QP1**)]
2. $\Diamond \forall p \Box \alpha \supset \Diamond \Box \alpha$ [⊢ $\alpha \supset \beta$ ⊢$_{KB}$ implies ⊢ $\Diamond \alpha \supset \Diamond \beta$ in 1]
3. $\Diamond \forall p \Box \alpha \supset \alpha$ [⊢$_{KB}$ $\Diamond \Box \alpha \supset \alpha$ in 2]
4. $\Diamond \forall p \Box \alpha \supset \forall p \alpha$ [$\alpha \supset \beta$ ⊢$_{QP}$ $\alpha \supset \forall p \beta$ (p not free in α) in 3]
5. $\forall p \Box \alpha \supset \Box \forall p \alpha$ [⊢$_{KB}$ $\Diamond \alpha \supset \beta$ implies ⊢$_{KB}$ $\alpha \supset \Box \beta$ in 4]

---

[6]The proof presupposes the thesis $\forall p \alpha \supset \exists p \alpha$, which follows from $\forall p \alpha \supset \alpha$ and $\alpha \supset \exists p \alpha$.
[7]In order to see how the substitution rule extended to quantifiers can be derived, note that from $\exists q(q \supset q)$ by existential instantiation (E$_\exists$) we obtain $(p \vee r) \supset (p \vee r)$, from which by double existential quantification one obtains $\exists p \exists r((p \vee r) \supset (p \vee r))$. One may actually prove that, for every proof in which substitution is applied to quantified formulas, the same result may be reached without substitution simply by the rules of **QP**.

The formulas which are converse of (**BF**) and (**BF'**), i.e., $\Box \forall p \alpha \supset \forall p \Box \alpha$ and $\exists p \Diamond \alpha \supset \Diamond \exists p \alpha$, are theses of the weak system $K\pi$ (see Exercise 9.6), so (**BF**) and (**BF'**) are actually equivalent to their variants in which $\equiv$ takes the place of $\supset$.

The semantics for modal systems with propositional quantifiers may be given in terms of a special kind of frames which we already know, the *general frames* of Chapter 5.[8] We recall that a general frame is a triple $\mathcal{F} = \langle W, R, \Pi \rangle$ in which W and R are the same as in the standard frames and Π is a non-empty collection of sets of worlds, i.e., a subset of the power set of W closed under certain operations. In other words:

(i) $W \neq \emptyset$

(ii) $R \subseteq W \times W$

(iii) $\Pi \subseteq \wp(W)$

As we identify propositions with sets of worlds (namely with the sets of worlds in which they are true), the set Π is then the domain of propositions to which our formulas make reference.[9]

Definition 9.1.1 *A $K\pi^*$-model is a 4-tuple $\mathcal{M} = \langle W, R, \Pi, V \rangle$, where W, R, Π have the properties (i),(ii),(iii) above and furthermore:*

(iv) *Π is Boolean (i.e., is closed under complementation and union).*

(v) *V is a function from the set Var to Π.*

A formula, intuitively, denotes a set of worlds, i.e., a proposition; we have to restrict our attention to models which in a certain sense preserve propositions:

Definition 9.1.2 *A model $\mathcal{M} = \langle W, R, \Pi, V \rangle$ is closed under formulas iff we require, instead of (iv) in the Definition 9.1.1 that, for every wff α and every model $\mathcal{M}'$, $\{w \in W : \mathcal{M}', w \vDash \alpha\} \in \Pi$.*

Definition 9.1.3 *A $K\pi$-model is a 4-tuple $\mathcal{M} = \langle W, R, \Pi, V \rangle$, where W, R, Π have the properties (i),(ii),(iii),(v) above and furthermore:*

(iv') *Π is closed under formulas.*

---

[8] See Definition 5.1.2 and remarks above it.
[9] The standard frames for modal logics, i.e., the frames $\langle W, R \rangle$, may be defined as general frames in which $\Pi = \wp(W)$, i.e., Π is exactly the power set of W.

9.1. PROPOSITIONAL QUANTIFIERS

In what follows, we shall restrict our interest to systems which are extensions of $\mathbf{K}\pi$.

The definition of truth of α at a world w of a $\mathbf{K}\pi$-model $\mathcal{M} = \langle W, R, \Pi, V \rangle$ is the same as the one already given for standard modal logic (see Section 3.2), with a supplementary clause for quantified formulas, extending the clauses given on page 56:

8. $\mathcal{M}, w \vDash \forall p_i \alpha$ iff $\mathcal{M}', w \vDash \alpha$ for every $\mathbf{K}\pi$-model $\mathcal{M}' = \langle W, R, \Pi, V' \rangle$ such that $V'(p_j) = V(p_j)$ for every $j \neq i$.

A wff α will be said to be $\mathbf{K}\pi$-valid iff α is true at every world of every $\mathbf{K}\pi$-model. If **S** is a normal modal system, the properties of the accessibility relation in $\mathbf{S}\pi$-models will be the same required for standard propositional **S**-models.

It is straightforward to prove that all theses of $\mathbf{K}\pi$ are $\mathbf{K}\pi$-valid (see Exercise 9.3).

Given that (**BF**) is a "bridge" formula (i.e., it describes a particular interaction between modal operators and quantifiers) a legitimate question concerns its semantical interpretation. The answer is that (**BF**) describes the invariance of the domain Π of the propositions with respect to every world of the frame. To make this point clear, let us introduce a slightly different definition of a model:

Definition 9.1.4 *An indexed π-model is a 5-tuple $\langle W, R, \Pi, f, V \rangle$, where f is a function assigning to each world $w, w', \cdots$ in W some non-empty subset $\Pi^w, \Pi^{w'} \cdots$ of Π.*

Intuitively, Π^w denotes the propositional domain of the world w, i.e., the set of propositions conceivable at w. The definition of $\mathcal{M}, w \vDash \alpha$ should now be extended by adding the proviso that in $\mathcal{M}$ and all its variants $\mathcal{M}'$ the functions $V, V', \cdots$ take their values in Π^w.

The above defined $\mathbf{K}\pi$-models are now the special indexed models with *constant domains*, i.e., are indexed π-models such that $\Pi^w = \Pi^{w'}$ for every $w, w' \in W$. We may now prove what follows:

Proposition 9.1.5 *(BF) is true at every world w of an indexed π-model $\mathcal{M}$ if and only if $\Pi^{w'} = \Pi^{w''}$ for every w' and w'' of W.*

Proof: Let us suppose $\Pi^{w'} = \Pi^{w''}$ for every $w', w'' \in W$ of a given indexed π-model $\mathcal{M}$ and let us suppose that, for some arbitrary $w \in W$, $v(\forall p \Box \alpha, w) = 1$, for v a valuation in an explicit model (see Definition 3.2.2). Let V' be any alternative assignment to atomic variables which is coincident with V except for the assignment to p, and let v' be the corresponding valuation in explicit models. Then, given the supposition for every v' as defined above, we have $v'(\Box\alpha[p/\beta],w) = 1$ (β free for p). So $v'(\alpha[p/\beta],w') = 1$ at every w' such that wRw'. We have then, by the truth conditions for quantified formulas, $v(\forall p\alpha, w') = 1$. But in such case $v(\Box\forall p\alpha, w) = 1$. Hence $v(\forall p\Box\alpha \supset \Box\forall p\alpha, w) = 1$ in $\mathcal{M}$ for arbitrary w.

Conversely, let us consider an indexed π-model $\mathcal{M} = \langle W, R, \Pi, f, v \rangle$ such that for at least a couple of worlds w', w'', $\Pi^{w'} \neq \Pi^{w''}$. Let us define a model $\mathcal{M}$ with the following properties: $W = \{w', w''\}$, $R = \{\langle w', w''\rangle\}$, $\Pi^{w'} = \{\{w'\}\}$, $\Pi^{w''} = \{\{w'\}, \{w''\}\}$. Specifically, let V be an assignment such that, for some α, $v(\alpha, w') = v(\alpha, w'') = 1$. Given that $w'Rw''$ and $v(\alpha, w'') = 1$, $v(\Box\alpha, w') = 1$, and furthermore $v'(\Box\alpha, w') = 1$ for every alternative assignment v' (since in $\Pi^{w'}$ there is only one proposition); hence $v(\forall p\Box\alpha, w') = 1$. But in the world w'' of $\mathcal{M}$, $v(\forall p\alpha, w'') = 0$; with respect to the domain $\Pi^{w''} = \{\{w'\}, \{w''\}\}$, even if $v(\alpha, w'') = 1$, there is an alternative assignment v' such that $v'(\alpha, w'') = 0$. So, given that $w'Rw''$, $v(\Box\forall p\alpha, w') = 0$. Consequently, for some α, in the given model $\mathcal{M}$, $v(\forall p\Box\alpha \supset \Box\forall p\alpha, w') = 0$. So there is an indexed π-model $\mathcal{M}$ with a non-constant domain which falsifies (**BF**). ♠

9.2 Necessary and contingent identities

The transition from the semantics of propositional quantifiers to the semantics of individual quantifiers obviously implies replacing domains of propositions indexed by worlds $\Pi^w, \Pi^{w'}, \cdots$ with domains of individuals indexed by worlds $D^w, D^{w'}, \cdots$ and providing a suitable definition of the valuation function. If the Barcan equivalence $\forall p\Box\alpha \equiv \Box\forall p\alpha$ expresses the invariance of the domain of propositions across possible worlds, $\forall x\Box\alpha \equiv \Box\forall x\alpha$ expresses the invariance of the domain of individuals across possible worlds. Unfortunately, while the invariance of the domains appears to be highly plausible for propositions and is normally assumed for them without discussion, it cannot be considered a plausible assumption for individuals. In particular, if worlds are identified with instants (as we did in treating with temporal models in Chapter 6), a consequence of the Barcan formula would be that variables take their values over eternal or indestructible objects. In the modal semantics for quantified formulas, then, we must be prepared to abandon the requirement of constant domain.

The equivalences derived from the Barcan formula deserve attention as they express the equivalence between some formulas in which modalities

9.2. NECESSARY AND CONTINGENT IDENTITIES

are in the scope of a quantifier, such as $\forall x \Box \alpha$ (note that α may contain x free) and formulas in which the quantifiers are in the scope of a modal operator, such as $\Box \forall x \alpha$. The former formulas are called, in the tradition of Western philosophy, *de re* modalities, while the the latter are called *de dicto* modalities. The Barcan formulas (**BF**) and (**BF'**) conjoined with their converses clearly do not state that every *de re* formula is equivalent to some *de dicto* formula – the equivalence may be shown to not hold in non-trivial modal systems[10] – but surely they state that a class of paradigmatic examples of *de re* formulas can be eliminated in favor of some equivalent *de dicto* formula. Unfortunately, the Barcan formulas belong to the quantificational variants of the strong modal system **KB** and its extensions (including **S5**) and are far from being considered desirable modal truths.

The possibility of quantifying over modal formulas which takes place in *de re* formulas has been a source of technical and conceptual difficulties. Let us consider for instance a *de re* statement such as "something is necessarily mortal", formalized by $\exists x \Box P(x)$, and let us suppose that to any world w is associated a domain D^w, which is the set of individuals existing in w. From the above outlined semantics, we have that $v(\exists x \Box P(x), w) = 1$ iff, for some object of the domain D^w named by a, $v(\Box P(a), w) = 1$. This implies that there is an object a in the world w such that a has the property P in all worlds accessible to w. This means that a belongs to every domain $D^{w'}$ (when wRw') and that the property P is an invariant property of a in every $D^{w'}$. The idea that some properties may invariantly characterize real objects across possible worlds has been at the the center of a philosophical controversy. According to W. V. O. Quine, for instance, *de re* modalities imply rejecting the idea of necessity as a language-dependent notion and give legitimacy to the distinction between properties which the prescientific tradition qualified as "essential" (necessary) as opposed to "accidental" (contingent) properties (*"aristotelian essentialism"*). According to this view, Socrates has the necessary property of being mortal, but not the necessary property of being bearded, since there exists a possible world where he is shaved without loosing his identity.[11]

Quine's argument presupposes what is called an objectual interpretation of quantifiers, i.e., the idea that "for every x, $P(x)$" means – loosely speaking – "for every object a of the domain inside which x takes its values, a has the property P". But an alternative reading of quantifiers is provided by the so-called substitutional view, according to which "for every x, $P(x)$" means

---

[10] See Hughes and Cresswell in [HC96] pp. 251–253.
[11] See W. V. O. Quine [vOQ66].

that, for every term t which can be substituted to x, $P(t)$ is a true statement. If we endorse this conception, then a *de re* statement saying that something is possibly mortal ($\exists x \Diamond P(x)$) simply means that some substitution instance of "x is possibly mortal" is true. In this interpretation the Barcan equivalence $\Diamond \exists x P(x) \equiv \exists x \Diamond P(x)$ looks acceptable, at least in the standard reading of $\Box$ and $\Diamond$.

The difficulty emphasized by Quine is better understood if we extend **QL** not only with the axioms of some modal system, say **K**, but also with the axioms for the special two-place relation called identity. Let us remark that, in second order logic, identity could be introduced by definition as follows:

$$x = y \stackrel{\text{Def}}{\equiv} \forall P (P(x) \equiv P(y))$$

so expressing the two Leibnizian principles (of indiscernibility of identicals and identity of indiscernibles) in compressed form.

With the resources of first-order logic, we may axiomatize identity by introducing the two following axiom schemas:

(**I1**) $x = x$.

(**I2**) $x = y \supset (\alpha \supset \beta)$, where α and β differ only in that α has free x in zero or more places where β has free y.

With the resources of modal language with identity, we will have the possibility of performing a proof as the following:

1. the Morning Star = the Evening Star   [Hyp.]
2. the Morning Star = the Morning Star   [(**I1**)]
3. $\Box$(the Morning Star = the Morning Star)   [(**Nec**) in 2]
4. $\Box$(the Morning Star = the Evening Star)   [3, (**I2**) in 1]
5. $\exists x \Box (x = $ The Evening Star)   [**QL2** in 4]

The wff of line (5) is a *de re* modality. The object which makes the formula true is an object which in all possible worlds has the property of being identical to the Evening Star. But it is puzzling to say which object is thus described, given that the identity stated in line (1) is only contingently true – it might be false in some different possible world – and the same should be said of the identity "Venus = the Evening Star".

There are several ways to treat the above problem – sometimes called "Frege's puzzle" – which has both logical and philosophical implications. A first strategy consists in denying legitimacy to contingent identities. In other words, the proposal consists in forbidding that the same term may

9.2. NECESSARY AND CONTINGENT IDENTITIES

denote different objects in different worlds: as it is usually said, every term must be treated as a *rigid designator*. With a further step, we may also forbid that different terms denote the same objects in different worlds. As a result, we should introduce the following two axioms for identity, where t and t' stand for arbitrary terms:

(**RI1**) $t = t' \supset \Box(t = t')$

(**RI2**) $t \neq t' \supset \Box(t \neq t')$

In every quantificational normal system (**RI1**) is actually redundant in presence of (**I2**) –in fact a consequence of (**I2**) is $x = y \supset (\Box(x = x) \supset \Box(x = y))$, hence $x = y \supset \Box(x = y)$ by **PC**, (**I1**) and (**Nec**).

In some strong normal systems, (**RI2**), which is equivalent to $\Diamond(t = t') \supset (t = t')$, is also derivable (see Exercise 9.5). So one could claim that (**RI1**), (**RI2**) or both describe logical truths involving identity. If this were true, we should conclude that there is something wrong in speaking of contingent identities. But this conclusion has been judged counterintuitive given that it seems that there is nothing necessary in a true identity such as "George Bush = the president of USA in 2007". A possible consideration on this subject is that the problem is not raised by identity, but by definite descriptions as "the president of USA" or "the Morning Star": one could, in fact, think that they are not rigid designators, while proper names are. This distinction is however questionable if we look at such identities involving proper names such as "George Sand = Andine Aurore Lucile Dupin": there is nothing necessary, one could argue, in the fact that the writer Dupin had the pen-name George Sand rather than some other pen-name.

A straightforward way to admit contingent identities is to modify Axiom (**I2**) in the following way, which amounts to making a restriction on substitution of identicals, by somehow "forbidding" substitutions within modal operators:

(**I2c**) $x = y \supset (\alpha \supset \beta)$, where α and β differ only in that α has x free, i.e., not occurring in the scope of a modal operator in zero or more places where β has y free.

Before subscribing to this drastic revision of Leibniz's law, however, it is useful to examine two theories which gave rise to interesting technical developments in this connection: *counterpart theory* and the theory of *intensional objects*.

Counterpart theory[12] has been anticipated by some suggestions of Leibniz's philosophy. If every individual is seen as the sum of its properties, both simple and relational, any change of relations among individuals which occurs in passing from a world to another world yields a change in the nature of the individuals themselves. Hence in the worlds y accessible to x we find, in the best case, individuals who are similar to the individuals in x, but not identical to them: the most similar to the individuals in x will be called *counterparts* of the individuals in x.

The *essence* of an individual is the set of properties which an individual shares with its counterparts. An individual may have more than one counterpart in another world (for instance "splitting" himself into a couple of twins), while various individuals may have the same counterpart or no counterpart at all in another world.

This new way of looking at individuals implies that the domains associated to different worlds are not only different but disjoint. Obviously, in this new framework the Barcan Formula turns out to be refuted, and it is easy to also produce counterexamples to $x = y \supset \Box(x = y)$. However, counterpart theory also provides counterexamples to innocent logical truths. In this framework, in fact, suppose that $\Box(aSb)$ in w means "in all worlds w' accessible to w all the counterparts of a are connected by S to all the counterparts of b". However, if we have that w sees w' but b has no counterpart in w', $\Box(aSb)$ turns out to be vacuously true at w while $\Box \exists x(aSx)$ is false at w'. In fact, it may happen that there is a world w' at which no x such that aSx has a counterpart.

Given that $\Box(aSb \supset \exists x(aSx))$ is an obvious modal thesis, the fact that $\Box(aSb) \supset \Box \exists x(aSx)$ turns out to be invalid implies that axiom (**K**) is invalid. Counterpart theory then takes us far away from the family of normal modal logics.[13]

Recent elaborations of counterpart theory abandoned the metaphysical apparatus introduced by Lewis and, accepting the basic principle that the domains are disjoint, treat the counterpart relation as an unqualified binary relation among individuals.

Let h be a function which associates to every individual a the world $h(a)$ in which a lives and let C be the counterpart relation among individuals of different worlds. Then we may establish an association between the counterpart relation and the accessibility relation R in this way: $aCb \supset$

---
[12]Formulated by David Lewis in 1969 (cf. [Lew69]).
[13]For this criticism see A. Hazen [Haz79]. Various revisions of Lewis' theory have been introduced to correct this defect (M. Ramachandran, A. Plantinga, G. Forbes). Unfortunately they have other flaws, as for instance the fact that $\Box(x = x)$ turns out to be invalid.

9.2. NECESSARY AND CONTINGENT IDENTITIES

$h(a)Rh(b)$. The counterpart relation may be generalized to a plurality of counterpart relations between individuals. This new approach solves some technical problems of the original theory, but the intuitive interpretations of such abstract relations among individuals and the connected relations among worlds need case-by-case qualification.[14]

As a matter of fact, the original counterpart theory has several critical features of its own beyond the ones which have been already evidenced. One asks, for instance, if beyond having counterparts of Venus in different worlds, we have counterparts of the Morning Star or the Evening Star. The idea that a distinction exists between such objects and the planet Venus seems to be implicit in the fact that they have different properties. One may say, for instance, that the Morning Star has been believed to be different from the Evening Star for centuries, but not that Venus has been believed to be different from Venus for centuries. They are actually different individual concepts or, as is customarily said, different *intensional* objects. This suggests that we may introduce a distinction between two domains (the domain D of the standard objects and the domain I of intensional objects) in the semantics. The predicate variables now will always refer to strings of objects in D, but we will have, beyond $x, y, z, \cdots$ (which take values in D), other variables $f, g, h, \cdots$ which take values in I. To accomplish this enrichment, we may introduce two kinds of quantifiers, one for standard objects and one for intensional objects.

This duplication of quantifiers is plausible, but unfortunately is not the only cloning of quantifiers which seems to be plausible. As we will see in Section 9.5, in some multimodal logics it could be useful to distinguish between the domain of actual objects and the domain of possible objects, and match this distinction with a distinction between quantifiers for actual and possible objects. Furthermore, it turns out that most normal systems extended with axioms for intensional objects turn out to be not axiomatizable systems.[15] This unwanted complication suggests that intensional objects should be dispensed in a reasonably simple theory.

It is an illusion to think that some solution of Frege's puzzle in the framework of first-order modal logic may receive a definitive consensus by both logicians and philosophers. However, two points may be stressed here:

1. The need for so-called non rigid designators is sensible in doxastic-epistemic contexts, but not in other modal and multimodal contexts

---

[14]A basic reference for this approach is T. Brauner and S. Ghilardi [BG07].
[15]See Hughes and Cresswell in [HC68] pp. 335–336.

(for instance, they have no interesting role in logics of mathematical provability).

2. In quantified normal modal systems with identity, the axioms leading to the necessity of identity are philosophically defensible, for instance, by eliminating non-rigid designators from the class of terms via appropriate definitions (for example, in the style of the well-known Russell's Theory of Descriptions).

In the next section, we will take for granted axioms (**RI1**) and (**RI2**) for identity, but we will see that, independently from Frege's puzzle, the semantics for first-order modal systems is a source of technically intriguing problems.

9.3 The problem of completeness in first-order modal logic

As already remarked (cf. Proposition 9.1.5), systems whose semantics asks for invariant domains will have among theorems every formula equivalent to (**BF**).

Given the importance of identity for intensional reasoning, our basis will be **QL** extended with the axioms for identity (**I1**) and (**I2**) and the couple of axioms (**RI1**) and (**RI2**). Such systems, still extended with (**BF**), will be named **QK$^=$+(BF)**, **QKT$^=$+(BF)**, **QS4$^=$+(BF)** etc. More generally, **QS$^=$+(BF)** will be the name of any modal system which consists of **S** extended with **QL**, identity axioms and (**BF**).

We may now define a **QS$^=$+(BF)**-model in the following way:

Definition 9.3.1 *A **QS$^=$+(BF)**-model is a 4-tuple $\mathcal{M} = \langle W, R, D, V \rangle$, where W and R are as in the modal frames defined for S, $D \neq \emptyset$ and V is a value assignment which is so defined:*

1. *For any term t, $V(t)$ is defined as in clauses (i) and (ii) of page 242.*

2. *For any n-place predicate letter Q^n and any $w \in W$, $V(Q^n)$ is a set of $n + 1$-tuples $\langle a_1, \cdots, a_n, w \rangle$ where $a_1, \cdots, a_n \in D$.*

3. *For every w, w' in W, $\langle V(t), w \rangle = \langle V(t), w' \rangle$.*

9.3. COMPLETENESS IN FIRST-ORDER MODAL LOGIC

The notion of truth of α at a world w of $\mathcal{M}$ is then defined in the following way:

4. $\mathcal{M}, w \models Q(t_1, \cdots, t_n)$ if $\langle V(t_1), \cdots, V(t_n), w \rangle \in V(Q)$. otherwise $\mathcal{M}, w \not\models Q(t_1, \cdots, t_n)$.

5. $\mathcal{M}, w \models t_1 = t_2$ iff $\langle V(t_1), w \rangle = \langle V(t_2), w \rangle$.

6. $\mathcal{M}, w \models \bot$, $\mathcal{M}, w \models \alpha \supset \beta$, $\mathcal{M}, w \models \Box \alpha$ are defined as in propositional modal logic.

7. $\mathcal{M}, w \models \forall y \alpha$ iff, for every $\mathcal{M}' = \langle W, R, D, V' \rangle$ such that V' differs from V at most for the assignment to the variable y, $\mathcal{M}', w \models \alpha$.

Remark 9.3.2 *Notice that, by an obvious modification of the notation for alternative interpretation $\mathcal{M}^{x/d}$ on page 243, we could equivalently put at item (7) "$\mathcal{M}, w \models \forall y \alpha$ iff for every $\mathcal{M}^{y/d}$, $\mathcal{M}^{y/d}, w \models \alpha$."*

A wff α is said to be **QS$^=$+(BF)**-valid if and only if $\mathcal{M}, w \models \alpha$ for every world w of every **QS$^=$+(BF)**-model $\mathcal{M}$.

The formulation of the definition in terms of the explicit value assignment v is a rephrasing of clauses (4)–(7).

The proof of soundness of any **QS$^=$+(BF)** is routine (Exercise 9.4). In particular, it should be clear that conditions (4) and (6) above jointly grant the validity of both (**RI1**) and (**RI2**). As far as completeness is concerned, as no one of these systems is decidable, it is unavoidable to resort to a non-constructive completeness proof.

We recall that the standard (Henkin-style) method to prove completeness for first-order logic with identity (henceforth **QL$^=$**) is essentially the following.

Definition 9.3.3 *Let $\mathcal{M}$ be the collection of maximal consistent extensions $\Gamma_1, \Gamma_2, \cdots$ of $\mathbf{QL}^=$. The canonical model $\mathcal{M}^\Gamma = \langle D, V \rangle$ for $\mathbf{QL}^=$ is built, starting from some $\Gamma_c \in \mathcal{M}$, as follows. Let V be an assignment function which is defined in such a way that $V(t)$ is the equivalence class of the terms which $\Gamma_c \in \mathcal{M}$ says to be identical with t: $V(t) = \{t' : t = t' \in \Gamma_c\}$. The domain D is the set of all classes $V(t)$ for every term t. V is defined, for every n-place predicate Q, as follows:*

(i) $\langle d_1, \cdots, d_n \rangle$ *is an element of $V(Q)$ if and only if $Q(t_1, \cdots, t_n) \in \Gamma_c$*

By applying the identity principles, it is easily proven that, for all atomic statements of form $Q(t_1, \cdots, t_n)$, $\Gamma_c \models Q(t_1, \cdots, t_n)$ iff $Q(t_1, \cdots, t_n) \in \Gamma_c$.

The Henkin proof consists in generalizing the preceding relation to arbitrary statements by induction on the construction of wffs. The inductive hypothesis is of course that, for α with arbitrary length:

(IH) $\mathcal{M}^\Gamma, \Gamma_c \vDash \alpha$ iff $\alpha \in \Gamma_c$

The interesting step is given by the case in which α is $\forall x \beta(x)$:

$$\mathcal{M}^\Gamma, \Gamma_c \vDash \forall x \beta(x) \text{ iff } \forall x \beta(x) \in \Gamma_c.$$

In order to perform this step, it is necessary to impose some restrictions on canonical models, which are not needed in the propositional case. In fact, maximal consistent sets must enjoy a new property named ω-completeness:

(ωC) A maximal consistent set Γ formulas is ω-complete if and only if, for every $\alpha \in \Gamma$, if $\alpha[x/y] \in \Gamma$ for every variable y, then $\forall x \alpha \in \Gamma$.

Other alternative and equivalent ways to define ω-completeness are the following:

(ωC1) If $\Gamma \vDash P(t)$ for every term t, then $\Gamma \vDash \forall x P(x)$ for every variable x;

(ωC2) If $\Gamma \cup \{\neg \forall x P(x)\}$ is consistent, then, for some term t, $\Gamma \cup \{\neg P(t)\}$ is consistent.

It is essential to prove the following lemma about ω-completeness:

Lemma 9.3.4 *If Γ is a consistent set of wffs of $QL^=$, then there is a maximal consistent extension of Γ which is ω-complete.*

Proof: We may suppose that all the wffs of $QL^=$ are disposed in some given order. Let us suppose that we start from a consistent set Γ and then build a cumulative sequence of sets $\Gamma_0, \Gamma_1, \cdots$ where Γ_0 is exactly Γ.

Let us assume that $\forall x P(x)$ is the $i+1$-th wff in the order. Then we stipulate that, if Γ_i is consistent, then Γ_{i+1} is $\Gamma_i \cup \{\neg \forall x P(x)\} \cup \{\neg P(t)\}$, where t is a term which does not occur in any one of the wffs of the set Γ_i, neither in $\forall x P(x)$.

We may prove that the set Γ_{i+1}, which is built in this way, is still consistent. In fact, if $\Gamma_{i+1} = \Gamma_i \cup \{\neg \forall x P(x), \neg P(t)\}$, by contradiction, were an inconsistent set, we should have $\Gamma_i \cup \neg \forall x P(x) \vdash P(t)$. As t by construction does not occur in Γ_i, neither in $\neg \forall x P(x)$ nor in $\forall x P(x)$, we may apply the rule (**R1**) so as to obtain $P(t) \vdash \forall x P(x)$ and conclude $\Gamma_i \cup \{\neg \forall x P(x)\} \vdash \forall x P(x)$. But this implies that $\Gamma_i \cup \{\neg \forall x P(x)\}$ is inconsistent, against our hypothesis.

9.3. COMPLETENESS IN FIRST-ORDER MODAL LOGIC

Consider now the following set of formulas defined as:

$$\Gamma_{max} = \bigcup_{i=0}^{\infty} \Gamma_i$$

Clearly, Γ_{max} is the required maximal consistent extension of Γ. Furthermore, from the definition of ω-completeness (ω**C2**), it follows that Γ_{max} is an ω-complete maximal consistent extension of Γ. ♠

We are finally in conditions to prove the most important step of the inductive argument, part of which is left as an exercise. The symbols $w, w', w'', \cdots$ will now be used to denote maximal consistent omega-complete extension of **QL**$^=$. The canonical model $\mathcal{M}^w$ on **QL**$^=$ is now built starting from some set w in W.

Lemma 9.3.5 *Let W be the collection of all maximal consistent and ω-complete extensions of **QL**$^=$. Then, for $w \in W$, $\mathcal{M}^w, w \vDash \forall x P(x)$ iff $\forall x P(x) \in w$.*

Proof: (Sketch) The steps of the double-inductive proof are the following (recall the definition of $\mathcal{M}^{x/d}$ on page 243): $\mathcal{M}^w, w \vDash \forall x P(x)$ iff

1. For every d of the domain D, $\mathcal{M}^{w\ x/d}, w \vDash P(x)$, iff

2. For every term t, $\mathcal{M}^{w\ x/V(t)}, w \vDash P(x)$, iff

3. For every term t, $\mathcal{M}^w, w \vDash P(t)$, [16] iff

4. For every term t, $P(t) \in w$ (from (3) by induction hypothesis), iff

5. $\forall x P(x) \in w$ (from (4) by the ω-completeness of w and (**R1**))

♠

We must now consider what happens in extending this method of proof to the construction of the canonical model for quantified modal logics. One might think that nothing forbids building such a model along the lines used for propositional modal logics: in other words identifying ω-complete extensions of the reference system with the worlds of the canonical model. There is apparently no problem in defining a canonical model for some modal logic **QS**$^=$+(**BF**) by starting from a couple $\langle W, R \rangle$, where:

---
[16] The equivalence between line (2) and line (3) is proved by induction on the length of $P(x)$ (Exercise 9.10).

(i) $W = \{w, w', w'', \cdots\}$ is the set of the maximal consistent extensions of $\mathbf{QS}^=+(\mathbf{BF})$ which contain $t = t'$ whenever $t = t'$ belongs to w in W.

(ii) wRw' iff $Den(w) \subseteq w'$, i.e, if $\Box\alpha \in w$, then $\alpha \in w'$ (as in Definition 4.2.6).

A seemingly obvious idea is then to define the canonical model for a quantified modal logic $\mathbf{QS}^=+(\mathbf{BF})$ in the style used for the propositional case, which now would be a 4-tuple of form $\langle W, R, D, V \rangle$. The Barcan Formula (**BF**), as already remarked, corresponds to the idea that D is unique and is independent from the worlds in W. Given the outlined construction, it would be natural to think that the general theorem for canonical models could be proved along the lines of Chapter 4.

Unfortunately, this strategy runs up against a difficulty which is generated by the problem of granting joint satisfaction to the following two conditions:

(i) If w' is a maximal consistent extension such that wRw' (so $Den(w) \subseteq w'$) w' contains every β such that $\Box\beta \in w$.

(ii) w' is ω-complete in the defined sense.

It is true, in fact, that we can consistently grant each (i) and (ii) separately, but this does not mean that we can consistently jointly grant (i) and (ii). Suppose in fact that in w we have $\Box(t = t)$, $\Box(t' = t')$ etc. (i.e., all the infinite substitution instances of $\Box(x = x)$). Then all the infinite arguments of the modal operators of this form will be in w, but in this case proving ω-completeness becomes a puzzling task. Let us recall in fact that to grant ω-completeness, we have to know that some t is a term that does not occur in any of the wffs of the consistent set which is built up to that point of the construction. But satisfying this requirement becomes impossible if we have to compare t with an infinite number of terms.

In order to solve the problem, we have to perform a deviation which passes through the proof of the following four lemmas, which make an essential use of (**BF**).[17] The key idea is that avoiding maximal consistent extensions containing extra terms we can preserve ω-completeness.

Lemma 9.3.6 *If w is an ω-complete set, then $w \cup \Gamma_0$ is also ω-complete, provided that Γ_0 is a finite set of wffs.*

---

[17] As a matter of fact, this strategy for a solution cannot be generalized to arbitrary quantificational modal systems. It does not work for all systems lacking (**BF**), or characterized by frames whose accessibility relations are not preserved under subsets.

9.3. COMPLETENESS IN FIRST-ORDER MODAL LOGIC

Proof: Let us suppose that $w \cup \Gamma_0 \vdash P(t)$ for all the terms t and let $\bigwedge \Gamma_0$ be the conjunction of the elements of Γ_0. It follows that $w \vdash \bigwedge \Gamma_0 \supset P(t)$ for all the terms t. As w is ω-complete, $w \vdash \forall x (\bigwedge \Gamma_0 \supset P(x))$ for every arbitrary choice of the variables. If x does not occur in $\bigwedge \Gamma_0$, we have by (**R1**) $w \vdash \bigwedge \Gamma_0 \supset \forall x P(x)$, so $w \cup \Gamma_0 \vdash \forall x P(x)$. By applying the laws of **QL**, we can replace the variable x of $\forall x P(x)$ for any other variable. It follows, then, that whenever we have that $w \cup \Gamma_0 \vdash P(t)$ for all terms t, we always have $w \cup \Gamma_0 \vdash \forall x P(x)$. It is thus proved that $w \cup \Gamma_0$ is ω-complete. ♦

Lemma 9.3.7 *Any ω-complete set w may be extended to a maximal consistent ω-complete set $\overline{w}$ whose formulas contain the same symbols of the formulas occurring in w.*

Proof: Let us build a maximal consistent extension of w by using a variant of the method described in Chapter 4 (and also used in Lemma 9.3.4). Let $\Gamma_{i+1} = \Gamma_i \cup \{\neg \forall x P(x)\}$ be a consistent set. As $\Gamma_i \cup \{\neg \forall x P(x)\}$ is consistent, it follows, by the ω-completeness in formulation (ω**C2**) (see Section 9.2), that $\Gamma_i \cup \{\neg P(t)\}$ is also consistent for at least one term t (note that t could already be in w). Hence $\neg P(t)$ and $\neg \forall x P(x)$ may be consistently added to Γ_i. The result is a finitary extension of Γ_i and, by Lemma 9.3.6, if Γ_i is ω-complete, so also is Γ_{i+1}. By a standard argument, it may be then verified that the union $\overline{w}$ of the Γ'_is is an ω-complete extension of w. ♦

Lemma 9.3.8 *If w is an ω-complete maximal consistent extension which contains $\neg \Box \beta$, then $w^* = Den(w) \cup \{\neg \beta\}$ is consistent and ω-complete.*

Proof: The method to prove that w^* is consistent is an extension of the one used in modal propositional logic (cf. Section 4.2), and we concentrate on the first-order cases only. By Lemma 9.3.6, w^* is ω-complete provided that so is $Den(w) = \{\alpha : \Box \alpha \in w\}$. Let us now assume that $Den(w) \vdash P(t)$ for every term t. Thanks to rule (**DR1**), which belongs to **K** and to every normal system, $w \vdash \Box P(t)$ for every term t; and as w is ω-complete, it follows that $w \vdash \forall x \Box P(x)$. By the Barcan Formula, it follows that $w \vdash \Box \forall x P(x)$. Since w is maximal, $\Box \forall x P(x) \in w$, so $\forall x P(x) \in \{\alpha : \Box \alpha \in w\}$. It follows then that $Den(w) \vdash \forall x P(x)$. Thus $Den(w)$ is ω-complete and so is w^*. ♦

Lemma 9.3.9 *If w is an ω-complete maximal consistent set which contains $\neg \Box \beta$ then $w^* = Den(w) \cup \{\neg \beta\}$ may be extended to a ω-complete maximal consistent set in which the same symbols of w^* occur.*

Proof: By Lemma 9.3.8, w^* is consistent and ω-complete. By Lemma 9.3.7, it may be extended to a ω-complete maximal consistent set containing the same symbols. ♠

Given Lemmas 9.3.6–9.3.9, we have the tools to prove the required completeness theorem. Let **QS$^=$+(BF)** be any normal modal logic as defined above. So, what follows can be proven.

Proposition 9.3.10 *α is QS$^=$+(BF)-consistent only if α holds in a QS$^=$+(BF)-model.*

Proof: Let $\mathcal{M}^\omega = \langle W, R, D, V \rangle$ be the **QS$^=$+(BF)**-canonical model. Let w be a ω-complete extension of **QS$^=$+(BF)**. W will be now the set of consistent ω-complete extensions w' of **QS$^=$+(BF)** which contain $t = t'$ only in the case in which $t = t'$ belongs to w. R is the same as in standard modal models, and D and V are described as in Definition 9.3.3.

The proof is by induction on the complexity of α. The basis of the inductive argument consists in the equivalence $\mathcal{M}^\omega, w \models P(t_1, \cdots, t_n)$ iff $P(t_1, \cdots, t_n) \in w$.

Induction works as in the propositional case for truth-functional formulas. For formulas of form $\forall x P(x)$, the argument has been already exposed in Lemma 9.3.5. It remains now to complete the argument considering wffs of form $\Box \alpha$. The argument goes the same as it does for the propositional case (see Section 4.2) except for a modification which we have to introduce to establish what follows:

(*) If $\neg \Box \alpha \in w$, there is an element w' of W such that $\neg \alpha \in w'$ and wRw'.

By Lemma 9.3.9, we know that there is a maximal consistent extension w' of $Den(w) \cup \{\neg \alpha\}$. We can prove that such w' is an element of W if we can prove that $t = t' \in w'$ iff $t = t' \in w$. Given that w' is a member of W, we already know that $t = t' \in w'$ iff $t = t' \in w$. Let us remark that, if $t = t' \in w'$, this implies by (**RI1**) $\Box(t = t') \in w'$, so $t = t' \in w$. If $t = t' \notin w'$, then $t \neq t' \in w'$, hence $\Box(t \neq t') \in w$ by (**RI2**).

It follows that w' contains exactly the same identities of w and, for this reason, it is an element of W. Since $Den(w)$ is a subset of w', we conclude that wRw'. This concludes the proof of (*). ♠

9.4 Inclusive domains and arbitrary domains

If we drop the requirement that the domain D on which the variables are interpreted is necessarily unique for any given model, we are left with various alternatives, among which two at least seem appealing. The first is to let the domains vary independently from the possible worlds to which they are associated; the other is to define some kind of dependence among the relation between domains and the relation between worlds. The most obvious connection which can be established is of course the one between inclusion of domains and accessibility.

Let us begin with the second class of systems. We shall use the name **QS$^=$** (where **S** is any normal modal system) for any system of quantified modal logic with identity which is as the above studied systems **QS$^=$+(BF)** except for the omission of **(BF)**.

Recalling what we have called indexed π-models on page 249, we define:

Definition 9.4.1 *A QS$^=$-model is a 5-tuple $\langle W, R, D, f, V \rangle$, where W, R, D are as for systems with (BF), f is a function such that, for every world w in W, $f(w) = D_w$ (where, for every w, $D_w \subseteq D$): in other words f assigns to every world w a subset D_w of the general domain D.*

As anticipated, the correlation is established by the following Preservation Condition:

(Pres) if wRw', then $D_w \subseteq D_{w'}$.

$V(t)$ and $V(P)$ are defined as in models for logics with **(BF)**, taking their values in D. But now, it may happen that some of the $V(t_1), \cdots, V(t_n)$ such that $t_1, \cdots, t_n$ occur in the formula under evaluation are in D but not in D_w, i.e., they are not in the domain associated to w to which the value assignment is relativized. In such a case, we say that the relation $\vDash$ is *undefined*.

If α is atomic (i.e., it has the form $P(t_1, \ldots, t_n)$), we have that, for any $w \in W$, if every one of the $V(t_1), \cdots, V(t_n) \in D_w$, then:

1. $\mathcal{M}, w \vDash \alpha$ if $\langle V(t_1), \cdots, V(t_n), w \rangle \in V(P)$

2. $\mathcal{M}, w \nvDash \alpha$ if $\langle V(t_1), \cdots, V(t_n), w \rangle \notin V(P)$

Otherwise, $\vDash$ is undefined.

As truth-functional wffs are concerned, the definition runs as in standard models (under the proviso that $\vDash$ is defined for all subformulas). For quantified wffs, again under the condition that every one of the $V(t_1), \cdots, V(t_n) \in D_w$, we have what follows:

(Quant) For any α and w:

(a) $M, w \models \forall x\alpha$ if for every M' which is like M except for having in place of V a V' which is like V, except for assigning to x a different member of D_w, it holds $M', w \models \alpha$.

(b) $M, w \not\models \forall x\alpha$ if for every M' as before, $M', w \not\models \alpha$.

Otherwise, $\models$ is undefined.

For wffs of form $\Box \alpha$, the truth conditions are as in standard propositional modal logic, provided that $\models$ is defined, where $M, w \models \Box \alpha$ is defined iff $M, w' \models \alpha$ is defined for every w' such that wRw' (as in Definition 9.4.1).

A wff will be said $\mathbf{QS}^=$-valid when $M, w \models \alpha$ for every $\mathbf{QS}^=$-model in which $\models$ is defined.

It is routine work to prove that the axioms of $\mathbf{QS}^=$ are valid for this semantics, for every $\mathbf{S}$, and that the rules preserve validity. One may also show that the Barcan Formula (BF) is falsified by this semantics by going back, *mutatis mutandis*, to the argument already developed for the logics with propositional quantifiers (see Exercise 9.4). Simply note that in the given argument the two domains Π^w and $\Pi^{w'}$ were such that $\Pi^w \neq \Pi^{w'}$ but $\Pi^w \subseteq \Pi^{w'}$, so that the relevant model satisfies the Preservation Condition.

It is to be noted that, as we already saw, (BF) turns out to be a theorem in $\mathbf{QS5}^=$ and in the weaker $\mathbf{QKB}^=$. The semantical analysis enlightens this syntactical property of such systems. It is enough to remark that in both KB- and S5-models R is symmetric, so wRw' implies $w'Rw$. From the Preservation Condition, this means that for every w and w', $D_w \subseteq D_{w'}$ implies $D_{w'} \subseteq D_w$, so $D_w = D_{w'}$. Therefore symmetry implies the invariance of all the domains indexed by possible worlds, hence the omnipresence of a unique domain. We may then conclude that (BF) holds in any symmetric frame of the defined kind.

We have to consider now the last suggested option, namely, the idea of defining models with arbitrary domains. This semantics results from the preceding one by dropping the Preservation Condition and simply retaining the proviso that every domain D_w associated to w is always a subset of a unique domain D.[18]

The definition of the value assignment V here is the same as the one given for models with constant domains as far as non quantified formulas are concerned. But now the free variables take values in D, and $V(P)$ is always a set of ordered $n + 1$-tuples whose members are all in D, even if

---

[18] This semantics has been first formulated in S. Kripke [Kri63b].

9.4. INCLUSIVE DOMAINS AND ARBITRARY DOMAINS

they need not be in the D_w indexed by the reference world w. In other words: $M, w \vDash P(x)$ if and only if $\langle V(x), w \rangle \in V(P)$, so that $\vDash$ is always defined.

The evaluation of the quantified wffs takes the same form of **(Quant)** of page 264, since quantified variables are again relativized to indexed subdomains of D, with the difference that $\vDash$ is always defined.

A remarkable consequence of this semantics is that it validates a set of theorems which is anomalous from the viewpoint of standard quantificational logic. It is easy to see, in fact, that the wff $\forall x P(x) \supset P(y)$ turns out to be invalid, so that the axiom **(QL2)** is also invalid. In fact, it may be seen that $M, w \vDash \forall x P(x)$, while $M, w \nvDash P(y)$ as the free variables take their values in D, while the universal quantifier is relativized to the indexed domain. It is easy to verify, however, that the universal closure of $\forall x P(x) \supset P(y)$, i.e., $\forall y(\forall x P(x) \supset P(y))$, turns out to be valid. So in the axiomatic basis for this system (see Exercise 9.14), the classical axiom **(QL1)** must be substituted by its universal closure.

A second interesting feature of systems equipped with this semantics concerns the fact that **(BF)** is refuted by the associated semantics even when the base system is **S5**. We may see in fact that there is a falsifying model for **(BF)** in which R is an equivalence relation. For the refutation, it is enough to consider a model M with the following properties:

- $W = \{w_1, w_2\}$
- $w_1 R w_1, w_1 R w_2, w_2 R w_1, w_2 R w_2$
- $D = \{u_1, u_2\}$, $D_1 = \{u_1\}$ and $D_2 = \{u_1, u_2\}$
- $V(x) = u_1$
- $V(P) = \{\langle u_1, w_1 \rangle, \langle u_1, w_2 \rangle\}$

Thanks to the clause (7) of the definition of **QS⁼**-model (Section 9.3) $M, w_1 \vDash \Box P(x)$. Since u_1 is the only element of D_1, we have $M', w_1 \vDash \Box P(x)$ for every M' which assigns to x some member of D_1; so also $M, w_1 \vDash \forall x \Box P(x)$. But $\langle u_2, w_2 \rangle \notin V(P)$, hence $M, w_2 \nvDash \forall x P(x)$, so the consequent of **(BF)** $\Box \forall x P(x)$ turns out to be false at w_1.

The relation R of this falsifying model is reflexive, symmetric and transitive: the model considered has then the properties of an **S5**-model. It follows that **(BF)** is not a theorem of such a system, and *a fortiori* also fails in all the weaker systems.

Another remarkable difference which characterizes such systems concerns the converse of Barcan Formula, which turns out to be invalid in all such systems. In fact, consider a model M where W and $V(x)$ are as before but with the following features:

- $D_1 = \{u_1, u_2\}$
- $D_2 = \{u_1\}$
- $V(P) = \{\langle u_1, w_1 \rangle, \langle u_2, w_1 \rangle, \langle u_1, w_2 \rangle\}$
- R is a universal relation

Then $M', w_1 \vDash P(x)$, for every V' assigning to x some element of D_1, and $M, w_1 \vDash \forall x P(x)$. In an analogous way, $M, w_2 \vDash \forall x P(x)$. Therefore $M, w_1 \vDash \Box \forall x P(x)$. But $M, w_1 \nvDash \forall x \Box P(x)$, as $M', w_1 \nvDash \Box P(x)$, where M' is like M except for the alternative assignment $V'(x, w_1) = \{u_2\}$. So $\Box \forall x P(x) \supset \forall x \Box P(x)$ is invalid for this semantics and is underivable in all axiomatic systems of this family.

A philosophical conclusion should be drawn. In the last modelization, as domains may vary in arbitrary ways, something that does not exist in a world w – say Pegasus – might exist in another imaginary world w'. Vice versa, what does exist at w might be nonexisting at some other world w'. This semantics makes the requirement that terms should stand for rigid designators especially problematic. As far as descriptions are concerned, in order to treat, for instance, "the president of U.S.A in 2007" as a rigid designator, we should introduce the restriction that it always refers to the same person at all worlds in which such a person exists, while "The Morning Star" refers to a certain planet at all worlds in which such planet exists. On the other hand, we have to stress that the more generic syntagma "the president of U.S.A." is not a descriptive way to refer to a single object, but is a predicate, and should be treated as such.

As a general comment about modality and quantification, we have to remark that the various strategies which have been devised to associate a semantics to quantified modal logics give origin to several difficulties in proving completeness results for such logics. It suffices to say that if **S** is a complete propositional system, the first order system **QL+S+(BF)** (without axioms for identity) needs not in general be a complete system. For instance, it has been proved that **QL+S4+(BF)+(McK)** is an incomplete system even if **S4+(McK)** is not such. The same holds for the system **S4.2** (i.e., **S4** + the converse of (**McK**)): even if **S4.2** and also **QL+S4.2** are complete systems, **QL+S4.2+(BF)** is not so. As a matter of fact, it has been proved that incompleteness in quantificational modal logic is not a sporadic phenomenon but appears to maintain some structural dependence on the features of relational frames themselves. For instance, it has been proved that, if **S** extends **S4** but is weaker than **S5** and some theorem of **S** is refuted in some finite connected frame, then **QL + S** is an incomplete system. Such results strongly suggest that a progressive direction of inquiry in quantified modal logic could go towards finding technical and conceptual resources stronger than the ones offered by relational semantics.

9.5 Quantification and multimodalities

When we depart from monomodal languages and pass to multimodal languages, the situation is complicated by the fact that, in principle, we face the possibility of introducing a constant domain semantics for some modal operators of the language, while for others it may make sense to adopt a different choice. The Barcan Formula, for instance, might be a valid axiom for one of the modal operators of the language but invalid for other operators.

The case of tense logic offers a meaningful instance. It makes sense, in fact, to think of the future as the "field of the possible", while the past would be the "field of the real". Past entities might be thought of as ordered by inclusion while the realm of the future might be seen as a unique domain of possible entities.

However, one could notice that other interesting distinctions could be drawn by distinguishing not different operators, but different quantifiers. Instead of introducing distinct domains, say one for actual entities and another one for possible entities, we may introduce two kinds of quantifiers, one for actual objects and another for possible objects, the so-called *possibilia*. The symbols which we will employ are $\forall^a$ (from which $\exists^a$ is defined in the expected way) and $\forall$, respectively.

We may thus axiomatize a system $\mathbf{DQ}^=$ for double quantification and identity as follows:

(DQ0) The set of all truth-functional tautologies

(DQ1) $\forall x(\alpha \supset \beta) \supset (\forall x\alpha \supset \forall x\beta)$

(DQ2) $\forall^a x(\alpha \supset \beta) \supset (\forall^a x\alpha \supset \forall^a x\beta)$

(DQ3) $\alpha \supset \forall x\alpha$, x not free in α

(DQ4) $\forall x\alpha \supset \forall^a x\alpha$

(DQ5) $\forall x\alpha \supset \alpha[x/t]$, provided t is free for x in α

(DQ6) $\forall x(x = x)$

(DQ7) $\forall^a x \exists^a y(y = x)$

(DQ8) $x = y \supset (\alpha \supset \beta)$, where α and β differ only in that α has a free x in zero or more places where β has a free y

To the given basis, we add the axioms for the **PF**-systems examined in Chapter 6, the weakest of which is $\mathbf{K}_t$. A frame for such a logic will be a 6-tuple $\langle T, \overleftarrow{R}, \overrightarrow{R}, U, P, f \rangle$ where, intuitively, T is a set of instants, $\overleftarrow{R}$ and $\overrightarrow{R}$ are converse relations, U is the set of actual objects, P is the set of *possibilia* and f is as introduced in Definition 9.4.1 with the difference that, for every instant t, $f(t) = U_t$, where $U_t \subseteq U$. While *possibilia* are invariant with respect to possible worlds, this does not hold for actual objects. Moreover, axiom (**DQ4**) implies that each actual object is *a fortiori* possible. We will have then, for every t in T:

1. $U_t \subseteq P$

The truth conditions for non-quantified formulas are analogous to the ones we already know for $\mathbf{QS}^=$.

For quantified wffs, we have the following clauses:

2. $\mathcal{M}, t \vDash \forall x \alpha$ iff $\mathcal{M}', t \vDash \alpha$ for every model $\mathcal{M}'$ where V' assigns to all variables except x the same elements in P which are assigned to them by V.

3. $\mathcal{M}, t \vDash \forall^a x \alpha$ iff $\mathcal{M}', t \vDash \alpha$ for every model $\mathcal{M}'$ where V' assigns to all variables except x the same elements in U which are assigned to them by V.

Intuitively, then, we can now distinguish two senses of "exists", one referring to *possibilia* and one referring to entities which have been, are or will be actual. In principle, we might also be willing to accept (**BF**) in the variant $\Box \forall x P(x) \supset \forall x \Box P(x)$ but not in the variant $\Box \forall^a x P(x) \supset \forall^a x \Box P(x)$, which appears to be less intuitive.

The question remains untreated as to whether or not the existential quantifier $\exists$ is indeed suitable to the notion of "possibly existent". In fact, by applying the principle (**DQ5**), we have the validity of the law:

$$x = x \supset \exists y(y = x)$$

But an instance of this could be "(the round square = the round square) $\supset \exists y(y=\text{the round square})$", from which one should conclude that the round

9.5. QUANTIFICATION AND MULTIMODALITIES

square belongs to the set of possible beings. This step could be blocked if we put on axiom (**DQ5**) the restriction that the quantified objects should be really existent objects (as it happens in logics which have been called *free logics*). But what is real existence? We cannot accept the idea that real existence could be defined by introducing an "existence predicate" in this way:

$$Ex \stackrel{\text{Def}}{=} \exists y(x = y)$$

since this would imply that everything exists, being $\exists y(x = y)$ a logical truth. An alternative viewpoint is the idea that what is characteristic of real existent beings, as opposed to imaginary or abstract existent beings, is not to be identical with something but to be endowed with contingent properties. On the contrary, it seems that imaginary or unexisting objects have non-contingent properties: either impossible properties (as the round square) or necessary properties, such as in the case of Pegasus, the property of "being the winged horse ridden by Bellerophon". This notion of real existence could then be formally represented by a second-order quantification in the following way:

$$E!a \stackrel{\text{Def}}{=} \exists Q(Qa \land \nabla Qa)$$

In a temporal logic context, of course, $\Diamond, \Box, \nabla$, could also be defined in terms of $\diamondsuit$ and $\diamondsuit$, as it has been seen in Chapter 6, so giving a temporal meaning to real existence (what exists has some non-eternal properties). In this way, we have to again take into consideration the deep interconnection between second-order language and modal language, already evidenced in correspondence theory.

But having at our disposal a multimodal language gives us the opportunity to look in a different way at the relationship between quantifiers and modal operators: in fact, all or some quantifiers could be seen, themselves, as modal operators of a special kind.

To begin with, we could introduce the concept of *existential modalities*. If $\alpha, \beta, \gamma, \cdots$ are variables for wffs, it is possible to introduce an operator $\Box^+$ such that $\Box^+\alpha$ may be intended to say that *everything* has the properties indicated in α. $\Diamond^+\alpha$ will be read as stating that *something* has the properties indicated in α. Part of what is expressible in the language of quantification theory may be then translated into modal language and vice versa.

This basic idea may be developed by using again the ideas which are at the root of possible worlds semantics. Let us recall that a value assignment V to all the atomic variables of the language may be thought of as a possible world. How may we now conceive an accessibility relation between assignments?

In the first place let us define an identity between assignments by putting $V \equiv_i V'$ iff for all $j \neq i$, $V(x_j) = V'(x_j)$. We may say then that $\equiv_i$ is an accessibility relation between two assignments V and V' when they are coincident except at the i-th place. Then the truth conditions for the existential quantifier, where x_i is an arbitrary variable, may then be adjusted as follows, where V is an assignment belonging to an interpretation $\mathcal{M}$:

(∃) $\mathcal{M}, V \vDash \exists x_i \alpha$ iff there is an assignment V' to the variables such that $\mathcal{M}, V' \vDash \alpha$, and $V \equiv_i V'$.

By an appeal to modal notation, $\exists x_i \alpha$ behaves then as $\Diamond_i \alpha$, and $\forall x_i \alpha$ behaves as $\Box_i \alpha$, where $\Diamond_i$ and $\Box_i$ are indexed modal operators.

We conclude then that all standard quantification theory could in principle be reconstructed as a multimodal logic whose language has an infinite number of primitive modal operators. *A fortiori*, any system of first-order modal logic could be reformulated as a special multimodal system.

This way to approach quantificational logic opens progressive perspectives which, although complex, are thoroughly technically treatable. The investigation to be pursued in this direction is at the beginning, but we cannot avoid observing that, after creating an impressive number of new varieties of logical systems, modal logic evidences its amazing versatility by working as a unifying tool between the modal and non-modal sides of the science of logic.

9.6 Exercises

1. Prove rules (E∀) (I∀), (E∃) and (I∃) as derived rules of **QL**. Prove that (E∃) is equivalent to $\dfrac{\exists \alpha[x] \supset \gamma}{\alpha[x] \supset \gamma}$ (x not free in γ).

2. Add the following to the definition of a **K**π-model: Π is the power set of W. Show that this condition corresponds to the idea that each world is describable, expressed by the following wff: $\exists p(p \land \forall q(q \supset \Box(p \supset q)))$.

3. Show the soundness of **K**π with respect to the class of all **K**π-models.

4. Show the soundness of **QK**$^=$+(**BF**) with respect to the semantics given in Section 9.3.

5. Prove that $\Diamond(x = y) \supset (x = y)$ (i.e. (**I2**)) is a theorem of **QL**+ **KT** + (**F**) (see Remark 3.2.15) + (**I1**).

9.7. FURTHER READING

6. Prove in $\mathbf{K}\pi$ the formula $\Box\forall p\alpha \supset \forall p\Box\alpha$, which is the converse of the Barcan formula.

7. Define an $\mathbf{S4}\pi$-model $\langle W, R, \Pi, V\rangle$ (with R transitive) that falsifies the Barcan formula for propositional quantifiers.

8. Give a rigorous proof that $\exists p(\Box p \wedge \neg p)$ is true on a $\mathbf{K}\pi$-frame iff R is irreflexive. Show also that $\exists p(\neg\Box p \wedge p)$ is true on a frame iff for every world in the frame there is at least one accessible world.

9. Show that rule (**RI2**) is derivable in every extension of $\mathbf{KB+QL+(R1)}$.

10. Show, by induction on length of terms, that for each term t, $v(P(t)) = 1$ iff $v(P(x)[x/t]) = 1$ (cf. Lemma 9.3.5).

11. Show that condition ($\omega\mathbf{C}$) on page 258 is equivalent to conditions ($\omega\mathbf{C1}$) and ($\omega\mathbf{C2}$).

12. Show that $\mathbf{QK4}^=+(\mathbf{BF})$ is sound and complete with respect to the class of frames $\langle W, R, D\rangle$ defined in Section 9.2 where R is transitive.

13. Eliminate from $\mathbf{QK}^=+(\mathbf{BF})$ the rules (**RI1**) and (**RI2**), and prove completeness for this reduced system.

14. Give a rigorous proof that the semantics with variable domains validates $\forall y(\forall yP(y) \supset P(y))$ but does not validate $\forall xP(x) \supset P(y)$.

9.7 Further reading

Propositional quantifiers have been associated with modal logic ever since C. I. Lewis and C. H. Langford in [LL32], where they are introduced to fill the expressive gap between modal logic and quantificational logic and, more specifically, to express the independence between the material conditional and strict implication. An important systematic treatment is the Ph.D. thesis of K. Fine, later abridged in [Fin70].

Essential references on counterpart theory are D. Lewis [Lew69], A. Hazen [Haz79] and T. Brauner and S. Ghilardi [BG07]. On counterpart theory and *de re* modalities, see A. Plantinga [Pla74].

For the incompleteness of the first-order extension of $\mathbf{S4} + (\mathbf{McK})$ and $\mathbf{S4.2}$, see G. E. Hughes and M. J. Cresswell [HC96] pp. 265 ff. The discovery of the inadequacy of relational semantics for quantified modal logic is an important, and in a sense revolutionary, result mainly due to S. Ghilardi (for a recent survey see T. Brauner and S. Ghilardi [BG07]).

The literature about quantified modal logic has undergone an uncontrolled growth from S. Kripke [Kri63a] onwards, and even compiling an essential bibliography is problematic. For a recent survey on intensional logic, see M. C. Fitting [Fit06]. For so-called free logics (logics free from existential presuppositions), see E. Bencivenga [Ben83] (reprinted as [Ben84]).

The technique used to prove completeness in the present chapter is inspired by the work of R. Thomason in [Tho70], recovered by J. Garson in [Gar84]. The double quantification was introduced by N. B. Cocchiarella in [Coc66a], while the reduction of quantifiers to modal operators, suggested by G. H. von Wright in [vW57], was more recently developed by M. Marx and Y. Venema (see [MV97]). For other important references see S. Kuhn [Kuh80] and N. Alechina [Ale95].

Bibliography

[AB05] S. Artemov and L. Beklemishev. Provability logic. In D. Gabbay and F. Guenthner, editors, *Handbook of Philosophical Logic*, volume 13, pages 229–403. Springer, Amsterdam, 2005.

[Ale95] N. Alechina. *Modal Quantifiers*. PhD thesis, Institute for Logic, Language and Computation, University of Amsterdam, Amsterdam, 1995.

[Aum76] R. J. Aumann. Agreeing to disagree. *The Annals of Statistics*, 14(6):1236–1239, 1976.

[B05] J.-Y. Béziau. Paraconsistent logics from a modal point of view. *Journal of Applied Logic*, 3:7–14, 2005.

[BdRV01] P. Blackburn, M. de Rijke, and Y. Venema. *Modal Logic*. Cambridge University Press, Cambridge, 2001.

[Ben83] E. Bencivenga. Free logics. In D. Gabbay and F. Guenthner, editors, *Handbook of Philosophical Logic*, volume 3, pages 373–426. Dordrecht, Reidel, 1983.

[Ben84] E. Bencivenga. *Free Logics*. Bibliopolis, Naples, 1984.

[BG07] T. Brauner and S. Ghilardi. First-order modal logic. In P. Blackburn, J. van Benthem, and F. Wolter, editors, *Handbook of Modal Logic*, pages 549–620. Elsevier, Amsterdam, 2007.

[BGM98] M. Baldoni, L. Giordano, and A. Martelli. A tableau calculus for multimodal logics and some (un)decidability results. In H. de Swart, editor, *Proceedings of Tableaux 98–International Conference on Tableaux Methods*, pages 44–59. Springer, Berlin, 1998.

[Boc61] I. M. Bochenski. *A History of Formal Logic*. University of Notre Dame Press, Indiana, 1961.

[Boo79]   G. Boolos. *The Unprovability of Consistency: An Essay in Modal Logic*. Cambridge University Press, Cambridge, 1979.

[Boo93]   G. Boolos. *The Logic of Provability*. Cambridge University Press, Cambridge, 1993.

[BS85]   G. Boolos and G. Sambin. An incomplete system of modal logic. *Journal of Philosophical logic*, 14:351–358, 1985.

[BS04]   B. Brogaard and J. Salerno. Fitch's paradox of knowability. In Edward N. Zalta, editor, *Stanford Encyclopedia of Philosophy*, volume http://plato.stanford.edu. Center for the Study of Language and Information–Stanford University, Stanford, 2004.

[BT99]   P. Blackburn and M. Tzakova. Hybrid languages and temporal logics. *Logic Journal of the IGPL*, 7:27–54, 1999.

[Bul66]   R. A. Bull. That all normal extension of **S4.3** have the finite model property. *Zeitschrift für mathematische Logik und Grundlagen der Mathematik*, 12:609–616, 1966.

[Bul68]   R. A. Bull. An algebraic study of tense logic with linear time. *The Journal of Symbolic Logic*, 33:27–38, 1968.

[Bur84]   J. P. Burgess. Basic tense logic. In D. Gabbay and F. Guenthner, editors, *Handbook of Philosophical Logic*, volume 2, pages 89–133. Dordrecht, Reidel, 1984.

[BvB07]   P. Blackburn and J. van Benthem. Modal logic; a semantic perspective. In P. Blackburn, J. van Benthem, and F. Wolter, editors, *Handbook of Modal Logic*, pages 1–84. Elsevier, Amsterdam, 2007.

[Car47]   R. Carnap. *Meaning and Necessity*. University of Chicago Press, Chicago, IL, 1947.

[Cat88]   L. Catach. Normal multimodal logic. In T. Mitchell and R. Smith, editors, *Proceedings of AAAI'88– Seventh National Conference on Artificial Intelligence*, pages 491–495. The AAAI Press, Menlo Park, CA, 1988.

[Cat89]   L. Catach. *Les Logiques Multimodales*. PhD thesis, Univérsité de Paris VI, France, 1989.

[Cat91]   L. Catach. Tableaux: a general theorem prover for modal logics. *Journal of Automated Reasoning*, 7(4):489–510, 1991.

BIBLIOGRAPHY

[CC07] W. A. Carnielli and M. E. Coniglio. Combining logics. In Edward N. Zalta, editor, *Stanford Encyclopedia of Philosophy*, volume http://plato.stanford.edu. Center for the Study of Language and Information–Stanford University, Stanford, 2007.

[CCG+07] W. A. Carnielli, M. E. Coniglio, D. Gabbay, P. Gouveia, and C. Sernadas. *Analysis and Synthesis of Logics: How to Cut and Paste Reasoning Systems*. Springer, Amsterdam, 2007.

[CCM07] W. A. Carnielli, M. E. Coniglio, and J. Marcos. Logics of formal inconsistency. In D. Gabbay and F. Guenthner, editors, *Handbook of Philosophical Logic*, volume 14, pages 1–93. Springer, Amsterdam, 2007.

[CD97] W. A. Carnielli and I. M. L. D'Ottaviano. Translations between logical systems: a manifesto. *Logic et Analyse*, 40(157):67–81, 1997.

[Che80] B. F. Chellas. *Modal Logic: An Introduction*. Cambridge University Press, Cambridge, 1980.

[Chi63] R. M. Chisholm. The logic of knowing. *Journal of Philosophy*, 60:773–795, 1963.

[CM02] W. A. Carnielli and J. Marcos. A taxonomy of C-systems. In W. A. Carnielli, M. E. Coniglio, and I. M. L. D'Ottaviano, editors, *Paraconsistency - The Logical Way to the Inconsistent*, volume 228 of *Lecture Notes in Pure and Applied Mathematics*, pages 1–94. Marcel Dekker, New York, 2002.

[Coc66a] N. B. Cocchiarella. A logic of actual and possible objects. *The Journal of Symbolic Logic*, 31:688–689, 1966.

[Coc66b] N. B. Cocchiarella. *Tense Logic: A Study of Temporal Reference*. PhD thesis, U.C.L.A, LA, 1966.

[Con71] J. H. Conway. *Regular Algebras and Finite Machines*. Chapman & Hall, London, 1971.

[Coo71] S. Cook. The complexity of theorem proving procedures. In M. A. Harrison, R. B. Banerji, and J. D. Ullman, editors, *Proceedings of the Third Annual ACM Symposium on Theory of Computing*, pages 151–158. Shaker Heights, Ohio, 1971.

[Cop02]   J. Copeland. The genesis of possible worlds semantics. *Journal of Philosophical logic*, 31:99–137, 2002.

[Cre88]   M. J. Cresswell. Necessity and contingency. *Studia Logica*, 47:145–149, 1988.

[CZ97]   A. Chagrov and M. Zakharyaschev. *Modal Logic*. Clarendon Press, Oxford, 1997.

[Dug40]   J. Dugundji. Note on a property of matrices for Lewis and Langford's calculi of propositions. *The Journal of Symbolic Logic*, 5:150–151, 1940.

[EC00]   R. L. Epstein and W. A. Carnielli. *Computability: Computable Funtions, Logic, and the Foundations of Mathematics, with Computability and Undecidability–A Timeline*. Wadsworth/Thomson Learning, Belmont, CA, 2nd edition, 2000.

[EdC79]   P. Enjalbert and L. Fariñas del Cerro. Modal resolution in clausal form. *Theorethical Computer Science*, 61(1):1–33, 1979.

[Fey65]   R. Feys. *Modal Logic*. Nauwelaerts, Louvain, 1965.

[FH02]   J. M. Font and P. Hájek. On lukasiewicz's four–valued modal logic. *Studia Logica*, 26:157–182, 2002.

[FHMV95]   R. Fagin, J. Y. Halpern, Y. Moses, and M. Vardi. *Reasoning About Knowledge*. MIT Press, Cambridge, 1995.

[FHV92]   R. Fagin, J. Y. Halpern, and M. Y. Vardi. What can machines know? On the properties of knowledge in distributed systems. *Journal of the ACM*, 39(2):328–376, 1992.

[Fin70]   K. Fine. Propositional quantifiers in modal logic. *Theoria*, 36: 336–346, 1970.

[Fit63]   F. Fitch. A logical analysis of some value concepts. *Journal of Symbolic Logic*, 28:135–142, 1963.

[Fit83]   M. C. Fitting. *Proof Methods for Modal and Tense Logics*. Dordrecht, Reidel, 1983.

[Fit91]   M. C. Fitting. Many-valued modal logics I. *Fundamenta Informaticae*, 15:235–254, 1991.

[Fit92]	M. C. Fitting. Many–valued modal logics II. *Fundamenta Informaticae*, 17:55–73, 1992.
[Fit06]	M. C. Fitting. Intensional logic. In Edward N. Zalta, editor, *Stanford Encyclopedia of Philosophy*, volume http://plato.stanford.edu. Center for the Study of Language and Information–Stanford University, Stanford, 2006.
[Gab76]	D. Gabbay. *Investigations in Modal and Tense Logics, with Applications to Problems in Linguistics and Philosophy*. Dordrecht, Reidel, 1976.
[Gär73]	P. Gärdenfors. On the extensions of **S5**. *Notre Dame Journal of Formal Logic*, 14:277–280, 1973.
[Gar84]	J. W. Garson. Quantification in modal logic. In D. Gabbay and F. Guenthner, editors, *Handbook of Philosophical Logic*, volume 2, pages 249–307. Dordrecht, Reidel, 1984.
[Gär88]	P. Gärdenfors. *Knowledge in Flux: Modeling the Dynamics*. MIT Press, Cambridge, 1988.
[GHR94]	D. Gabbay, I. Hodkinson, and M. Reynolds. *Temporal Logic: Mathematical Foundations and Computational Aspects*. Oxford University Press, Oxford, 1994.
[GKWZ03]	D. Gabbay, A. Kurucz, F. Wolter, and M. Zakharyaschev. *Many-Dimensional Modal Logics: Theory and Applications*. North Holland, Amsterdam, 2003.
[Göd32]	K. Gödel. Zum intuitionistischen aussagenkalkül. *Anzeiger der Akademie der Wissenschaften in Wien*, 69:65–66, 1932. mathematisch-naturwissenschaftliche Klasse.
[Gol75]	R. Goldblatt. Solution to a completeness problem of Lemmon and Scott. *Notre Dame Journal of Formal Logic*, 16:405–408, 1975.
[Gol93]	R. Goldblatt. *The Mathematics of Modality*. CLS Publications, Stanford, CA, 1993.
[Grz67]	A. Grzegorczyk. Some relational systems and the associated topological spaces. *Fundamenta Mathematicæ*, 60:223–231, 1967.

[GT75]   R. I. Goldblatt and S. K. Thomason. Axiomatic classes in propositional modal logic. In J. N. Crossley, editor, *Algebra and Logic*, volume 450 of *Lecture Notes in Mathematics*, pages 163–173. Springer, Berlin, 1975.

[Haa74] S. Haack. *Deviant Logic*. Cambridge University Press, Cambridge, 1974.

[Har79] D. Harel. *First-Order Dynamic Logic*, volume 68 of *Lecture Notes in Computer Science*. Springer, Berlin, 1979.

[Haz79] A. Hazen. Counterpart-theoretic semantics for modal logic. *The Journal of Philosophy*, 76(6):319–338, 1979.

[HC68]   G. E. Hughes and M. J. Cresswell. *An Introduction to Modal Logic*. Methuen, London, 1968.

[HC84]   G. E. Hughes and M. J. Cresswell. *A Companion to Modal Logic*. Methuen, London, 1984.

[HC86]   G. E. Hughes and M. J. Cresswell. A companion to modal logic: some corrections. *Logique et Analyse*, 29:41–51, 1986.

[HC96]   G. E. Hughes and M. J. Cresswell. *A New Introduction to Modal Logic*. Methuen, London, 1996.

[HF98]   P. W. Humphreys and J. Fetzer. *The New Theory of Reference, Kripke, Marcus, and Its Origins*, volume 270 of *Synthese Library*. Kluwer, Dordrecht, 1998.

[HG73]   B. Hansson and P. Gärdenfors. A guide to intensional semantics. In *Modality, Morality and Other Problems of Sense and Nonsense: Essays Dedicated to Sören Halldén*, pages 151–167. CWK Gleerup Bokfürlag, Lund, 1973.

[HH91]   C. W. Harvey and J. Hintikka. Modalization and modalities. In T. Seebohm, D. Follesdal, and J. N. Mohanty, editors, *Phenomenology and the Formal Sciences*, pages 59–77. Kluwer, Amsterdam, 1991.

[Hin62] J. Hintikka. *Knowledge and Belief*. Cornell University Press, Ithaca, NY, 1962.

[Hin63] J. Hintikka. The modes of modality. *Acta Phylosophica Fennica*, 16:65–81, 1963.

[HKT00]   D. Harel, D. Kozen, and J. Tiuryn. *Dynamic Logic*. MIT Press, Cambridge, 2000.

[HM85]   J. Y. Halpern and Y. Moses. A guide to the modal logics of knowledge and belief. In A. K. Joshi, editor, *Proceedings of the 9th International Joint Conference on Artificial Intelligence (IJCAI 85)*, pages 480–490. Morgan Kaufmann, Los Angeles, 1985.

[HM92]   J. Y. Halpern and Y. Moses. A guide to completeness and complexity for modal logics of knowledge and belief. *Artificial Intelligence*, 54:319–379, 1992.

[Hod01]   W. Hodges. Logic and games. In Edward N. Zalta, editor, *Stanford Encyclopedia of Philosophy*, volume http://plato.stanford.edu. Center for the Study of Language and Information–Stanford University, Stanford, 2001.

[Hum81]   L. Humberstone. Relative necessity revisited. *Reports on Mathematical Logic*, 13:33–42, 1981.

[Jan90]   R. Jansana. *Una Introducciòn a la Lògica Modal*. Tecnos, Madrid, 1990.

[JdJ98]   G. Japaridze and D. de Jongh. The logic of Provability. In S. R. Buss, editor, *Handbook of Proof Theory*, volume 137 of *Studies in Logic*, pages 475–546. Elsevier, Amsterdam, 1998.

[Kal35]   L. Kalmár. Über die Axiomatisierbarkeit des Aussagenkalküls. *Acta Scientiarum Mathematicarum*, 7:222–243, 1935.

[Kam68]   H. Kamp. *Tense Logic and the Theory of Linear Order*. PhD thesis, U.C.L.A, LA, 1968.

[Kau60]   R. Kauppi. Über die Leibnizsche Logik mit besonderer Berücksichtigung des Problems der Intension und der Extension. *Acta Philosophica Fennica*, 12:1–279, 1960.

[KK62]   W. Kneale and M. Kneale. *The Development of Logic*. Clarendon Press, Oxford, 1962.

[KM64]   D. Kalish and R. Montague. *Logic. Techniques of Formal Reasoning*. Harcourt, Brace & World, New York, 1964.

[Koz79]   D. Kozen. On the representation of dynamic algebras. Technical Report RC7898, IBM Thomas J. Watson Research Center, October 1979.

[Koz80]   D. Kozen. A representation theorem for models of *-free PDL. In *Automata, Languages and Programming - Lecture Notes in Computer Science*, Lecture Notes in Computer Science, pages 351–362. Springer, Berlin, 1980.

[Kra93]   M. Kracht. How completeness and correspondence theory got married. In M. de Rijke, editor, *Diamonds and Defaults*. Kluwer, Dordrecht, 1993.

[Kri59]   S. Kripke. A completeness theorem in modal logic. *The Journal of Symbolic Logic*, 24:1–14, 1959.

[Kri63a] S. Kripke. Semantic analysis of modal logic I, normal propositional calculi. *Zeitschrift für mathematische Logik und Grundlagen der Mathematik*, 9:67–96, 1963.

[Kri63b] S. Kripke. Semantical considerations on modal logic. *Acta Philosophica Fennica*, 16:83–94, 1963.

[Kri65]   S. Kripke. Semantical analysis of intuitionistic logic I. In J. N. Crossley and M. A. E. Dummett, editors, *Formal Systems and Recursive Functions–Proceedings of the Eighth Logic Colloquium Oxford, July 1963*, pages 92–130. North-Holland, Amsterdam, 1965.

[Krö87]   F. Kröger. *Temporal Logic of Programs*. Springer, New York, 1987.

[Kuh80]   S. Kuhn. Quantifiers as modal operators. *Studia Logica*, 39: 145–158, 1980.

[Kuh89]   S. Kuhn. The domino relation: flattening a two-dimensional logic. *Journal of Philosophical Logic*, 18:173–195, 1989.

[Lem66a] E. J. Lemmon. Algebraic semantics for modal logics I. *The Journal of Symbolic Logic*, 31(1):44–65, 1966.

[Lem66b] E. J. Lemmon. Algebraic semantics for modal logics II. *The Journal of Symbolic Logic*, 31(2):191–218, 1966.

[Lem66c] E. J. Lemmon. A note on Halldén incompleteness. *Notre Dame Journal of Formal Logic*, 7:296–300, 1966.

[Len78]   W. Lenzen. Recent work on epistemic logic. *Acta Philosophica Fennica*, 30:1–219, 1978.

[Len79]   W. Lenzen. Epstemologische betrachtungen zu S4, S5. *Erkenntnis*, 14:33–56, 1979.

[Len80]   W. Lenzen. *Glauben, Wissen und Wahrscheinlichkeit*. Springer, Vienna, 1980.

[Lew69]   D. Lewis. *Convention: A Philosophical Study*. Harvard University Press, Cambridge, MA, 1969.

[Lin71]   L. Linsky. *Reference and Modality*. Oxford Readings in Philosophy, Oxford University Press, Oxford, 1971.

[LL32]   C. I. Lewis and C. H. Langford. *Symbolic Logic*. The Appleton-Century Company, New York, 1932. reprinted in paperback by Dover Publications, New York, 1951.

[Löb55]   M. H. Löb. Solution of a problem of Leon Henkin. *The Journal of Symbolic Logic*, 20(2):115–118, 1955.

[Lor55]   P. Lorenzen. *Einführung in die operative Logik und Mathematik*. Springer, Berlin, 1955.

[LS77]   E. J. Lemmon and D. Scott. *An Introduction to Modal Logic*. Blackwell, Oxford, 1977.

[Łuk70]   J. Łukasiewicz. A system of modal logic. In L. Borkowski, editor, *Jan Łukasiewicz's Selected Works*. North-Holland, Amsterdam, 1970.

[Mag82]   R. Magari. Primi risultati sulla varietà di Boolos. *Bollettino della Unione Matematica Italiana*, 6:359–367, 1982.

[Mak66]   D. Makinson. On some completeness theorems in modal logic. *Zeitschrift für mathematische Logik und Grundlagen der Mathematik*, 12:379–384, 1966.

[Mak69]   D. Makinson. A normal modal calculus between **T** and **S4** without the finite model property. *The Journal of Symbolic Logic*, 34:35–38, 1969.

[Mak70]   D. Makinson. A generalization of the concept of a relational model for modal logic. *Theoria*, 36:331–335, 1970.

[Man75] A. B. Manaster. *Completeness, Compactness and Undecidability. An Introduction to Mathematical Logic*. Prentice Hall, Englewood Cliffs, NJ, 1975.

[Mar89] N. M. Martin. *Systems of Logic*. Cambridge University Press, Cambridge, 1989.

[Mat89] B. Mates. *The Philosophy of Leibniz: Metaphysics and Language*. Oxford University Press, New York, 1989.

[McG06] C. McGinnis. Tableau systems for some paraconsistent modal logics. *Electronic Notes in Theoretical Computer Science*, 143: 141–157, 2006.

[Men64] E. Mendelson. *Introduction to Mathematical Logic*. van Nostrand, Princeton, NJ, 1964.

[Mey03] J. J. Meyer. Modal epistemic and doxastic logic. In D. Gabbay and F. Guenthner, editors, *Handbook of Philosophical Logic*, volume 10, pages 1–38. Kluwer, Dordrecht, 2003.

[MH69] J. McCarthy and P. Hayes. Some philosophical problems from the standpoint of Artificial Intelligence. *Machine Intelligence*, 4:463–502, 1969.

[Mon70] R. Montague. Universal grammar. *Theoria*, 36:373–398, 1970.

[MP92] Z. Manna and A. Pnueli. *The Temporal Logic of Reactive and Concurrent Systems: Specification*. Springer, Berlin, 1992.

[MT48] J. C. C. McKinsey and A. Tarski. Some theorems about the sentential calculi of Lewis and Heyting. *The Journal of Symbolic Logic*, 13:1–15, 1948.

[Mug01] M. Mugnai. *Introduzione alla Filosofia di Leibniz*. Einaudi, Turin, 2001.

[MV97] M. Marx and Y. Venema. *Multi-Dimensional Modal Logic*. Kluwer, Dordrecht, 1997.

[OM57a] M. Ohnishi and K. Matsumoto. Corrections to our paper 'Gentzen method in modal calculi I'. *Osaka Mathematical Journal*, 10:147, 1957.

[OM57b]	M. Ohnishi and K. Matsumoto. Gentzen method in modal calculi I. *Osaka Mathematical Journal*, 9:113–130, 1957.
[OM57c]	M. Ohnishi and K. Matsumoto. Gentzen method in modal calculi II. *Osaka Mathematical Journal*, 11:115–120, 1957.
[Par81]	D. Park. Concurrency and automata on infinite sequences. In P. Deussen, editor, *Theoretical Computer Science*, volume 104 of *Lecture Notes in Computer Science*, pages 167–183. Springer, Berlin, 1981.
[Pel00]	F. J. Pelletier. A history of natural deduction and elementary logic textbooks. In J. Woods and B. Brown, editors, *Logical Consequence: Rival Approaches*, volume 1, pages 105–138. Hermes Science Publications, Oxford, 2000.
[Pie06]	A.-V. Pietarinen. Peirce's conributions to possible–worlds semantics. *Studia Logica*, 82:345–369, 2006.
[Piz74]	C. Pizzi. *La Logica del Tempo*. Boringhieri, Turin, 1974.
[Pla74]	A. Plantinga. *The Nature of Necessity*. Clarendon Press, Oxford, 1974.
[Pop94]	S. Popkorn. *First Steps in Modal Logic*. Cambridge University Press, Cambridge, 1994.
[Pra65]	D. Prawitz. *Natural Deduction. A Proof-Theoretical Study*. Almquist and Wiksell, Stockholm, 1965. Second edition by Dover Publications, 2006.
[Pra76]	V. R. Pratt. Semantical considerations on Floyd-Hoare logic. In *Proceedings of the 17th Annual IEEE Symposium on Foundations of Computer Science*, pages 109–121. IEEE, 1976.
[Pra92]	V. R. Pratt. Origins of the calculus of binary relations. In *Proceedings of the 7th Annual IEEE Symposium on Logic in Computer Science*, pages 248–254. IEEE, 1992.
[Pri57]	A. N. Prior. *Time and Modality*. Oxford University Press, Oxford, 1957.
[Pri67]	A. N. Prior. *Past, Present and Future*. Clarendon Press, Oxford, 1967.

[Pri68]   A. N. Prior. *Papers on Time and Tense*. Clarendon Press, Oxford 1968.

[Rau79]   W. Rautenberg. *Klassische und nichtklassische Aussagenlogik*. Friedr. Vieweg & Sohn, Braunschweig, 1979.

[Rei47]   H. Reichenbach. *Elements of Symbolic Logic*. Macmillan, New York, 1947. Reprinted by Dover, 1980.

[Ren70]   M. K. Rennie. Models for multiply modal systems. *Zeitschrift für mathematische Logik und Grundlagen der Mathematik*, 16:175–186, 1970.

[RU71]   N. Rescher and A. Urquhart. *Temporal Logic*. Springer, Vienna, 1971.

[Ryb97]   V. V. Rybakov. *Admissibility of Logical Inference Rules*. Elsevier Science, Amsterdam, 1997.

[Sah75]   H. Sahlqvist. Completeness and correspondence in the first and second order semantics for modal logic. In S. Kanger, editor, *Proceedings of 3th Scandinavian Logic Symposium*, pages 110–143. North-Holland, Amsterdam, 1975.

[Sch60]   T. Schelling. *The Strategy of Conflict*. Harvard University Press, Cambridge, MA, 1960.

[Sch68]   K. Schütte. *Vollständige Systeme modaler und intuitionistischer Logik*. Springer, Berlin, 1968. Vol. 42 of Ergebnisse der Mathematik und ihrer Grenzgebiete.

[Sch89]   G. F. Schumm. Some compactness results for modal logic. *Notre Dame Journal of Formal Logic*, 30:285–290, 1989.

[Sch01]   G. Schurz. Rudolf carnap's modal logic. In W. Stelzner and M. Stöckler, editors, *Zwischen traditioneller un moderner Logik. Nichtklassische Ansätze*, pages 365–380. Mentis, Paderborn, 2001.

[Sco70]   D. Scott. Advice on modal logic. In K. Lambert, editor, *Philosophical Problems in Logic. Some Recent Developments*, pages 143–173. Reidel, Dordrecht, 1970.

[Scr51]   S. J. Scroggs. Extensions of the Lewis system **S5**. *The Journal of Symbolic Logic*, 16:112–120, 1951.

[Seg67] K. Segerberg. Some modal logics based on a three-valued logic. *Theoria*, 33:53–71, 1967.

[Seg68] K. Segerberg. On the logic of tomorrow. *Theoria*, 31:199–217, 1968.

[Seg70] K. Segerberg. Modal logics with linear alternative relations. *Theoria*, 36:301–322, 1970.

[Seg71] K. Segerberg. *An Essay in Classical Modal Logic*. University of Uppsala, Uppsala, 1971.

[Seg73] K. Segerberg. Two-dimensional modal logic. *Journal of Philosophical Logic*, 2:77–96, 1973.

[Seg77] K. Segerberg. A completeness theorem in the modal logic of programs. *Notices of the American Mathematical Society*, 4(6):1–552, 1977.

[Seg82] K. Segerberg. *Classical Propositional Operators: An Exercise in the Foundations of Logic*. Clarendon Press, Oxford 1982.

[Seg95] K. Segerberg. Belief revision from the point of view of doxastic logic. *Logic Journal of the IGPL*, 3(4):535–553, 1995.

[Smo85] C. Smorynski. *Self-Reference and Modal Logic*. Springer, New York, 1985.

[Smu68] R. M. Smullyan. *First-Order Logic*. Springer, New York, 1968.

[Sol76] R. Solovay. Provability interpretations of modal logic. *Israel Journal of Mathematics*, 25:287–304, 1976.

[Tap84] B. Tapscott. Correcting the tableaux procedure for S4. *Notre Dame Journal of Formal Logic*, 25:241–249, 1984.

[Tar41] A. Tarski. On the calculus of relations. *The Journal of Symbolic Logic*, 6(3):73–89, 1941.

[Tho70] R. H. Thomason. Some completeness results in modal predicate calculus. In K. Lambert, editor, *Philosophical Problems in Logic*, pages 56–76, Reidel, Dordrecht, 1970.

[Tho72] S. K. Thomason. Noncompactness in propositional modal logic. *The Journal of Symbolic Logic*, 37:716–720, 1972.

[Tho74]   S. K. Thomason. An incompleteness theorem in modal logic. *The Journal of Symbolic Logic*, 40:30–34, 1974.

[Urq81]   A. Urquhart. Decidability and finite model property. *The Journal of Philosophical Logic*, 10:367–370, 1981.

[vB76]   J. van Benthem. *Modal Correspondence Theory*. PhD thesis, University of Amsterdam, Amsterdam, 1976.

[vB78]   J. van Benthem. Two simple incomplete logics. *Theoria*, 44:25–37, 1978.

[vB83a]   J. van Benthem. *The Logic of Time. A Model Theoretic Investigation in to the Varieties of Temporal Discourse*. Reidel, Dordrecht, 1983.

[vB83b]   J. van Benthem. *Modal Logic and Classical Logic*. Bibliopolis, Naples, 1983.

[vB84]   J. van Benthem. Correspondence theory. In D. Gabbay and F. Guenthner, editors, *Handbook of Philosophical Logic*, volume 2, pages 167–247. Reidel, Dordrecht, 1984.

[Ven93]   Y. Venema. *Many-Dimensional Modal Logic*. PhD thesis, Department of Mathematics, University of Amsterdam, Amsterdam, 1993.

[vOQ66]   W. v. O. Quine. *The ways of Paradox and Other Essays*. Random House, New York, 1966.

[vW57]   G. H. von Wright. *Logical Studies*. Routledge & Kegan, London, 1957.

[vW65]   G. H. von Wright. And next. *Acta Philosophica Fennica*, 18:293–304, 1965.

[vW82]   G. H. von Wright. *Wittgenstein*. Blackwell, Oxford, 1982.

[Waj33]   M. Wajsberg. Ein erweiteter Klassenkalkul. *Monatshefte für Mathematik und Physik*, 40:113–126, 1933.

[Wil92]   T. Williamson. On intuitionistic modal epistemic logic. *Journal of Philosophical Logic*, 21:63–89, 1992.

[Wil00]   T. Williamson. *Knowledge and Its Limits*. Oxford University Press, Oxford, 2000.

[Wit01] L. Wittgenstein. *Tractatus Logico-Philosophicus*. Routledge, London, 2001. David Pears and Brian McGuinness, translators.

[Xu88] M. Xu. On some US-tense logics. *Journal of Philosophical Logic*, 17(2):181–202, 1988.

[Zem73] J. J. Zeman. *Modal Logic*. Clarendon Press, Oxford 1973.

Index of names

Alchourrón, C., 204

Alechina, N., 272
Aristotle, 3, 25, 27, 162
Artemov, S., 138
Aumann, R. J., 204

Béziau, J.-Y., 239
Baldoni, M., 238
Beklemishev, L., 138
Bencivenga, E., 272
Bernardi, C., 138
Blackburn, P., 70, 86, 130, 176, 181
Bochenski, I. M., 48
Boole, G., 2, 22
Boolos, G., 116, 138, 238
Brauner, T., 255, 271, 272
Brogaard, B., 203
Bull, R. A., 139, 181
Burgess, J. P., 181

Carnap, R., 48, 54, 55, 83, 85
Carnielli, W. A., 11, 23, 41, 239
Catach, L., 235, 238
Chagrov, A., 86, 116
Chellas, B. F., 93, 116
Cocchiarella, N. B., 157, 180, 272
Coniglio, M. E., 11, 239
Conway, J. H., 214
Cook, S., 23
Copeland, J., 85
Cresswell, M. J., 48, 85, 115, 116, 138, 251, 255, 271

D'Ottaviano, I. M. L., 41
de Jongh, D., 116
De Morgan, A., 238
de Rijke, M., 86, 130, 181
Diodorus Cronus, 162
Dugundji, J., 50, 84

Enjalbert, P., 238
Epstein, R. L., 23

Fagin, R., 203, 204
Fariñas del Cerro, L., 238
Fetzer, J., 24
Feys, R., 48
Fine, K., 271
Fitch, F., 187, 203
Fitting, M. C., 85, 239, 272
Font, J. M., 85
Forbes, G., 254

Gärdenfors, P., 140
Gödel, K., 50, 107, 133
Gabbay, D., 181, 239
Garson, J., 272
Gentzen, G., 23
Gettier, E., 185
Ghilardi, S., 255, 271, 272
Giordano, L., 238
Goldbach, C., 186
Goldblatt, R., 68, 86, 116
Goranko, V., 137
Gouveia, P., 239
Grzegorczyk, A., 116
Gärdenfors, P., 139, 204

Hájek, P., 85
Haack, S., 24
Halpern, J. Y., 203, 204
Hansson, B., 140
Harel, D., 238
Harvey, C. W., 239
Hayes, P., 203
Hazen, A., 254, 271
Heyting, A., 50
Hintikka, J., 85, 86, 203, 239
Hodges, W., 86

289

Hodkinson, I., 181
Hughes, G. E., 48, 85, 115, 116, 138, 251, 255, 271
Humberstone, L., 140
Humphreys, P. W., 24

Jansana, R., 139
Japaridze, G., 116

Kalish, D., 23
Kalmár, L., 23
Kamp, H., 180
Kanger, S., 85
Kauppi, R., 85
Kneale, M., 48
Kneale, W., 48
Kozen, D., 238
Kracht, M., 238
Kripke, S., 32, 72, 85, 115, 116, 264, 272
Kröger, F., 181
Kuhn, S., 140, 272
Kurucz, A., 239

Löb, M. H., 107
Langford, C. H., 31, 48–50, 54, 271
Leibniz, G. W., 53, 85, 254
Lemmon, E. J., 99, 116, 139
Lemmon, J., 139
Lenzen, W., 203
Lewis, C. I., 31, 48–50, 54
Lewis, D., 204, 254, 271
Linsky, L, 24
Lorenzen, P., 86, 138

Magari, R., 138
Makinson, D., 116, 131, 139, 204
Manaster, A. B., 23
Manna, Z., 181
Marcos, J., 11
Martelli, A., 238
Martin, N. M., 48
Marx, M., 272
Mates, B., 85
Matsumoto, K., 48
McCall, H., 2
McCarthy, J., 203
McGinnis, C., 239
McKinsey, J. C. C., 139
Mendelson, E., 23
Meyer, J. J., 204

Montagna, F., 138
Montague, R., 23, 85, 139
Moses, Y., 204
Mugnai, M., 85

Nietzsche, F., 158

Ohnishi, M., 48

Park, D., 86
Peirce, C. S., 85, 238
Pelletier, F. J., 23
Pietarinen, A.-V., 85
Pizzi, C., 181
Plantinga, A., 254, 271
Plato, 183, 184, 203
Pnueli, A., 181
Popkorn, S., 116
Pratt, V. R., 238
Prawitz, D., 23
Prior, A. N., 158, 174, 180

Quine, W. v. O., 24, 251, 252

Ramachandran, M., 254
Rautenberg, W., 139
Reichenbach, H., 177, 181
Rennie, M. K., 238
Rescher, N., 157, 180
Reynolds, M., 181
Russell, B., 23, 185
Rybakov, V. V., 139

Sahlqvist, H., 116, 235, 238
Salerno, J., 203
Sambin, G., 138
Schelling, T., 204
Schröder, E., 238
Schumm, G. F., 139
Schurz, G., 85
Schütte, K., 48
Scott, D., 99, 116, 139, 157, 167, 211, 237
Scroggs, S. J., 139
Segerberg, K., 24, 86, 116, 139, 140, 180, 204, 237, 239
Sernadas, C., 239
Smorynski, C., 116, 212, 239
Smullyan, R., 23
Socrates, 183, 203
Solovay, R., 116, 239

INDEX OF NAMES

Tapscott, T., 85
Tarski, A., 139, 238
Theaetetus, 183
Thomason, R., 272
Thomason, S. K., 85, 138
Tiuryn, J., 238
Tzakova, M., 181

Urquhart, A., 139, 180
Ursini, A., 138

Valentini, S., 138
van Benthem, J., 70, 86, 117, 118, 138, 176, 181, 237

Vardi, M. Y., 203, 204
Venema, Y., 86, 130, 140, 181, 272
von Wright, G. H., 54, 166, 170, 180, 272

Wajsberg, M., 84
Whitehead, A., 23
Williamson, T., 203
Wittgenstein, L., 54
Wolter, F., 239

Xu, M., 181

Zakharyaschev, M., 86, 116, 239
Zeman, J. J., 48

Index of notation

Axioms and formulas
   Alt_n, 135
   D_n, 134
   $G(\varphi, \psi)$, 219
   $G(a, b, \varphi)$, 218
   $G(a, b, c, d)$, 214
   $G^{k,l,m,n}$, 92
   $\nabla\alpha$, 27
   $\triangle\alpha$, 27
   (@1), 177
   (@2), 177
   ($4K_i$), 195
   (**4**), 37
   ($5K_i$), 195
   (**5**), 37
   (**A1.1**) to (**A5.2**), 175
   (**AN1**) to (**AN4**), 166
   (**AN2'**), 168
   (**AO**), 187
   (**Ax1**) to (**Ax3**), 6
   (**BA1**) to (**BA2**), 145
   (**BF**) and (**BF'**), 247
   (**Ban**), 32
   (**B**$_i$), 201
   (**B**), 37
   (**CB**$_i$), 201
   (**CKr**), 198
   (**C**), 198
   (**D'**), 30
   (**D1**), 102
   (**DK1**) to (**DK2**), 199
   (**DQ0**) to (**DQ8**), 267
   (**Dum**), 165
   (**D**), 30
   (**F**), 62
   (**G1**), 92
   (**GL**$_p$), 160
   (**GL**), 42
   (**Grz**), 116
   (**H**), 117
   (**I1**) to (**I2**), 252
   (**I2c**), 253
   (**Int**), 173
   (**K**K_i), 186
   (**KA**), 184
   (**KP**), 187
   (**K**$^\square$), 142
   (**K**$_\square$), 144
   (**K**$_\boxplus$), 144
   (**K**), 35
   (**LP**), 142
   (**Lin'**), 103
   (**Lin**), 103
   (**MM1**) to (**MM4**), 213
   (**MV**), 118
   (**McK**$_f$), 160
   (**McK**), 67
   (**Mk**), 131
   (**QL1**) to (**QL2**), 243
   (**QP1**$^+$), 247
   (**QP1**) to (**QP2**), 246
   (**RI1**) to (**RI2**), 253
   (**T'**), 29
   (**T***), 30
   (**TL1**) to (**TL5**), 170
   (**Triv**), 33
   (**T**K_i), 195
   (**T**), 29
   (**US.1**) to (**US.4**), 172
   (**VB**), 117
   (**Ver**), 33
   (**W1**) to (**W2**), 167
   AT1 to **AT4**, 45
   $K_{[a]}$, 213

Complexity
   NP-complete, 23
   NP-problem, 23

Connectives and operators
  C, 198
  C^k, 197
  E, 197, 269
  E^k, 197
  Q_a, 245
  Q_b, 245
  $\bot$, 5
  $\cup$, 207
  $\equiv$, 5
  $\exists$, 242
  $\exists^a$, 267
  $\forall$, 241
  $\forall^a$, 267
  $\neg$, 5
  $\odot$, 207
  $\dashv$, 31
  $\dashv_a$, 222
  $\supset$, 5
  $\top$, 5
  $\vee$, 5
  $\wedge$, 5

Logics and systems
  US-logic, 172
  $G^{\langle\varphi,\psi\rangle}$, 213
  $G^{\langle a,b,\varphi\rangle}$, 213
  $G^{\langle a,b,c,d\rangle}$, 213
  $(T^\square)$, 142
  AN, 166
  Ban, 34
  CKT^m, 198
  CK^m, 198
  CL, 157
  CR, 157
  $CS4^m$, 198
  $CS5^m$, 198
  DK^m, 199
  $DQ^=$, 267
  $DS4^m$, 200
  $DS5^m$, 200
  DT^m, 200
  $KD45^m$, 201
  KD, 38
  KG1, 93
  KGL, 87, 108
  KGrz, 116
  KT4=S4, 37
  KTB4=KT5=KDB4=KDB5=S5, 37
  KTB=B, 37
  $KT^\square$, 142
  KT^m, 187, 194
  KT=T, 37
  KVB, 117
  K, 32, 36
  $KG^{k,l,m,n}$, 93
  $K\pi$, 248
  $K^{\square\square}$, 145
  K^m, 194
  K_b, 157
  K_t, 144
  $K_t + \Diamond\alpha \supset \Diamond\Diamond\alpha$, 159
  $K_t^{\circ\circ}$, 161
  K+(H), 117
  L_n, 135
  Mk, 131
  PCr, 158
  PC, 4
  $PC^\triangle$, 30
  $PC^\square$, 26
  $PC^\square$+(Ban), 33
  $PC^\square$+(Triv), 33
  $PC^\square$+(T), 29
  $PC^\square$+(Ver), 33
  PL, 158
  QL, 41, 241
  $QL^=$, 257
  QL_1, 57
  QL+S+(BF), 266
  $QS^=$, 263
  $QS^=$+(BF), 256
  S1 to S5, 31
  S4.1, 67
  S4.3, 102
  $S4^m$, 194
  $S5^m$, 194
  SL, 157
  $S\pi^*$, 247
  TLUS, 172
  TL, 170
  Triv, 34
  Ver, 34
  W, 167

Maps
  F^φ, 228
  ρ, 224
  ρ_S, 226
  e-transform, 108
  f, 41, 163, 165

INDEX OF NOTATION

f^{-1}, 57
g, 165
Miscellaneous
$0^\exists$, 73
$0^\forall$, 73
$1^\exists$, 73
$1^\forall$, 73
Φ, 207
Φ_0, 207
Θ, 207
$\alpha[p/\beta]$, 6
α^*, 12
(**MSDM**), 36
(**SDM**), 8
(**IH**), 258
(**Pres**), 263
(**Quant**), 264
PA, 107
$\mathbf{S}^\Phi$, 213
S-tableau, 72
Modalities
ATN, 170
K_i, 185
O, 211
P, 211
S, 171
T, 167
T_S, 167
U, 171
Y, 167
Y_S, 171
$[a]$, 208
@, 177
$\boxdot$, 141
$\boxminus$, 28
$\diamondsuit$, 144
$\diamondsuit'$, 172
$\blacklozenge$, 144
$\blacklozenge'$, 172
$\langle a \rangle$, 208
$\overline{\square}$, 170
$\boxplus$, 144
$\boxplus'$, 172
$\boxdot$, 144
$\boxdot'$, 172
$\widehat{\diamondsuit}$, 163
$\widehat{\square}$, 163
$\widehat{\widehat{\diamondsuit}}$, 163

$\widehat{\widehat{\square}}$, 163
Models and frames
$\mathcal{F}$, 60
$\mathcal{G}$, 119
$\mathcal{M}$, 60
Models
π-model
indexed, 249
$\mathbf{K}\pi$-model, 248
$\mathbf{K}\pi^*$-model, 248
Operations on relations and sets
$\Rightarrow$, 223
R^{-1}, 223
$\cap$, 223
$\cup$, 223
$\odot$, 223
$\oplus$, 223
$\overline{R}$, 223
$\triangleright$, 223
$\wp$, 51
Properties
$\omega\mathbf{C1}$, 258
$\omega\mathbf{C2}$, 258
$\omega\mathbf{C}$, 258
(fmp), 126
Relations
R^+, 132
R^0, 62
$R^\square$, 143
R^n, 62
$\overleftarrow{R}$, 145, 178
$\overrightarrow{R}$, 145, 178
$(\mathbf{T}_n)$, 62
Rules
(**CKr**), 198
(**CR1**), 148
(**CR2**), 148
(**DKm**), 199
(**DR1**), 38
(**DR2**), 38
(**Eq**), 7
($\mathbf{E}_\exists$), 243
($\mathbf{E}_\forall$), 243
($\mathbf{I}_\exists$), 243
($\mathbf{I}_\forall$), 243
(**MP**), 7
(**NecR**), 36
($\mathbf{Nec}_{K_i}$), 186

Rules (*Continued*)
    (**Nec**$_{T_S}$), 167
    (**Nec**$_{\mathbb{D}^n}$), 175
    (**Nec**$_{\mathbb{D}}$), 145
    (**Nec**$_{\mathbb{P}^n}$), 175
    (**Nec**$_{\mathbb{P}}$), 145
    (**Nec**), 35
    (**R1**), 243
    (**RT**$_S\overline{\square}$), 171
    (**UQ**), 246
    (**US**), 6
    Nec$_{[a]}$, 213
Sets of formulas
    $Den(w)$, 95
    $Den_a(w)$, 225
    $Poss(w)$, 98
    $Poss_a(w)$, 236
    WFF, 34

Syntactical and semantical consequence
    $\Gamma \vDash_{\mathfrak{F}} \alpha$, 88
    $\Gamma \vdash \alpha$, 10
    $\Gamma \vdash_S \alpha$, 36
    $\mathcal{F}, w \vDash \alpha$, 60
    $\mathcal{F} \vDash \alpha$, 60
    $\mathcal{M}, w \vDash \alpha$, 56
    $\mathcal{M} \vDash \alpha$, 56
    $\mathfrak{F} \vDash \alpha$, 88
    $\mathfrak{M}$, 89

Index of subjects

PF-logics, 144
Absorption Theorem, 45

Accessibility relation, 60
    as a binary predicate, 64
    of the canonical frame, 99
Actual objects, 267
Admissible
    model, 119
    set, 119
Algebra
    Boolean, 123
        with operators, 123
    modal, 123
    topological Boolean, 125
Algebraic semantics, 123
Anti-Omniscience Principle, 187
Aristotelian essentialism, 251
Aristotle's square, 25
Arithmetical provability
    logic of (**KGL**), 107
Arity, 242
Atomicity, 67
Axiom schema, 5
Axiomatic basis for **PC**, 6
Axiomatization
    non-separable, 220
    separable, 220

Banal, Banalization, 33
Barcan formula, 247
    converse of, 251
Belief
    conjunctive, 201
    strong, 201
    weak, 201
Bimodal logics, 141
Bisimulation
    between frames, 69
    between models, 69
    contraction, 70
    invariance, 70
Boolean algebra
    with operators, 123
Bridge axiom, 167
Bull's Theorem, 134

Canonical
    frame(s), 95
        of **KGL**, 113
        of **S4.3**, 103
    system, 99
Canonicity, 87, 107, 117
Circuit rules, 148
Collapse of modalities, 33
Compactness
    semantic
        modal, 88
        syntactical, 11
Completeness
    algebraic, 125
    arithmetical, 133
    constructive, 89
    Henkin's method, 95
    of **CKT**^{m}, 198
    of **CK**^{m}, 198
    of **CS4**^{m}, 198
    of **CS5**^{m}, 198
    of **DKT**^{m}, 200
    of **DK**^{m}, 200
    of **DS4**^{m}, 200
    of **DS5**^{m}, 200
    of **G**$^{\langle a,b,\varphi \rangle}$, 233
    of **G**$^{\langle a,b,c,d \rangle}$, 227
    of **K45**, 102
    of **K4**, 102
    of **K5**, 102
    of **KB4**, 102

Completeness (*Continued*)
    of **KB**, 102
    of **KD45**, 102
    of **KD45**^{m}, 201
    of **KD4**, 102
    of **KD5**, 102
    of **KDB**, 102
    of **KD**, 102
    of **KT4=S4**, 102
    of **KTB**, 102
    of **KT**^{m}, 196
    of **KT=KDT**, 102
    of **K**^{m}, 195
    of **PC**, 12
    of **QS**$^=$+(**BF**), 262
    of **S4**^{m}, 196
    of **S5**, 102
    of **S5**^{m}, 196
    of first-order logic, 257
    strong, 13
    structural, 138
    w.r.t. general frames ($\mathcal{G}$-completeness), 121
    w.r.t. a class of frames ($\mathcal{F}$-completeness), 105, 106
    w.r.t. a class of models ($\mathcal{M}$-completeness), 104, 105
Conditional Elimination, 7
Conjunctive Normal Form (CNF), 17
Conjunctive Normal Form
    theorem reduction to, 17
Connectives
    complete set of, 21
    Sheffer, 21
Consistency, 11
Contingency, 27
Contingent, 25
Contingent identities, 253
Contradictory statements, 25
Contraposition, 7
Contrary
    statements, 25
Correspondece Theory, 65
Counterpart theory, 253
Craig interpolation property, 122

De Morgan's Laws I and II, 7
Decidability
    and finite model property, 126
    of **K45**, 130
    of **K4**, 130
    of **KB4**, 130
    of **KB**, 130
    of **KD45**, 130
    of **KD4**, 130
    of **KDB**, 130
    of **KD**, 130
    of **KT4=S4**, 130
    of **KTB4=KT5=S5**, 130
    of **KTB**, 130
    of **KT**, 130
    of **PC**, 14
Deducible (derivable), 6
Deduction
    in **S**, 6
    of α from Γ, 6
Definite descriptions, 253
Degenerated
    modal system, 37
Denecessitation, 47
    multimodal, 226
Deontic logics, 209
Deontic temporal logics, 211
Derived (meta)rule, 8
Determinism, 158
Diagram, 72
    chain, 72
    degree, 75
Diodorean
    axiom, 108
    fragment, 163
    necessity, 162
    possibility, 162
Disjunction Introduction (meta-rule of), 8
Disjunction property
    semantical, 111
    syntactical, 111
Disjunctive Expansion, 7
Distribution of
    $\vee$ over $\wedge$, 7
    $\wedge$ over $\vee$, 7
Double Negation, 7
Doxastic logic, 183
Dugundji's formula, 50, 134
Dugundji's Theorem, 52
Dummett's formula (*Dum*), 165
Dynamic algebra, 238
Dynamic logic, 210, 238

INDEX OF SUBJECTS

Epistemic
    state, 190
        satisfiability in, 191
Epistemic logic, 183
    in Atificial Intelligence, 184
Epistemic temporal logics, 211
Epistemic-doxastic logics, 209
Equational logic, 123
Essence
    of an individual, 254
Excluded Middle, 7
Existence Postulate, 54

Filtration, 126
Filtration Theorem, 127
Finite frame property, 129
Finite model property, 126
    holds for **KVB**, 131
First-order
    definability, 67
    logic, 57, 252
        translation into, 65
    modal logic, 255
    model, 66
Fitch's Theorem, 246
Formation operators, 207
Formula
    affirmative, 217
    characteristic of a digram, 89
    complexity of, 5
    denecessitated set of, 95
    input, 14
    length of, 5
    modal degree of (dg), 28
    negative, 217
    possibilitated set of, 98
    subformula of a, 35
        immediate, 35
    well-formed, 5
        set of (WFF), 34
Formula schema, 5
Frame(s)
    p-morphism between, 69
    diamond property, 93
    canonical
        of a system, 95
    characterized by, 88
    Church-Rosser, 93
    cohesive, 115
    elementary class, 68
    finite
        for **KGL**, 109
        for a system, 88
    general, 119, 248
    generated, 110
        strongly, 110
    incestual, 93
    model over, 60
    multi-agent, 190
    multi-relational, 224
    neighborhood, 140
    relational, 60
    subframe of, 110
        generated by, 112
    temporal, 145
    true at a world in a, 60
    truth on a, 66
    validity on a, 60
Free logics, 269
Frege's puzzle, 252
Fundamental Theorem
    for Affirmative Systems, 230
    for Basilar Systems, 226
    of Canonical Models, 98

Gödel-Löb formula, 107
Gettier's problem, 185
Goldblatt-Thomason Theorem, 68

Halldén-completeness, 122
Halldén-incompleteness, 122
Halting clause, 208
Hybrid logics, 177

Identity, 7
Importation-Exportation, 7
Incompleteness
    Halldén-, 121
    in quantificational modal logic, 266
    of the system **KVB**, 117
    w.r.t. a class of frames
        ($\mathcal{F}$-incompleteness), 117
Indexed Π-model, 249
Induction
    axioms, 219
    on complexity of formulas, 5
    on complexity of proof, 5
Intensional object, 253
Interpolation formula, 22

Interpretation
    alternative, 243
    of **QL**, 242
Introspection axiom
    mutual negative, 219
    negative, 195
    positive, 195

Kamp's Theorem, 181
King's puzzle, 198
Kleene algebra, 214
Knowledge
    as justified true belief, 185
    by acquaintance, 185
    by description, 185
    common, 197
    distributed, 199
    implicity, 196
    minimal logic of, 189
Knowledge Axiom, 184

Laws
    □◊-interchange, 27
    Absorption, 45
Leibniz's law, 67, 253
Leibnizian principle
    identity of indiscernibles, 252
    indiscernibility of identicals, 252
Lindenbaum algebra, 124
Lindenbaum's Lemma, 96

Matrix
    characteristic, 50
    for Dugundji's formula, 51
Maximal consistent extension, 95
McKinsey axiom (**Mck**), 67
McKinsey property, 68
Megaric
    fragment, 163
Modal
    algebra, 123
        regular, 123
    degree, 27
    function, 27
    language, 34
Modal logic
    degenerated system, 37
    non-normal system of, 36
    normal system of, 32, 35

Modal parameter, 207
    atomic, 207
        identity, 207
        null, 207
Modal Syntactical Deduction Metatheorem (**MSDM**), 36
Modalities
    alethic, 30
    de dicto, 3
    de re, 3
    deontic, 30, 211
    Diodorean, 163
    existential, 269
    extensionalist view, 54
    global (or universal), 86
Modality (sequence of modal operators), 43
Model
    canonical, 95
        of a system, 95
    Carnapian
        explicit, 55
        implicit, 55
        valid in a, 56
    closed under formulas, 248
    multi-agent, 190
    of **QL**, 242
    Relational
        p-morphism between, 69
        validity in a, 60
    Relational (Kripke model), 59
    submodel of, 110
        generated by, 112
    temporal, 145
    with constant domain, 249
Model for, 51, 60
Modus Ponens
    rule of (**MP**), 7
Monadic fragment of **QL**, 84
Monomodal logics, 141
Monotonicity (meta-rule of), 8
Multi-frame, 224
Multi-model, 225
Multimodal
    language, 172, 207
    logics, 141

Natural Deduction Calculus, 9
Necessary identities, 253

INDEX OF SUBJECTS

Necessitation Rule, 32
    multimodal, 213
    variants of, 145
Necessity, 3
    logical, 142
    physical, 142
Nominals, 177
Non-canonicity, 107
    of **KGL**, 113
Non-standard logics, 1
    intuitionistic, 1
    paraconsistent, 1, 11

Omniscience, 186
Organon, 3, 25

Paradox of knowability, 187
Partial order
    strict, 110
Peano arithmetic (**PA**), 107
Permutation of Antecedents, 7
Physical modalities
    logic of, 141
Possibilia, 267
Possibilitation, 47
    multimodal, 226
Possibility, 3
Possible objects, 267
Possible worlds, 53
Post-completeness, 13, 18, 80
Pragmatic contradiction, 245
Principle of Knowability, 187
Principle of Plenitude, 162
Probability theory, 2
Processes
    identity, 208
    parallel, 208
    serial, 208
Proof, 6
    length of, 6
Proof by Cases, 7
Properties
    accidental (contingent), 25
    actual (real), 25
    essential (necessary), 25
    pontential (possible), 25
Propositional constants, 245
Propositional function, 2
Pseudo-Scotus, 7

Quantifier(s)
    existential, 242
    for actual objects, 267
    objectual interpretation, 251
    propositional, 244
    scope of, 242
    substitutional interpretation, 251
    universal, 241

Rectification under a valuation, 12
Reduction Theorem, 44
Relation
    n-dense, 62
    asymmetric, 68
    converse, 145
    dense, 62
    empty, 61, 223
    euclidean, 61
    identity (diagonal), 223
    intransitive, 68, 173
    irreflexive, 68
    linear (Linearity), 103
    of contrariety, 30
    of subalternance, 30
    plausibility, 189
    reflexive, 61
    serial, 61
    symetric, 61
    transitive, 61
    trichotomic (Trichotomy or Connectedness), 103
    universal, 223
    weakly connected, 103
    well-covered, 109
Replacement of Proved Equivalents (**Eq**), 7
Rigid designator, 253

Sahlqvist's monomodal systems, 102
Sahlqvist's schema, 102
Scrogg's Theorem, 134
Second-order
    definability, 67
    logic, 241, 243, 244
    quantification, 67
Secondary Modus Ponens (meta-rule of), 8

Semantics
    bidimensional, 140
    Kripke, 72
    neighborhood, 139
    relational, 59
Sillogism
    categorical, 25
    modal, 25
Simplification, 7
Soundness, 10
    strong, 11, 13
Stop rule, 77
Strict implication, 31
    and multimodalities, 220
Subcontrary statements, 26
Substitution, 6
    of identicals
        restriction on, 253
    simultaneous, 6
Syntactical compacteness, 11
Syntactical Deduction Metatheorem (**SDM**), 8
System
    affirmative, 213
    basic elementary, 68
    basilar, 213
    Catach-Sahlqvist, 213
    elementary, 68
    extension of, 18
        proper, 18
    heterogeneous, 206
    homogeneous, 206
    multimodal
        separable, 221
    Post-complete, 18

Tableau
    alternatives, 14
    method, 15, 85, 238
    relational, 72
        for **KT**, 87
        for **K**, 87
        for $\mathbf{K}^\square$, 143
        for $\mathbf{K}^{\square\square}$, 146
        for $\mathbf{K}_t$, 148
    semantic, 14
    sequential, 14
Tautology, 9

Temporal logic, 145, 210
    as multimodal logic, 144
    of programs, 170
Tense logics, 144
    metric, 175
Term, 242
    as a rigid designator, 253
    free for, 243
Theaetetus, 183, 184, 203
Theorem of an arbitrary system, 6
Time
    ascending linear, 210
    branching, 157
        in the future, 157
        in the past, 157
    circular, 158
    descending linear, 210
    discrete, 167
    linear, 157
Topological logics, 180
Transitive algebras, 125
Transitivity, 7
    meta-rule of, 8
Translation
    arithmetical, 133
    from **KTB** into $\mathbf{K}_t$, 165
    from $\mathbf{K}_t$ into **KTB**, 165
    from **QL** into **S5**, 57
    from **S5** into **QL**, 41
    function, 41
    standard, 65
    strong, 41
Trivial, Trivialization, 33

Uniform substitution
    rule of (**US**), 6
Unraveling, 70

Valid consequence
    with respect to a class
        of frames, 88
Validity
    for a system **S** (**S**-validity), 72
    of a **QL** formula, 243
Variable(s)
    atomic or propositional, 5
    bound, 242
    free, 242
    set of (Var), 34

INDEX OF SUBJECTS

Verificationist Principle, 187
Verum, 33

Weakening, 7
Well-coveredness, 109
Well-formed formula
    (wff), 5

World
    accessible from, 75
    as index, 223
    reference, 56
    semi-terminal, 64
    terminal, 62
    true at a, 56

GPSR Compliance
The European Union's (EU) General Product Safety Regulation (GPSR) is a set of rules that requires consumer products to be safe and our obligations to ensure this.

If you have any concerns about our products, you can contact us on

ProductSafety@springernature.com

In case Publisher is established outside the EU, the EU authorized representative is:

Springer Nature Customer Service Center GmbH
Europaplatz 3
69115 Heidelberg, Germany

www.ingramcontent.com/pod-product-compliance
Ingram Content Group UK Ltd.
Pitfield, Milton Keynes, MK11 3LW, UK
UKHW022230230426
12048UKWH00016BA/1172